A Basilisk Glance

Robert Templer is a journalist, writer, lecturer and policy consultant. Educated at the University of Cambridge and the University of London, he is the author of the highly acclaimed *Shadows and Wind: A View of Modern Vietnam* (Penguin, 1999).

A BASILISK
GLANCE
Poisoners from Plato to Putin

ROBERT
TEMPLER

First published in the United Kingdom by Bui Jones 2024
Copyright © Robert Templer

Robert Templer has asserted his right under the Copyright, Designs and
Patents Act 1988 to be identified as the author of this work

Cover (detail) Portrait of Simonetta di Vespucci by Piero di Cosimo
Set in Times New Roman

BUIJONES

buijones.com
Bui Jones Limited Company Number 14823240

Printed and bound in Great Britain by Clays Ltd, Elcograf S.p.A.
A CIP catalogue record for this book is available from the British Library

ISBN 978-1-7394243-4-3 Paperback
ISBN 978-1-7394243-7-4 ebook

MIX
Paper | Supporting
responsible forestry
FSC® C018072
www.fsc.org

Bui Jones books are printed on
paper from responsible sources.

*For Catherine and Henry
and in memory of Deborah Rogers and David Miller*

CONTENTS

INTRODUCTION: THE EMPTY SAND

*All it is, is taking a drink... to go to sleep. That's
what death is, sleep.*
>Jim Jones, from a recording of the mass
suicide at Jonestown, 18 November 1978

At the edge of Evergreen Cemetery in the dry hills overlooking
the Californian city of Oakland is a small granite gravestone, barely
more than a foot square. It is inscribed: "In memory of the victims
of the Jonestown tragedy." There are no names of the dead, no sense
even of the number of people buried there. A few feet away, another
memorial stone set flat in the dusty ground gives some idea of that
tragedy. It was put there by Fred Lewis, an African American man
from San Francisco. He had come home from his work as a butcher
at Petrini's Grocery on a Saturday in August 1978 to find his family
gone. All that was left in his home was a single bed and a black-and-
white television. His wife, Doris, aged forty-six when she left; and his
children; Lisa, sixteen; Karen, nineteen; Freddie Jr., thirteen; Barry,
eleven; Adrienne, nine; Cassandra, eight; and Alecha, seven, had gone
to Guyana where they died in the worst recorded mass suicide and
murder by poison. Another nineteen members of Lewis's family also
died there.

In the apocalyptic end to the Peoples Temple, a new Christian
movement founded in Indiana in 1955, 907 people died from
potassium cyanide mixed with grape Flavor Aid. Among them were
276 children. More than 240 of those children lie unidentified under
the small marker in Evergreen Cemetery.

The deaths ended a movement that had evolved from a church run by a charismatic preacher, Jim Jones, into the dream of a socialist utopia in the jungle of Guyana and finally into a nightmare of obliteration. Jones had started out in Indianapolis as the pastor for a Disciples of Christ church, leading a racially integrated flock in a city that was notorious for the popularity of the Indiana Ku Klux Klan. Increasingly afflicted by racism in Indiana and worried about the threat of nuclear war, he moved his church in the 1970s to Northern California, as he had read in *Esquire* magazine that residents of Ukiah in the Redwood Valley might survive an atomic holocaust.

This small, mixed-race church grew rapidly as people were drawn to Jones's utopianism, charisma and sense of community. Although later it would appear more a mix of Marxism, mind control and the Book of Revelation, it started off with an appealingly idealistic message. Jones promised racial equality and economic opportunity to draw in his congregation, mostly African Americans. He set up churches in San Francisco and Los Angeles, cities struggling with racial tensions, inequality and poverty. In those charged times, he proved himself an adept political player, even chairing the important city housing commission in San Francisco. He was often photographed with senior Democratic Party politicians, including President Jimmy Carter, California Governor Jerry Brown and San Francisco Mayor George Moscone. By the mid-1970s, he was an important figure in local politics, lending his support to several civic schemes, including a plan to build a suicide prevention fence across the Golden Gate Bridge. Moscone relied on members of the Temple to canvas for votes.

But as the Temple grew, so did the confrontation between Jones and a group known as the Concerned Relatives. In 1974, Jones sent the first group of Temple members to Guyana to start work on a new community based around agricultural self-sufficiency. It was to be cut out of remote jungle in the former British colony. In his mind, Temple members would be safe there from nuclear attack and from the increasingly intrusive concerns of their families, with whom they had often cut all ties. The relocation further inflamed the anxieties of the families who felt they had lost loved ones to a cult. The Temple,

Jones himself and the group's suspect finances came under increasing scrutiny. In 1977 an article in *New West* magazine exposed some of the violent punishments doled out to those who violated the preacher's arbitrary rules. One sixteen-year-old was beaten across the buttocks seventy-five times in front of a congregation of hundreds for kissing another woman. Meanwhile, Jones was sleeping with some of the men and women in his inner circle.

Jones moved to Guyana in 1977, just hours before the magazine article was on the newsstands. By then his vision had turned into a nightmare of recriminations and defections. He had been staging "White Nights", the name given to rehearsals of mass suicide, for some time, inculcating his flock with ideas of persecution. Annie Moore, a young nurse and passionate devotee of Jones, wrote in a letter several months before the end:

> The only course of action open to us was a mass suicide for the glory of socialism. We were told we would be tortured by mercenaries if we were taken alive. Everyone, including the children, was told to line up. As we passed through the line, we were given a small glass of red liquid to drink. We were told that the liquid contained poison and that we would die within forty-five minutes. When the time came when we should have dropped dead, Reverend Jones explained that the poison was not real and that we had just been through a loyalty test. He warned us that the time was not far off when it would become necessary to die by our own hands.

By the end of 1978, Jonestown was struggling. The initial ardour to create a new society in this difficult environment was fading under the immense challenges of jungle life. Jones was facing numerous personal problems, including a worsening drug addiction and a custody battle over a child of one of his many lovers. On 14 November 1978, Leo Ryan, a Democratic congressman from San Mateo, flew to Guyana with a small team of people to investigate accusations that people were being abused and held against their will.

When Ryan reached Jonestown, Temple members had been coached on how to behave but a small group decided to leave with him. Jones was furious at these "defectors", as were other members of the group, one of whom tried to kill Ryan with a knife. The congressman and others left for an airstrip, fearing they were about to be killed. As they boarded their planes, a tractor pulled up, towing a trailer of armed men from the Temple. They opened fire, killing the congressman and four others as they cowered behind the wheels of a chartered Guyana Airlines plane.

After the attack on Ryan, Jones and his confidantes began the final "White Night", telling people that they were about to be murdered by Guyanese troops who had surrounded the compound, although this was not the case. Temple members, mostly poor African American families from Indiana and Northern California, gathered by a tin-roofed building hung with a sign slightly misquoting the philosopher George Santayana: "Those who cannot remember history are condemned to repeat it." Jones began a rambling attempt to explain what was happening. A recording captured the confusion and fear. It lasts for around an hour and contains the last discussions between Jones and his flock as they killed their children and took their own lives. Several people can be heard protesting, begging Jones to find an alternative such as moving the group to the Soviet Union. Jones repeatedly told them that it was over.

Jones was too addled by his consumption of drugs to organise the killings. That was done mostly by a group of women around him: Carolyn Layton and Maria Katsaris, who had both been his lovers; Harriet Tropp, who ran the radio room; and Annie Moore, who was also Layton's sister. Debbie Blakey, a member of the Peoples Temple for six years who left before the deaths, said later that they had been talking about suicide for years. What began in 1973 as a discussion within the inner circle soon grew to include plans for the deaths of the entire congregation. By 1978, Jones had lost control of the group. The suicides and murders were decided upon and carried out by a handful of mostly white men and women who by then were making the critical decisions at Jonestown. Moore debated various means of carrying out the killings. "I started out for revolutionary suicide, almost switched to fighting but stick to suicide now," Moore wrote. "I never thought people would line up to be killed but actually think a select group would have to kill the majority of the people secretly without the people knowing it." She went on to discuss the merits of pumping carbon monoxide into rooms while people were sleeping.

Towards the end, Jones was spinning stories of nuclear war between Russia and China and telling black Temple members that the US government was setting up concentration camps for them. While they lived in California, Temple members maintained some tenuous links to the outside world, but in Guyana they were entirely isolated, often ill and increasingly exhausted from arduous manual labour. Jones had always controlled his flock through sleep deprivation, isolation and even sexual violence. Later, survivors would describe themselves as being just too tired to think for themselves and too fearful to argue with Jones and those around him. The lengthy harangues over the loud speakers, frequent suicide drills and paranoid fantasies about attacks kept everyone in Jonestown in a state of intense anxiety. Their diet was meagre, often consisting of rice gruel with no protein. Despair and the powerful controlling hand of Jones created an environment in which the unthinkable became entirely plausible.

Moore's letters provide some evidence of the mind-set of the inner circle as the Temple spiralled towards the end. "The main reasons for suicide – to assure safety to the children and from the standpoint of history it would go down better and might stir others to become socialists or more active... I don't relish the idea of participating in killing the children and I don't think anyone else does but I will do it because I think I could be as compassionate as the next person about it and I don't hate children." The deaths of the children are shocking. Many of them should never have been there, having effectively been kidnapped by Temple staff. Some forty-five were taken there by people who were not their parents, at least thirty-six had been taken illegally, some bundled onto Pan Am jets to South America with their hands tied.

It is not known with certainty who mixed the poison or who served it. One of the few people who survived by escaping into the surrounding jungle said Moore and Joyce Touchette, both nurses, had administered the poison. It was a mix of tranquilisers such as Valium, Phenergan, chloral hydrate and the poison potassium cyanide. This cocktail was mixed with a sweet powdered drink to reduce the bitterness of the cyanide. On the tape, a woman's voice, calm and authoritative with none of the notes of hysteria in Jones's speech, can be heard saying:

> You have to move and the people who are standing there
> in the aisles, go stand in the radio room yard. Everybody
> get behind the table and back this way. Okay. There's
> nothing to worry about. Everybody keep calm and try
> and keep your children calm. And all those children that
> need help, let the little children in and reassure them.
> They're not crying out of pain. It's just a little bitter
> tasting. They're not crying out of any pain.

This woman has never been identified but one thing is clear: it was a group of women, several of them medical professionals, who organised the killings, mixed the poison in a large metal barrel and injected it down the throats of children and the elderly. Many may have been forcibly injected as syringes with broken or bent needles were found scattered around the site. On the tape, the voice of a man, presumably surrounded by dead children, is heard saying: "I would rather see them laying like that than to see them die like the Jews did."

Pathologists carried out autopsies on only a handful of bodies. Most were in an advanced state of putrefaction when they were flown to Dover Air Force Base in Delaware, home to a special mortuary that the US military uses to handle battle deaths. Nobody knows how many were forced to take the poison and how many did it voluntarily, but the children and many of the elderly were in no position to make the decision to kill themselves.

In the immediate aftermath of the killings, Jonestown evoked such horror that there was an enormous effort by Americans to dissociate themselves from the events. Secretary of State Cyrus Vance asked the government of Guyana to bury the bodies in a mass grave at Jonestown. The Guyanese prime minister insisted all the corpses be removed from the country. It took months to find a burial place as numerous cities, including almost all those linked to the Temple, refused to take them. The victims, most of whom had not been identified and had decayed horribly in the jungle heat so that their features were no longer recognisable, lacked death certificates. On these grounds, several states, including Delaware, where the bodies were stored, refused them burial.

There was a public outcry at the thought of the potential contagion that the bodies represented. Some believed they represented an actual threat and would poison the earth in which they might be buried. Others were concerned that a mass grave would become a place of pilgrimage for cults. Each attempt to find a place for a grave provoked similar concern of contagion and poisoning, both real and spiritual. Burial at sea was put forward as a way to dispose of the bodies without contaminating the soil of the United States. It was all part of an effort to distance Jonestown from the mainstream of religion in America, home to many new faiths like Scientology and the Unification Church. Americans came to terms with Jonestown by insisting that the victims were not sane, not Christian and not American.

The United States at that moment was turning away from many of the changes of the 1960s and 1970s. Ronald Reagan was emerging as a force, conservative Christian organisations held ever greater sway in politics and a backlash to progressive views was building. Reagan, running for office at the time, was asked about Jones and responded by linking him and his followers to the Democrats he had cultivated in San Francisco. For the right, Jonestown was just another sign of America gone very wrong.

Over the years many writers have tried to make sense of Jonestown, explaining the deaths through psychological and sociological analysis of Jones and his followers. The role of the conflict between the growing anti-cult movement and this increasingly insular group has been weighed, along with Jones's drug use and the social changes in America in the 1970s. The lives of the people who joined the group have been mined to provide rationales for why they should have followed this man. It has been examined through the lenses of race and class. It has puzzled many why so many poor black people followed a white man and his middle-class acolytes into the jungle. Jim Jones and his charisma have always been put at the heart of it, but there is a critical aspect to the group that has been less understood.

The women around Jones were far from passive and adoring devotees. They were the driving force of the Temple and were formidably organised, intelligent and capable. They ran the finances, manned the radio room, dealt with health crises, soothed Jones's raging paranoia and kept Temple members in line. Some were involved in sexual relationships with Jones, who had fathered children with several of them, but they also showed immense independence and passion in their work. Far from being brainwashed sexual slaves, as they were sometimes portrayed, several of them felt they had gained more power and autonomy at the Temple than ever before in their lives. But none of the extensive studies of Jonestown can answer a critical question – perhaps because it cannot be answered with any satisfaction. How did these smart and caring women, some of them nurses, many of them mothers, reach a state in which they could calmly poison hundreds of children?

JONESTOWN BECAME ONE of the most widely known events in the United States, matching a level of public awareness that had only been reached at that time by the attack on Pearl Harbor and the dropping of atomic bombs on Japan. It entered the culture almost immediately, ensuring a deepening of suspicion around anything that might be labelled a cult. It also threw up an enduring phrase – "to drink the Kool-Aid" – words that have come to mean a blind belief in a cause, even though it identifies another brand of powdered drink from the Jonestown poisoning. The phrase has become curiously detached from its tragic origins, a line of mockery directed at anyone who buys into any delusion, no matter how minor.

Those who died were written off as lost souls who had fallen into the grip of a madman. There was little sense that many of them had found fulfilment in a community and no understanding of why they might have killed themselves. The poisonings inspired a wave of conspiracy theories, many of which persist to this day; the usual cast of suspects appeared, including the CIA and the KGB. The deaths were a source of immense fascination; dozens of books were written about Jonestown. Movies and documentaries followed. Part of the fascination was that this was a deliberate mass poisoning; itself an extremely rare event but something that has haunted people since ancient times.

This book is the story of that dread, a fear that has defined and developed us as humans, persisting to this day. These fears have been focused on certain groups and expressed in manifest ways in culture and society. It examines why we assess the risks of poisoning in a different manner to other threats to life and limb. Fears endure even as they are taken over by new anxieties. But there are patterns, even repetitions.

Involuntary exposure, unfamiliarity and invisibility are three attributes of risk that provoke significantly greater fears. Not knowing if exposure has occurred inevitably raises anxieties further. When the risk is hard to understand, the long-term effects are unclear and the number of people exposed is difficult to predict, anguish only grows. Poison, a threat that often combines all these issues, touches a nerve and always has. Car accidents may kill vastly more people than deliberate poisonings, but the illusion of control and the sense that we understand the risk mean that cars, however deadly, are not often a source of dread.

The anxieties surrounding poisonings are a way of creating order among inchoate threats to health, fertility, stability, family and life itself. Myths are filled with poison stories and it is in these stories we find the earliest fixations with poisoners. Later, the idea of poison was attached to an ever-growing list of potential threats from those who represent outsiders: young fertile women with a threatening sexual allure; menopausal women who reminded people of the end of fertility and mortality; and Jews who were for millennia seen as a threat to the very survival of Christianity in Europe. In ancient Greece, doctors had been seen as potential killers, as medicine was as likely to kill as to cure until quite recently. Many of the enduring associations of poison were established very early in European history. With shifts from Galenic medicine to the ideas of Paracelsus in the sixteenth century, there was a new surge in the fear of science and its misuses. The fear of poison rebounding on its user, especially in time of war, appears at the same time as the first poisoned weapons. Unsatisfied with the dangers of real poisons, people have been adept at inventing many new and mostly imaginary threats.

Our fears of poison have produced a catalogue of misogynistic or discriminatory views of women, Jews, the elderly and other outsiders from mainstream European society. But while these groups frequently reappear as victims of excessive fear, the associations with poison change over time. In the Greek world, poison was associated with barbarians, gods and myths, but was still seen as a natural force. Poison was both curse and cure; a necessary part of the world for its role in the shaping of identities in opposition to darker forces. For Romans, poison was distinctly feminine and foreign; the antidote was a masculine and nationalistic culture that reinforced an identity constantly under threat as the empire grew and brought in more diverse peoples. In medieval Europe, poison was linked to Christian ideas of sin; this was the pervasive invisible force that threatened humankind. The antidotes were faith and redemption.

The Renaissance took poisons away from the supernatural and back to nature, but retained many of the associations of the medieval world, mostly relating to the supposedly toxic natures of women and Jews. In England, then a country racked with religious divisions and uncertain about its nationhood, poisons were inextricably associated with Italy and Catholicism. The antidotes in this case were Protestantism and Englishness, an idea that was mostly defined at the time as not being Italian. This was also an era of medical revolution as Galenic ideas of balanced humours within the individual were gradually replaced with ideas of diseases as specific entities that existed outside any one person and were to be cured with specific drugs. Harsh new chemical drugs produced new anxieties about doctors as poisoners.

In the late seventeenth century in Naples, then one of the largest, busiest cities in Europe, a woman was selling poisons to enable wealthy women to be rid of their tiresome husbands. Her name was Giulia Tofana and she would endure as a figure of fear for centuries. Her concoction, known as Acqua Tofana, or Acqua di Napoli, was most likely a solution of arsenic mixed perhaps with belladonna; it would have killed slowly and painfully. Its symptoms were close to those of cholera, dysentery and other diseases that ravaged the city.

In a different time and place, Tofana would have been called a witch. She was an elderly, unmarried woman selling potions and cosmetics; across Europe in the previous two centuries, tens of thousands of women like her had been tried and executed as witches or lynched by enraged mobs. Tofana, however, was accused of poisoning, not witchcraft. She was accused of a modern, almost scientific, crime and not the crime of magic but the fears she aroused were similar to those stirred by witchcraft. Judges and officials may have been trying to move towards supposedly more rational and secular forms of justice, but they would still torture an old woman into confessing. Tofana was said to have sold poisons that killed at least six hundred men. The frenzy of witch hunts might have been coming to an end, but old women were still a focus of fear and anxiety.

By the industrial age of the nineteenth century, poisons were still associated with women but they reflected new forces that were transforming family and social life. Poisoning was a crime that disrupted the carefully crafted image of Victorian family. The expansion of scientific evidence, the rise of a mass media and the poisonous environment of early industrialisation created a new era in the grip of toxic anxiety. Later, these fears would become dispersed and unfocused; the invisibility of poison was taken to an extreme in radiation, the ultimate poison of the twentieth century.

Poisons have been used as weapons of war for at least three millennia. Poisoned arrows and tainted wells were clouded in a moral ambivalence; no civilisation has been comfortable killing in this way, although that anxiety has not stopped some. The oldest rules of war relate to poisons. They were a dishonourable weapon only to be used on those regarded as somehow less than human. By the end of the twentieth century, the era of most-poisonous warfare, the use of chemical weapons was banned internationally.

This is also a history of the attempts to sow discord, entertain or divert by exploiting the fears that poisons arouse. These start close to home with the most ancient myths and stories examining and inculcating the fears men have of women, of the traps of domesticity, of outsiders and of the unknown and misunderstood. We still live in a world where we fear unseen forces – religious and political extremists – who view poisons as weapons to instil terror. The killing of a Russian man with a handgun on a London street would have aroused some concern; the murder of a man in a hotel with polonium-210, a rare radioactive poison, became a global sensation. Vladimir Putin has made the fear of poison ubiquitous: if someone can die in the quiet town of Salisbury from a rare nerve agent, the same threat could strike anyone anywhere.

Our fears of poison, due to its invisibility and its inescapable nature, provide us with a potent source of thrills; the poisoner has been a staple of drama since the first plays were written. Poison appears in many of Shakespeare's works and is central to the plots of two of the most popular, *Hamlet* and *Romeo and Juliet*. Restoration drama was obsessed with poisonings. Later they would pervade the works of Sir Arthur Conan Doyle and Agatha Christie, who turned her knowledge as a pharmacist to good use by using poisonings or overdoses in sixty-six of her novels. Now it is the rogue scientist, usually Russian, that dominates toxic fiction.

Even as our understanding of poisons and their mechanisms has grown, the fears remain unchallenged; the ability of most people to assess the risks of the world around them is little better now than in the Middle Ages. Our fear rests deep within us, perhaps entwined in some way in our very nature as omnivores required to be both adventurous in what we consume and sensitive to the possibility of harm. But poison exists more in the realm of culture than reality, more often encountered as a way to arouse anxiety than a means to kill. Almost every civilisation has viewed poison as a threat to its very existence and prescribed antidotes to that exaggerated threat. Poison is a weapon of fear, far more important for its impact on our imaginations than our bodies. It is a weapon of political intrigue, elitist and theatrical. It is a weapon of terror, invisible, unknown and little understood.

PART ONE

1. THE SOUL-HUNTING FOG

*And so, tragic discourse gave birth to one invention
after another and added daily increments to the
horrors of war.*
> Lucretius, *On the Nature of the Universe*

When Hercules killed the Hydra of Lerna, he dipped his arrows into the monster's venom and created the first poisoned weapons. The Roman natural historian Aelian would later write that Hercules got the idea from seeing wasps buzzing around the corpse of a viper, absorbing its poison to make their stings more potent. Hercules harnessed the poison of the Hydra and created weapons that would bring tragedy to him and his family even as they turned the tides of great wars.

Hercules first used the arrows against a group of centaurs. On his way to his Fourth Labour, Hercules visited Pholus, a friendly centaur. Pholus took Hercules into his cave and generously offered him roasted meats and wine. A nearby group of more feral centaurs, attracted by the scent of the wine, crashed into the cave. Hercules killed most of them with his poisoned arrows and chased the others. Chiron, another artistic and cultured centaur who had taught mankind about medicine, was wounded in the knee with an arrow. Racked with terrible pain, he decided to renounce his claim to immortality and die. Pholus also died when he pulled an arrow from another centaur and wounded himself with it. Already, toxic weapons were having unintended consequences.

Hercules then committed what proved to be a fatal violation of the rules of war. Another centaur, Nessus, attempted to abduct Hercules's wife, Deianeira. To stop Nessus, Hercules shot him in the back with a poisoned arrow. As he died, Nessus told Deianeira to take his cloak and store it in a chest. If Hercules strayed from his marriage vows, she could give him the garment and it would bring him back to her and ensure his undying love.

Years later, Hercules took a younger wife. Deianeira, desperate to win him back, sent him the cloak sealed in a casket with instructions that only he should wear it. He put it on to perform a sacrifice and immediately began sweating. His skin started to burn but he was unable to remove the robe. He plunged into a stream but that did nothing to alleviate the agony. Thermopylae, a hot spring near Mount Oeta, is said to be the site of his death from the poisoned cloak and the waters are still believed to cure skin diseases. Unable to bear the searing pain, Hercules called on his followers to burn him alive on a funeral pyre. Only Philoctetes, a renowned archer, was willing to follow the command. He became the heir to the poisoned arrows.

As he sailed to fight at Troy, Philoctetes stuck himself with one of the poisoned arrows, leaving a wound that festered so severely that his shipmates could not endure the stench or his constant cries of pain. He was abandoned on the island of Chryse, near Lemnos, while the others carried on to fight. As the conflict dragged on, they realised that they needed the arrows if they were to win and returned to the island to get them. The arrows would change the outcome of the war, even killing Paris, who, of course, had started the whole terrible conflict by kidnapping Helen. By then, the toxic arrows had set in motion a chain of tragedies that would harm all involved.

NEARLY EVERY WAR produces an advance in weapons and a nostalgia for earlier, simpler times. The Peloponnesian War of 431–404 BCE between Athens and Sparta was regarded as among the most brutal up to then. Almost immediately people wrote of an era when conflict was somehow less terrible. That same feeling was evident throughout the twentieth century when the brutality of modern warfare led to hankering for when armies fought in the open, when they adopted chivalrous techniques and did not resort to perilous and uncontrollable weapons. But the reality of warfare is that it has always been brutal and underhand. Biological and chemical warfare has been seen as an anomaly in the ancient world, a rare violation of a powerful poison taboo, but the historian Adrienne Mayor has found more than fifty references to the use of such weapons in ancient texts. Their prevalence in mythology cannot be considered a direct reflection of the extent of their use, but it does illustrate the degree of preoccupation with poisons and contagion going back thousands of years.

Records from Sumer, in present-day Iraq, show that there was a basic understanding of both poisons and contagion as early as 1700 BCE. It took little time to recognise that this might be useful in times of war. Hittites, who lived in what is now the far east of Turkey, sent plague victims into enemy lands to spread disease. From this, ancient people developed an array of chemical and biological weapons. From the eighth century BCE, when Homer detailed the use of poisons and potions in *The Odyssey*, to the second century CE, when Greek historians wrote of chemical and biological warfare, these weapons were seen in a variety of ways, sometimes as legitimate means of war, often as deeply reprehensible.

Poisoned arrows were likely the first toxic weapons – the word "toxic" comes from *toxon* or "bow" – and they evoked horror from the start. Odysseus was initially rebuffed in his search for poison for his arrows in Ephyra. Ilus refused him the poison, fearing that it would upset the "ever living gods". This has often been taken as a sign of a Greek taboo on the use of poison in warfare, but there is nothing to suggest that it was a major constraint. Greek and Roman writers from Galen to Dio Cassius often refer to poison arrows but almost always accuse nameless "barbarians" of using them. Socrates wrote that such "vile tricks and treachery" were shameful to a true warrior. Arrows were possibly dipped in different toxins including rhododendron sap, monkshood, sea urchin toxin and snake venom.

Writers from the time tend to view such arrows as deadly and terrifying but doctors discuss the treatment of wounds in a way that suggests they were not always fatal. None of the arrows would have had the power of those made with powerful neurotoxins derived from tropical vines.

The most feared archers were the Scythians, a nomadic people whose population stretched from the Black Sea to Mongolia from the fifth century BCE to around the first century CE. The Scythians claimed descent from Hercules, who had given them the poisoned arrows with which they would defeat the armies of the Persian king Darius I in the fifth century BCE and Alexander the Great in 331 BCE. For the Greeks, Scythians were colourful, dangerous barbarians, admired but preferably from a distance. They were fearsome horsemen who drank undiluted wine, smoked large quantities of marijuana and consumed a psychedelic drug made from the mushroom fly agaric, the iconic red-topped toadstool. Skilled archers, even on horseback, they moved with the ease of centaurs and were prized as mercenaries. Pliny the Elder described them firing arrows to shoot chunks of turquoise off the sides of cliffs. He may not have been exaggerating much. Archaeologists have found skulls with Scythian arrowheads embedded right between the eyes. Herodotus, writing in the fifth century BCE, describes the people with a mixture of horror and awe. "For eight-and-twenty years then, the Scythians were rulers of Asia, and by their unruliness and reckless behaviour everything was ruined."

The Scythians poisoned their arrows with a concoction based on viper venom, which retains its toxicity even when dried. Scythicon, as it was known, was described by several writers including Aristotle, Theophrastus and Aelian, and from their descriptions, we can guess at the recipe. Viper flesh was left to rot and then mixed with human blood. Animal dung or human faeces was added in a leather bag, which was buried and the contents allowed to decay further. Strabo, the first-century Greek geographer, commented that "even people who are not wounded by the poison projectiles, suffer from their terrible odour". The viper venom would have possibly retained some of its toxicity but the real danger came from the fact that even the slightest scratch might lead to a painful or even deadly infection.

On top of this, the arrows were barbed, meaning they could only be cut out of flesh in an excruciatingly painful manner. Arrow shafts preserved in the Siberian permafrost show that the Scythians also painted them with zigzags and diamond patterns so they resembled snakes. It was hardly the clean and quick death that was regarded as noble in a warrior culture. People would have suffered terribly. The treatment of wounds by cauterising them and then covering them in tar would have been screamingly painful. The Roman poet Ovid later wrote of the "double death" delivered by arrows that were hooked and dripping with poison.

Mass poisonings were also part of ancient warfare. The poisoning of wells and water supplies was first recorded in 590 BCE at the battle of Kirrha, a fortified city that controlled access to the oracle at Delphi. Those laying siege to the city diverted the river but the inhabitants survived on rain and well water. Realising they could not deprive them of water, the besiegers decided to poison it instead. They gathered hellebore, a commonly used plant also known as skunk cabbage, and added it to the dammed river before allowing the waters to flow again. The thirsty people of Kirrha gorged on the waters and suffered debilitating diarrhoea. The defenders left their posts and the city was overrun. Even in the early accounts there was a certain ambivalence about the use of this sort of warfare. The town was said to have been near where Hercules killed the centaur Nessus, whose body poisoned the waters in the surrounding area.

The trickery of poisons always evokes concern; their users are regarded as both admirably cunning and reprehensible. Along with the earliest use of these weapons came justifications that have been used ever since. Polyaenus, the author of *Stratagems of War*, a volume written for the Roman emperor Marcus Aurelius in the second century CE, twisted the story of Hercules and the drunken centaurs, saying that he had deliberately lured them with poisoned wine to be rid of them because of their chaotic behaviour. Polyaenus is among the first writers to justify the use of poisons by saying they might be used against inferior cultures. Barbarians of all stripes were inordinately fond of alcohol, Polyaenus advised, and as such could be poisoned with wine mixed with mandrake or aconite. The idea that using poison against one's own people was unforgivable, but the use could be justified against barbarians, however that was defined, would take a terrible hold in the Western mind.

A MURDERED FATHER, a dead friend, a suspected stepmother, rival half-siblings, two scorned women desperate to revive a dying love, two motives for murder, the threat of torture, a deathbed promise and a bottle of poison. The speech "Against a Stepmother", written by Antiphon, the Athenian orator, sometime between 450 and 411 BCE, contains in its few pages the full drama of poisoning and its myriad associations that would last for centuries. This short speech, an exercise written for a mock trial, captures what poison meant in these times.

The dead man, who is unnamed in the son's speech but is known to have married twice and had children with both women, was friends with a man called Philoneos. Philoneos had a concubine but he had tired of the woman and planned to send her off to a brothel, a terrible fate for an Athenian woman. The dead man's wife – the stepmother of the orator – learned about this and befriended the concubine. Warning the woman of her fate, the stepmother gave her what she said was a love philtre, a potion that would reawaken the feelings Philoneos had once felt for her when added to his wine.

The orator, who was prosecuting the case before the court, described what happened next:

> The two men poured out their libation; and then, taking in hand that which was their own destroyer, they drained their last draft. Philoneos dropped dead instantly. My father was seized with an illness from which he died in three weeks. For this the woman, who had acted under orders, has paid the penalty for her offence, in which she was an innocent accomplice: she was handed over to the public executioner after being broken on the

wheel. But the woman who was the real cause, who thought out and engineered the deed – she will pay the penalty now, if you and Heaven so decree.

The son's speech is sarcastic and vengeful; it is filled with an undiluted hatred of his stepmother, although he offers little evidence of her complicity in the crime. Her defence was that she was only offering Philoneos's concubine a love philtre and she did not know it would prove toxic. The son calls her "this Clytemnestra" but she might have described herself as more like Deianeira, the unsuspecting poisoner of Hercules.

In just a few pages, Antiphon illustrates most of the domestic anxieties about poison in classical Greece. Women and their use of love philtres were a source of fear. Scorned women were dangerous while stepmothers have always inhabited a special realm of suspicion. Men are unsuspecting innocents, vulnerable to the hidden threat of poison at any time. The secrecy arouses the greatest fear and is used by the son to appeal to the court for a guilty verdict and a harsh punishment. "People who are plotting the murder of their neighbours do not prepare their plans and make their preparations in the presence of witnesses. They do so with the greatest secrecy so that no other human soul will know. But the victims of their machinations know nothing until they are caught in the grip of menace."

The Greek word for "poison" was *pharmaka*, an ambiguous word that could mean both poison and remedy. It captures an enduring fact – that medicines can be poisons and poisons can be medicines. From Plato to Jacques Derrida, the ambiguity at the heart of this word has been a fascination. Plato used it to describe writing itself – a force that could be both harmful and curative. Medicine was a much more significant part of education than it is today, now that knowledge is isolated among a small professional group. Many Greeks had a significant understanding of what was then known about the body and medicine. Medical matters were a common reference in philosophy and ethics. In these discussions, the word *pharmaka* captured a world of drugs, everyday products used for treatments such as wine, olive oil or vinegar, as well as poisons and magical potions. The Chinese character for poison, *du*, captures the same idea of potency, for good or ill. Earlier versions varied slightly – one suggesting positive growth, one prohibiting growth. No word exists in English that covers both meanings.

POISONS, ACCORDING TO ancient Hindu beliefs, were an early effort at pest control that went awry. The Lord Brahma, creator of the universe in Hindu cosmology, was resting on a lotus leaf when he was pestered by a devil called Kaitabha, who was formed from the ear wax of the Lord Vishnu. Wishing to be rid of Kaitabha so he could continue his work, Brahma created poisons. The effect was rather more dramatic than he intended. Feeling abashed by his excessive reaction to an annoying sprite, Brahma tried to mitigate the effects of his invention by spreading poison throughout the animal, vegetable and mineral worlds. He also came up with the idea of antidotes, thoughtfully providing one for each type of poison. Other variations have the devil Kaitabha paired with another called Madhu. Both were killed by Vishnu after they annoyed him. Cast into the sea, the pair ended up forming soil but the toxic aspect of their irritating personalities meant that soil could not be eaten. Yet another origin myth has Vishnu churning an ocean of milk to create a nectar of immortality. In the first stage of making this nectar, it was poisonous but he tried it anyway. The drops that fell from his lips were licked up by scorpions and snakes, providing them with their venomous power.

The first recorded Indian expert on poisons was Kashyapa, a revered Vedic sage and doctor who lived in the sixth century BCE around the time of the birth of the Buddha. Kashyapa had vowed to protect a king who had been told he would die of a snake bite. Takshaka, the king of the serpents, was heading to kill the king when Kashyapa intercepted him. To prove how toxic he was, the snake bit a blossoming tree and turned it to ashes but Kashyapa restored it to life with his antidotes. Later the snake persuaded Kashyapa against using his medical skills to help the king and the curse did come to pass. Ancient Indian toxicology may be grounded in legends, but it had a profound influence on the development of the science. Indian knowledge reached Greece and informed the development of classical natural sciences. Indian works also influenced Arab scientists, providing much of the base for Islamic toxicology.

The most important early Indian work is *the Sushruta Samita* (or *Compendium of Sushruta*), an encyclopaedic collection of medical knowledge that probably dates from around the sixth century BCE. It shaped not only thousands of years of Indian views of poisons but also, via a translation into Arabic in the eighth century, Islamic medicine, and from there much of European scientific thought in the Renaissance. It is the first work to describe plastic surgery, caesarean sections, the need for sterilisation of instruments during operations and the use of intoxicants for anaesthesia. Sushruta included eight chapters on toxins, ranging from how to prevent foods from being poisoned to curing poisonings with the sound of drums. His recommendations for the uses of poison clashed with the *Manusmriti*, also known as the *Laws of Manu*, a work that laid out rules of war. "When he fights with his foes in battle, let him not strike with weapons concealed (in wood), nor with (such as are) barbed, poisoned, or the points of which are blazing with fire."

One of the earliest surviving texts of Greek toxicology is "Theriaca" by Nicander. This poem of nearly a thousand lines examines the nature of toxic animals. A shorter poem, "Alexipharmica", is a study of remedies and antidotes. The poems, both somewhat awkward and impenetrable to the modern ear, were most probably drawn from the writer Numenius, whose works on such subjects as diet, banqueting and fishing were lost but widely quoted by others. The survival of information from this time is something of a matter of chance, but Nicander's works became highly influential. After their rediscovery in the fifteenth century, the poems once again became an important source for scientists.

Nicander was a hereditary priest at the Shrine of Clarus, although he was originally from Colophon, "the snow-white town". Nicander has not been highly regarded as a poet or as a toxicologist: the poem was later dismissed as a "long, disordered and untrustworthy medley". But his work shaped the way people understood snakes and their effect on man for nearly two thousand years. "Alexipharmica" does give some sense of what was known about poisons at this time. Nicander mentions aconite, lead, hemlock and opium as poisons, mostly recommending the drinking of large quantities of olive oil as an emetic, a remedy that is still used.

The Greek pantheon abounded with healing deities. Demeter looked after women and children, Persephone cured rotten teeth and bad eyes, Genetyllis dealt with infertility, Leto was the surgeon of the gods, Hecate specialised in pediatrics and Athena could cure blindness. Asclepius was a demi-god who had been taught medicine by Chiron, the centaur killed by Hercules. Asclepius' daughters – Panacea, Meditrine and Hygeia – still figure in our vocabulary to this day. Images of Hygeia often have her feeding snakes or holding a dish of food for a coiled serpent at her side. Before the seventh century BCE, the daughters of Asclepius were shown administering cures on their own. After that time, they are reduced to the role of nurses, assisting their father as he treats his patients. Hygeia, the female advocate of cleanliness and preventative medicine, was shoved aside by the male doctor with his expensive cures in a sign of things to come. More than three hundred temples have been found dedicated to this family. Many have rooms in which patients slept so that gods could come to them and offer cures. Snakes were allowed to slither around the dormitories at night while the ill slept, expecting to wake free of illness.

Homer saw poisoning as the archetypal act of betrayal. In *The Odyssey*, women are closely associated with both medicine and poison. Helen of Troy is known now for her beauty and for launching a thousand ships but she was also a respected healer who had studied with Polydamna, a queen of Egypt and herself a celebrated physician. Helen helps Telemachus overcome his misery over his missing father, Odysseus, by "casting a drug into the wine whereof they drank, a drug to lull all pain and anger and bring forgetfulness of all sorrow". There are various suggestions as to what Helen prescribed including verbena, a root called *oinopia*, or opium. *The Odyssey* refers to a complex pharmacopoeia including such powerfully toxic drugs as mandrake, opium and hellebore.

From the earliest Linear B tablets of Mycenaean Crete in the sixteenth century BCE, there is evidence of an extensive knowledge of *pharmaka*. During the Homeric Age, from the eighth century BCE, knowledge of drugs flourished. Mentions of drugs and salves are found in Lyric poetry and in the works of Aristophanes and others from the fifth century BCE. There was a more systematic study of plants for their medicinal and poisonous actions. Theophrastus's *Historia Plantarum* and *De Causis Plantarum*, which date from around 300 BCE, are the earliest surviving works on botany. Theophrastus took much of his information from *rhizotomoi*, rural people who gathered plants for medicinal use. His work was a classification of plants rather than a discussion of their uses. There was no effort to develop a theory of toxicology and no explanation for the beneficial or harmful impact of plants.

Hippocrates, born in 460 BCE, and Aristotle, born around 384 BCE, moved medicine and knowledge about poisons away from the realm of mythology. Aristotle's wife, Pythias, worked with him on studying plants and animals and although he described her as his assistant, her contribution to his work may have been much greater than acknowledged. Women were slowly moved away from the centre of medical practice as their status declined in classical Greece. Suspicions were commonplace about their use of potions, notably the love philtres used to revive marriages or induce lust in a man. Antiphon's speech mentions that the stepmother was known to have used poisons before, although he offers no evidence for this. Any woman who could cure could also kill. The *pharmakon* was always double edged.

Most of the poisons that were known in the ancient world were the same set of toxins that would dominate medicine until the nineteenth century. Henbane and belladonna were prescribed as intoxicants to ease pain. Mandrake was used as a soporific but was known to be dangerously toxic. The most common poison used was hemlock, which is mentioned by both Lucretius and Pliny the Elder. Aconite was probably the swiftest of poisons available. Mixed with wine, it was said to be an antidote to the sting of a scorpion. Hellebore, another common poison, was a purgative and was taken for epilepsy. Yew trees were known to be poisonous; one could die simply sleeping under one or drinking from a vessel made from its wood, so it was believed. Salamanders were the deadliest of all animals; they could wipe out entire tribes, according to Pliny. If they climbed a tree, all the fruit in that tree would become toxic. It could have the same devastating effect if one fell into a well or if its foot touched the wood on which bread was baked. (In fact, most salamanders are not dangerous although one species, the European fire salamander, secretes a poisonous mucus from its skin.)

The toad was also said to be toxic. (Only a South American toad, *Ceratophrys ornata*, has a poisonous bite. Some frogs from that region excrete toxins through their skins.) Looking poisonous was enough to get an animal classified as poisonous, a trick adopted by a range of animals that mimic actually venomous species to avoid being eaten. Pliny added the blood of snakes and of gelded goats to the list of poisonous substances along with the flesh of the weasel and shrew mouse. Mineral poisons were less well understood. Lead was known to be a poison, as were arsenic compounds that were also used as medicines. The waters of several springs in Arcadia, Thessaly and Macedonia were believed to be fatal. One of the many stories of the death of Alexander the Great involves him being poisoned with water from the River Styx.

Dioscorides, writing in the first century CE, was the first Greek to attempt to draw some order out of the mass of knowledge about *pharmaka*. He classified drugs by the effect they had – stopping bleeding, for example. He also developed the idea of "drug affinity", which endures in homoeopathic medicine. This concept – essentially that certain plants that in some way resembled a disease or a body part could cure illness – required a detailed knowledge of the entire life cycle of the plant. For example, opium was said to cure certain illnesses relating to the head because the seed pod looks like a human head. Medicine would for a long time follow these ideas of transference and sympathetic healing. Sheep would be brought into a room with someone with a fever in the hope that the fever might be transferred to the animal. Red foods and herbs would combat a disease that produced red marks on the body such as smallpox. Saffron was a common cure for jaundice. Dioscorides's *Materia medica* contained some six hundred species of plants and includes a variety of drugs introduced into Greek and Roman medicine from Asia. It was an important medical work for centuries, in part because the classification of plants was an enormous achievement at a time when there were no botanical gardens or plant collections in museums.

The physician Galen, born in AD 130, emerged as the most enduring medical writer of the ancient world. Among his works were volumes on poisons and theriacs, universal antidotes that were said to have been discovered by Mithradates, the king of Pontus. Galen believed poisons represented a grave threat to mankind because people could run away from other dangers but poisons were stealthy. He lived a dramatic and peripatetic life, at one stage working as a doctor to gladiators in his hometown of Pergamon before fleeing an outbreak of plague there. He was an extraordinarily prolific author, although many of his works were lost when a fire burned the Temple of Peace on the Palatine Hill in Rome in AD 192. He ended up as a prominent court and military doctor in Rome, treating several outbreaks of plague that ravaged the army then fighting the Germanic tribes.

Galen had little regard for his own profession, describing doctors as ignorant and thieving, accounts echoed by others such as Pliny and Juvenal. This probably was as much to do with Roman hauteur as doctors were often slaves or Greeks. Many of the drugs were imported from far away, although Galen said most doctors were ignorant of those herbs available just outside Rome. The emperors had large stores of drugs for their own use and these were replenished as Galen travelled around the region.

His enduring legacy was to structure his medical thinking around the idea that all living objects were composed of the four elements: earth, air, water and fire. He had taken this idea from Hippocrates and Aristotle, but through his works it was to become the dominant Western way of thinking about the body for 1,500 more years. Health depended on the proper balance of the humours that corresponded to the elements – black bile, yellow bile, blood and phlegm. Diseases were not seen as something that existed outside the individual as we mostly-see them today, but were specific to a person and their unique imbalance of humours. These were in a constant state of flux and a perfect balance was almost impossible to achieve. Therapies tended to focus on correcting an imbalance by draining off an excess of a humour through purges, bleeding or vomiting. Moderation in all things was the key to maintaining balance. There were six areas in which people could control the humours themselves: breathing, nutrition, exercise, sleep, emotional balance and excretion.

Humours could be set off balance by outside forces. Invasion of the body or pollution were also possible causes of illness. Leprosy, for example, could be caused by moral failings. Such things as bad air, the alignment of the planets and even sin could influence the balance of the humours. Galen rejected Dioscorides's idea on drug affinity, believing instead that drugs influenced the balance within the body. Poisons were also thought to act in this way. A poison could enter the body in any number of ways and then disrupt the humoral system by being too cold or too hot. Those poisons had a narcotic effect; for example opium cooled the body too swiftly, while other substances heated it dangerously.

Alongside the Galenic system was a belief in magic and potions. The poison administered according to Antiphon's speech was not given to kill but was a love philtre, a drug to bring wandering men back to women. These philtres make many appearances in Greek literature and mythology and in almost all cases, it is women who administer them, often with tragic results. Love philtres were a way for women to trap men; they represented the fearful side of domesticity. This appears in *The Odyssey* in the figure of Circe. Her skill with potions and poisons was both appealing to men and terrifying, much like the confining aspects of marriage and sexuality. Odysseus and his men land on the island of Aiaia to stock up on food and water. He sends his men off to re-supply the ship but they meet Circe, who lures them to her stone house in the forest with the promise of music and food. The house was filled with her pet lions and wolves, themselves victims of Circe's magic potions that render them tame. The sailors, famished from their voyage, eat Circe's meal and turn into pigs.

Poison often comes mixed with something sweet, an archetypal male view of the desires and dangers that women represent. Circe serves Odysseus and his men food in a manner that elevates their masculinity. But by adding her "own vile pinch" to the wine and honey, she turns them into pigs and keeps them in a sty. Poison is not an honourable weapon. There is no open competition as there might have been in duelling or war, but instead only deceitfulness. Poison equalises by matching secret knowledge against masculine strength.

One sailor evades Circe's clutches and returns to Odysseus, who is furious that his men have been treated in this way. He sets off and with help from the god Hermes, fortifies himself with the antidote "moly", which prevents Circe's potions from working. At the point of a sword, she relents and reverses the spell on the sailors. She serves them a banquet and provides Odysseus with advice on avoiding the treacherous sirens and other problems that lie ahead on his journey. The serving of food is a key ritual in male and female interaction. The story of Odysseus and Circe has turns in which one and then the other is dominant. Odysseus arrives at Circe's home weak and hungry but is wary of entering. He kills a stag and feeds himself, becoming stronger and therefore more able to deal with the risk of domestic servitude under Circe's thumb.

In the ebb and flow of female and masculine power, poison is in the hands of women. The idea of a woman feeding a man something toxic still arouses intense fears. In the story of Adam and Eve, a woman gives a man a dangerous food, beguiling him with what it might offer. Even the snakes in many paintings of the scene have the faces of women, although there is nothing in the book of Genesis to suggest that the reptile was female. A central part of poisoning stories is often the idea of something good and positive made dangerous. The poisoning of food, sex and clothing are common themes in mythology, being reworked frequently up to the present day.

THE POISONS IMAGINED by Euripedes in his telling of the story of Medea, first performed in 431 BCE, are almost as horrifying as the play's famous denouement. Medea is described in the play as "Asiatic". She was from Colchis on what is now the Black Sea coast of Georgia and therefore not Greek. She concocts a poison that appears like a blend of the most potent modern nerve gas and napalm, using it to kill Glauke, the new teenage bride of Jason, her unfaithful husband and father of her children. Hidden in an embroidered gown and a gold coronet, the poisons leach into Glauke's body, as reported to Medea by the messenger who had delivered the poisoned gifts: "Then suddenly we saw a frightening thing. She changed/ Colour and staggered sideways, shaking in every limb." White froth spills from her lips, her eyeballs roll back and blood drains from her face. She recovers for a moment only to find that her coronet is now a band of fire that she cannot remove. "Save to her father she was unrecognisable/ Her eyes, her face, were one grotesque disfigurement/ Down from her head dripped blood mingled with flame; her flesh/ Attacked by the invisible fangs of poison, melted/ From the bare bone, like gum drops from a pine-tree's bark/ A ghastly sight." King Creon of Corinth reaches down to grasp his daughter and succumbs to the poison in her gown, sticking to his daughter so that when he tries to pull away in agony, his flesh peels from his bones.

Medea's story has been told many times, although the best-known version is by Euripides, who casts her as an outsider, poisoner, child murderer and the emblematic evil woman. She may have been wronged by Jason but her response is so extreme and brutal as to diminish the possibility of sympathy. In other versions, she is from a Greek royal family and was born in Thessaly, a place notorious for poisoners and healers. She is also often described as Circe's niece, establishing a link with another woman skilled in magic and potions. Jason is the main beneficiary of her knowledge, being able to withstand the tests set for him during his retrieval of the Golden Fleece only because of her abilities and advice. It is Medea who gives him an ointment to protect him from the fire-breathing oxen that he must use to plough a field, tells him how to defeat the army of warriors that sprout there and provides him with a potion to kill the sleepless dragon that guards the fleece. Despite all this help, Jason is ungrateful and callow, leaving Medea for King Creon's daughter to gain the royal power that has eluded him up until then. Until the Euripides version, she was not often blamed for killing her children; in one telling, after she poisons Creon, the people of Corinth kill her children in revenge.

In the 1930s, stories circulated in the United States about a woman who bought a new dress to attend a dance. At the dance, she felt faint, collapsed and died. A postmortem examination found that she had formaldehyde in her blood. On investigation, the gown was found to have been used before to dress a young woman who had died and been embalmed. The family decided it was too expensive to be buried in and returned it to the store, whereupon it was sold to the unsuspecting victim, or in some cases rented to her for the evening. The urban myth cropped up repeatedly over the next few decades, often taking on racist overtones as it was said that a black woman had worn the dress at her funeral. The centaur's cloak and Glauke's dress returned in modern form in the 1980s in a panic surrounding supposedly deadly temporary tattoos that were said to be impregnated with poisons or drugs. There was no real risk from these as even the glue had been FDA-approved. Most recently, a poisoned wedding dress made an appearance in an episode of the forensic procedural *CSI: NY*. The associations of women, jealousy and poisons are ancient and modern.

2. AN EVIL NOTORIETY

Come il basilisco soavamente fischinando nella sua caverna...
                      Niccolo Machiavelli, *The Prince*

Basilisks became more dangerous as each new description appeared in the works of Greek and Roman naturalists. The animals ruled alone over the empty desert as other snakes would flee at their sound, wrote Nicander of Colophon, the second-century BCE Greek doctor, in his poem "Theriaca", a compendium of antidotes and cures. Two centuries later Galen believed they could kill not just with their glance but also with their hissing. By the time Pliny the Elder was writing in the first century CE, they were able to kill plants with their breath and the vapours that emanated from their bodies. They could even burst open rocks, according to the Roman natural philosopher. These powers were put to good use by the people of Pergamon in Asia Minor who paid a fortune for the skin of a basilisk to be hung in the Temple of Apollo to drive out swallows and spiders that might defile the space.

Pliny was one of the few writers to give a detailed description of the basilisk. Citing some sixty Greek sources, most of them lost to us now, he described it as being only a foot long and adorned with a bright white marking. It moved like a caterpillar, raised in the middle. "It is believed that once, one was killed with a spear by a man on horseback and the infection rising through the spear killed not only the rider but the horse."

In 331 BCE, the men of Rome, then a burgeoning republic and the pre-eminent city of the Italian peninsula, were dying in alarming numbers, such that the moment would gain what the historian Livy called an "evil notoriety". Important men succumbed to an illness ravaging the city until a servant went to the magistrate and told him that women were poisoning their husbands. The servant said she could help the magistrates catch the women in the act of preparing poisons. Some twenty women, many of them from aristocratic families, were arrested, caught in the act of what they insisted was preparing medicines. To test whether they were medicines, they were forced to drink them and several died. In the end, 170 women were found to have joined in what is the first recorded mass criminal poisoning, an act of what Romans called veneficium, akin to witchcraft. Valerius Maximus, a collector of historical anecdotes in the first century CE, saw the murders as crimes of the weak and scurrilous. Women were not strong enough to confront men directly so they resorted to insidious crimes.

Livy, who was writing around the time of Christ in the first century CE, was sceptical of the whole account, suggesting that "the unhealthy weather" might have been to blame for the deaths. "I would gladly believe – and the authorities are not unanimous on the point – that it is a false story which states that those whose deaths made the year notorious for pestilence were really carried off by poison."

These killings were the first in a series of alleged mass poisonings that haunted Rome. In 186 BCE, a group involved in the licentious worship of Bacchus were accused of poisonings. Some two thousand people were executed, according to an account by Livy written three centuries after the events. But if one digs a little deeper than Livy's account of these orgies of sex and murder, the motives for the prosecution are revealed. The Roman writer describes the cult just after he writes of the immoral practices of soldiers returning from Anatolia in 187 BCE and so the assumption has always been that they were linked. But it seems more likely that this particular cult was introduced to Rome along with the thirty thousand prisoners taken at the Battle of Tarentum in 208 BCE. This town on the heel of Italy now known as Taranto was the home of a Bacchanalian cult that Pliny the Elder describes as involving the whole population in acts of drunkenness and everything that goes along with that. Tens of thousands of people from this area were brought back to Rome as slaves, a great many others came as refugees. Doubtless, they would have brought their religious rites with them, stirring up massive anxieties among the conservative residents of Rome.

Livy does not tell us where he thought the cult was from, only that it had been active for some time. It was based in a poor area of the city near the docks behind the Aventine Hill and was led by a woman from Campania. Nights of wine-fuelled rioting took over the area and the cult leaders threatened anyone who betrayed their secrets with poisoning. His sources for this are not clear. What is suspected, however, is that the people involved were poor, foreign, often enslaved and generally little understood by Romans. They were ripe for accusations of licentiousness and poisoning, just as later Romans interpreted the Eucharist in Christian rites as part of a plot to administer poisons. The aggressive response probably had less to do with real concerns about poisoning and more to do with a rise in nationalism at the time when the government was eager to limit foreign influences. Most of those accused and executed were slaves, Greeks and other foreigners who had no right to trials in front of Roman courts.

In 182 BCE, another series of killings took place, this time at the highest levels of Roman society, going against the previous pattern of mass killings mostly involving foreigners and slaves. The killings were blamed on the wife of a consul who was trying to ensure that her son by an earlier marriage rose to the powerful position held by her husband. She was put to death along with three thousand others, according to Livy. By the first century, the words "poisoner" and "adulteress" were considered almost synonymous, which might account for the apparent prevalence of poisoning in Roman society. Marcus Cato had said two centuries earlier that all adulteresses were poisoners as well. The satirist Juvenal provides the most scandalous material on poisonings in Rome, such that his commentary on the prevalence of mothers killing their children, wives poisoning husbands and everyone poisoning those who stood in the way of wealth and position have to be discounted somewhat. "If you want to be anyone nowadays," Juvenal wrote, "you must dare some crime that merits narrow *gyara* [exile on an island in the Aegean] or a gaol; honesty is praised and starves. It is to their crimes that men owe their pleasure grounds and high commands, their fine table and old silver goblets with goats standing out in relief."

For the honourable Roman, poisoning was something that entered their society from outside and was to be condemned in the strongest terms, even in times of war. Cicero cites the example of the consul Gaius Fabricius Luscinus, who returned a defector from the camp of their enemy as he had offered to poison his commander. "Thus they stamped with their disapproval the treacherous murder even of an enemy who was at once powerful, unprovoked, aggressive and successful." Romans, masculine, honourable and decent to a man, were never supposed to sink to such levels.

The first outsiders Romans associated with both medicine and poisoning were the Marsi, a nomadic people from the Abruzzi region in the mountains of central Italy with an exceptional knowledge of botany. The Marsi were also known as excellent soldiers and many served in the Roman army. It was said that the Romans never triumphed over them or without them. They sold herbs and medicines to Romans and performed as snake charmers. Galen consulted them on drugs. The position of the Marsi set a pattern for Roman views of medicines and poisons. Both the best doctors and the best poisoners came from somewhere else.

21

Following the earlier Greek model, poisons and infections were seen as part of the natural world rather than punishment by God. Seneca in his *Quaestiones Naturales* wrote that lightning contained a pestilential force and that this could have an impact on other substances. If lightning struck wine, for example, it froze it, but after it returned to its liquid state it was toxic and could kill or reduce those who drank it to madness. Lightning also left a foul smell behind in anything it touched, evidence of its toxic nature. Pliny the Elder's works are mostly about life – death rarely encroached upon his vision of the natural world and yet he often returns to the subject of poisons.

William Heberden, author of *An Essay on Mithridatium and Theriaca*, wrote in the eighteenth century that the ancients were "under perpetual alarms from an apprehension of poisons: They had probably seen the ill effect of a few substances on the human body and, like people in the dark, immediately made their dangers more and greater than they were." The ancient world did contain a great many dangers that we would now dismiss: visual and auditory poisons, toxic women and dangerous drums. Dying animals were also said to produce toxins and putrescence was the main mechanism by which poisons were formed in the world. Even the earth produced toxic airs that could break forth after earthquakes to produce epidemics.

Classical literature gives the impression of Rome as a murderous and dangerous world but there is a measure of exaggeration. It is impossible to know what the crime rate in Rome was; historians cannot even agree on the population of the city. Literary sources are the main basis for any projection of the prevalence of poisoning and they rarely give an accurate sense of anything other than the fears that existed within a society or could be aroused to entertain it. Most of the histories that survive from Roman times deal with court life and so it is there that we see what appears to be an alarming rate of poisonings. While it may have been seasons of pestilence that aroused men to accuse the women of Rome of poisoning, it does appear that the wives and mothers of the court may have been more inclined to resort to murder in this way.

Tacitus's *Annales*, the author's last work written before he died in 117 CE, covers the rule of four key Roman emperors – Tiberius, Caligula, Claudius and Nero – who together ruled from 14 CE to 68 CE. It starts with the death of Augustus Caesar, supposedly killed with poisoned figs given to him by his wife Livia. She had already poisoned Augustus's grandsons, Lucius and Gaius, to ensure that her son Tiberius came to the throne. Tiberius was also said by the writer Suetonius to have met his death through poison. Caligula was said to have had a vast collection of toxins that he used freely, even rigging sporting events by poisoning gladiators and rival horse racing teams. When Claudius succeeded him, he discarded the poisons in the sea, causing dead fish to wash up on the beach.

Locusta was believed to have been involved in the death of Claudius's son Britannicus. Emperor Nero hired her to find a poison, which was administered by the fourteen-year-old's tutors. The first attempt failed. The second attempt was subtler. He was offered food that had been tested by his tasters. Finding his drink to be too strong, he asked that it be diluted. Poisoned water was added to the cup and "he simultaneously lost both voice and breath". Nero reassured the other guests at the banquet that this was a result of the prince's epilepsy and not a matter of concern. The body was rapidly prepared for burial but turned black, which Romans believed was evidence of poisoning. Britannicus was covered with plaster of Paris but this washed off in the rain as his body was carried to the pyre, revealing to all that he had been murdered. Nero was said to have followed this up by attempting to poison his mother Agrippina but she survived, having fortified herself with antidotes. Other victims include Burros, the *praefectus praetorio* during Nero's reign. He suffered from a throat problem and died when his palate was smeared with a toxic medicine. Nero's aunt, Domitia, was also said to have been poisoned, although Suetonius suggests that it might have been an overdose of medicine.

Poisoning was believed to be so common that aristocrats had food tasters known as *praegustators*. These slaves were numerous enough to have formed their own association, headed by a *procurator praegustorum*. But even they could not be fully trusted. According to Tacitus, Agrippina used the Emperor Claudius's food taster in an attempt to kill him. She had employed Locusta, an expert on poisons, to find a toxin that would not act too quickly and raise suspicions, yet she did not want it to take too long to kill him for fear that he might realise what had happened and transfer power to an older son from an earlier marriage. The poison provided by Locusta was supposed to have deranged his mind to prevent any backtracking and was to be administered by the eunuch Halotus. Somehow the poison, mixed in a dish of mushrooms, failed to act. Agrippina moved to plan B. She ordered Xenophon, the emperor's doctor, to kill him. To encourage Claudius to vomit and purge himself of the poisoned mushrooms, Xenophon tickled the back of his throat with a poisoned feather and the emperor died not long afterwards.

Claudius died in October 54 CE. We know about the events surrounding his death from accounts by Tacitus, Suetonius and Dio Cassius, none of whom were born at the time. We do not know the sources for their accounts. Tacitus suggests that Claudius fell ill and then showed signs of recovery. Xenophon, a wealthy Greek doctor, was called and finished him off. Suetonius, a younger contemporary of Tacitus, is more restrained, suggesting that Claudius might have received the poisoned mushrooms from his food taster or his wife. He also suggests that Claudius appeared to recover and was then poisoned again, although he does not mention Xenophon. Dio Cassius, writing a century later, has Locusta providing a poison to Agrippina, who mixes it with some mushrooms. She eats from an untainted part of the plate and serves the poisoned mushrooms to her husband.

Romans enjoyed mushrooms even though they were aware that they could kill. Pliny describes entire households dying after eating them. Judging from their writing, they had little sense of which were poisonous and few useful guides as to which mushrooms to avoid. Location was regarded as more important than species in determining what might be dangerous; snake holes or rusty nails in the neighbourhood of a mushroom were regarded as signs of toxicity. Pliny describes favoured mushrooms, helping mycologists determine that Romans ate fungi from the genus *Amanita*. There are many edible species in this genus but it also includes the death cap mushroom, which is responsible for most sickness and deaths caused by fungi in Europe today. The phallus-shaped fungus appears in autumn, around the time that Claudius died. It also produces unusual symptoms. Unlike with most toxic fungi, the onset of poisoning is not immediately evident. Symptoms take time to appear and there is often a period of remission before a final bout of illness and eventual death.

Claudius, it seems, died from mushroom poisoning, as always claimed by his son Nero when he said that his father had become a god "after eating the food of the gods". Agrippina may have given it to him deliberately but it is doubtful she would have eaten from the same dish of mushrooms that killed Claudius. Romans most often braised mushrooms in olive oil – the resulting dish would have been thoroughly poisonous if one of the fungi was toxic. The likelihood is that Claudius died from consuming a dish of mushrooms that no one at the time knew was deadly.

TACITUS IS THE main source for many of the stories of the murderous Julio-Claudian dynasty and he aimed to show how the concentration of power in the hands of this family debased the politics of Rome. Poisoning was one of the lowest of all crimes and yet it was repeatedly used by the family to determine succession of men who held absolute power. Despite these imaginative histories, often written many years after the events and shaded by the reputations that developed around the rulers, it is hard to show that Roman aristocrats were at serious risk of being poisoned. Between Augustus coming to power in 27 BCE and the sixteen-year-old Romulus Augustus being overthrown by Germanic invaders in 476 CE, a period of just over half a millennium, Rome was ruled by seventy-seven emperors. From what we know, things became more dangerous as imperial rule went on; in the Early Empire, up until 193 CE, emperors lived an average of 53.4 years and generally ruled for a dozen years. In the nearly two hundred years of the Late Empire, those averages fell to forty-three years and a reign of six years.

Early on, most emperors appear to have died of natural causes, most likely infectious diseases such as malaria or typhoid, with only Claudius thought to have been poisoned, and we do not know if that was deliberate. One died after eating alpine cheese, most likely from food poisoning of some type rather than poison. Rulers of the Late Empire fared worse. Of the fifty-nine men who ruled in this period, twenty-nine died by the sword or dagger. Five died on the battlefield, one drowned, one was struck by lightning, one died in prison, and one each was hanged, strangled or stoned to death. Some fifteen died by natural causes and five committed suicide. One of the suicides poisoned himself and one died from the fumes from a brazier in his tent in what was likely a case of carbon monoxide poisoning.

Life expectancy at birth in Rome was low by modern standards. For the general population, it was just twenty-five years for women and 22.9 for men. For aristocrats, who were better fed, it was just 32.5 for women and 30.1 for men. This mostly indicates that large numbers of babies and children died; half of all infants died before the age of five. Men who lived to be twenty-five, the age at which they could join the Senate, could expect to live for thirty more years. Although emperors and aristocrats benefitted from better nutrition, they also lived in the cities of Rome and later Constantinople, hotbeds of pestilence due to high population densities and poor sanitation. They may not have been helped much by medicine, as much of the advice was either useless or positively harmful, such as the injunction for aristocratic women not to feed babies themselves for the first twenty days of life but to always use wet nurses. Mother's milk in the first month is much richer in protein and antibodies than the milk produced after this.

Emperors may have lived slightly longer than others but they reproduced at a slightly lower rate and they were more likely to die violently. Low birth rates were something of a curse in Rome, contributing to the eventual fall of the empire. Very few emperors had more than one son and many had no children. For some time, it was believed that this reduction in fertility might have been caused by lead poisoning among the aristocracy from household pipes and from grape syrup, the most common sweetener. Food and wine were often stored in lead-lined copper vessels or mixed with lead acetate, known as "sugar of lead" and both a sweetener and preservative. Adults can absorb about a tenth of soluble lead that passes through them, damaging nerves and inhibiting certain metabolic enzymes. Children may absorb up to half of any amount dissolved in food, making the problem significantly more dangerous.

By examining ice cores from Greenland, researchers have assessed the amount of lead processed by Greeks and Romans for the eight hundred years between around 500 BCE and 300 CE. Around four hundred tonnes ended up in the Arctic when precipitation brought down this air pollution. At their peak, Romans may have been using 80,000 tonnes of lead a year. Much of this was in plumbing – the very word derives from the Latin for "lead" – with one study showing that 12,000 tonnes were used in a single distribution system connecting an aqueduct in Lyon.

The idea of the Roman Empire collapsing due to inadvertent lead poisoning does not stand up to scrutiny. Romans were aware of the toxicity of lead – it had been noted by Xenophon in the fourth century BCE and by Lucretius in the first century BCE – but there are relatively few descriptions of symptoms or outbreaks, suggesting that it was too rare to have caused the downfall of the empire. The first clear recognition of chronic, rather than acute, lead poisoning is from the seventh century CE when Paul of Aegina describes it in some detail. Studies of lead in bones show that Romans had about half the amount found in the skeletons of modern Europeans. Current thinking is that only very high doses would have had a serious impact on fertility, reducing the population significantly. Lead poisoning does not appear to have been a problem for the wine-swilling, sugar syrup-drinking aristocrats with their dangerous plumbing, as amounts found in bones are similar across the empire, although higher in cities than the country.

Child mortality and monogamy limited the number of children born to Roman aristocrats. Those born out of wedlock did not figure in successions and little effort was made to legitimise them or include them in family life, as European aristocracies would later do with various forms of fictive kinship. A range of factors almost certainly hastened the transformation of the Roman Empire into Byzantium, moving its centre from Rome to Constantinople. Climate change, diseases – the Antonine pandemic of 165 CE, the two decades of the Plague of Cyprian from 250 CE and the terrible Justinianic plague of 541 CE – the problems of demographic recovery with low birth rates, and shifting food supplies all contributed. Lead poisoning was not likely a part of it.

Roman emperors were not at great risk of being poisoned. They mostly did not die young, nor did they have enough heirs to provoke the sorts of terrible fratricide seen among later Byzantine and Ottoman rulers. During what came to be known as the "Crisis of the Third Century", a succession of short-reigning emperors were murdered; in almost all cases, the sword sufficed. The historian Greg Woolf described the murder of emperors in terms of the board game Cluedo: "I accuse the Praetorian guard, in the palace, with the dagger." The killing of an emperor was best done in public by many people, in the manner of the death of Julius Caesar, to share the blame and avoid rumours that the ruler had survived, was now in hiding and would soon re-emerge.

IN BOTH GREEK and Latin, the word for poison is also used for potent herbs, both good and bad, and also for magic charms. The stems "vene-" in Latin and "pharma-" in Greek may have led us astray for centuries. All materials that deal with poison or with magic are thus somewhat difficult to translate as the meaning is never known for sure but can only be interpreted through the context. The ideas of magic and poison were probably not distinguished well in the minds of people anyway. A *venenum* might be swallowed by a victim or employed at a distance by someone practising magic. Tacitus describes the death of Germanicus as being caused by *veneficia*. He says that in different parts of his house, human remains that Tacitus describes as "half-burned human bodies dripping with gore" were found along with lead tablets with his name on it. Tacitus says that friends of Germanicus believed he had been poisoned but that his enemies denied this, saying that no traces of poisoning were seen on his body when it was displayed in the forum at Antioch after his death.

Poison was a common means of killing oneself in Rome. It was regarded as an honourable way out for those who had been conquered or faced justice. Livy said it was usual for kings to keep poison in their palaces against the uncertainties of fortune. Pliny also mentioned that those suffering from unendurable illnesses would be given poison. Martina, the provider of poisons who was suspected in the death of Germanicus and was being sent to Rome for trial, killed herself en route, having concealed some poison in the knot of her hair. Her jailers could find no indication of how she had killed herself, according to Tacitus. In most of the cases described by the historian, poisoning occurs at the same time as medical treatment, which may have led us into believing that poisonings were more common than they actually were. The blurring of categories between medicine, poison and magic charms means that these deaths may not all have been malicious.

DOCTORS WERE HEALERS and killers. Women used their knowledge of potions for both good and evil. Those who gathered and traded the numerous plants used for medicines were often reviled. The Psylli, Nasamones and Paleothebans were also reputed to be knowledgeable about drugs and also faced suspicions due to this. Early Christians refused them entry to the church on the grounds of their suspect occult learning.

The Roman historians who chronicled the behaviour of the elites suggested that knowledge of poison was something transmitted in specialised schools. Those like Locusta who developed a reputation for being skilled in the administering of poisons took on pupils. Physicians also became infamous for their knowledge of poisons, as were the travelling salesmen who peddled snake oil cures. Legally, poisoning was first marked out as a specific crime in the Lex Cornelia passed by the dictator Sulla in 81 BCE. This called for the punishment of not just those who used poisons but also anyone who made, sold, possessed or bought it, an extension of its remit that never applied to those who made or sold swords or daggers, for example. Later additions to the law punished women who administered drugs against sterility to banishment if they caused the death of the woman taking the drug. Pharmacists were banned from offering love philtres. For those of a higher rank, the sentence was banishment. Those lower on the social scale were thrown to wild beasts.

A question that troubled Romans was why poisons even existed. Much later this was also a concern for Christian theologians trying to justify why a beneficent God would allow the presence of poisons in so many places. Hindus justified their existence as part of an accident; overkill by a God enraged by the presence of evil.

Pliny explained that poisons were put there to help mankind: "It is out of compassion to us that (the earth) has ordained certain substances to be poisonous, in order that when we are weary of life, hunger, a mode of death the most foreign to the kind disposition of the earth, might not consume us." Instead poisons were available that left the body undefiled by wounds or nooses. Animals would avoid eating someone who had died by poison and therefore the body would return to the earth without damage of any kind. "He who has perished by his own hand is reserved for the earth." Pliny also believed that the Roman Empire was justified because it had conquered many lands where poisons flourished and had effectively cured them. He writes extensively about poisons but locates their origins almost always in countries that had fought Rome at some time; poisons were clearly something foreign and barbaric to him. Poisons were particularly associated with Egypt and North Africa.

From the earliest days of Rome, there were laws that controlled magic. These were not in any way absolute. Magic was permissible, in fact a part of daily life, as long as it did not interfere with anyone else or their property. Cato, a fierce critic of doctors and magicians, even provides readers with a magic ritual to cure sprains. But magic also evoked tomb robbers, sexual crimes and unfortunate foreign influences. For Roman rulers, magic and poisoning were linked to sedition, to threats to the very republic or empire itself and were thus to be treated in the harshest possible way. The *senatus consultum* against magic that was aimed at the Bacchanalian cult of 187 BCE was the starting point for a series of measures, mostly repeating the same ban against magic that would be reintroduced several times until the fall of the empire.

Apart from these statutes, the key laws against poisoning and magic were put into place under Emperor Justinian in the sixth century CE, a time of strong Christian influence that profoundly altered Roman law. It is unclear whether these laws were in place in earlier periods or whether they were new aspects of the Justinian code reflecting a strong Christian opposition to magic. One section is titled "Concerning magicians, astronomers and all such like". This chapter includes the injunction: "It is worse to destroy a man by 'medicine' than it is to slay him with a sword." It also says: "By the same law also are condemned to death the sorcerer – poisoners who slay men by their hateful arts as well as by 'medicine' as by magic spells or those who publicly sell hurtful 'medicine'." Even people who gave out medicines without evil intent could be punished. A woman was sent into exile for giving another woman a potion to use as a contraceptive that was later said to have killed her. Pharmacists who prescribed hemlock, salamander, aconite, mandrake or Spanish fly could also face punishment if anyone came to harm.

Poison and magic were closely entwined, linguistically and legally, in such a way that we will never be clear exactly how they were seen by the Romans. Much of what we know comes from a handful of historians writing long after the events in a manner calculated to entertain as much as anything. Views about Tacitus have shifted over the centuries; some have viewed his work as essentially satire while others see him as a restrained and rigorous historian. Much of what we know about poison is filtered through ideas of magic, a great source of anxiety for the Romans and later for Christians who saw it as a threat to their faith.

3. WHAT A SPLENDID RESULT OF REASON

And ate laste the feend, our enemy,
Putte in his thought that he shoulde poison beye,
With which he myghte sleen his felaws tweye;

And forth he gooth, no longer wolde he tarie
Into the toun, unto a pothecarie,
And preyde hym that he hym wolde selle
Some poison, that he myghte his rates quelle.
Geoffrey Chaucer, *The Ploughman's Tale*

Belief in the basilisk endured into the Middle Ages. It was mentioned in the Bible and therefore there was a widespread view that it must exist. It appears twice in the book of Isaiah. "Rejoice not thou, whole Palestina, all of thee, because the rod that smote thee is broken, for out of the serpent's root shall come a basilisk and his fruit shall be a fiery flying serpent." More ominous is the line: "And the suckling child shall play on the hole of an asp and the weaned child shall put his hand on the basilisk's den." In the later King James version, there is a third mention: "They hatch cockatrice' eggs and weave the spider's web; he that eateth of their eggs dieth and that which is crushed breaketh out into a viper."

In his twelfth-century guidebook for preachers,

Pantheologus, Peter of Cornwall, the prior of the Holy Trinity Church in Aldgate, described all the animals mentioned in the Bible, including the basilisk and the unicorn. Their appearance in sacred texts meant belief in their existence was strong but the lack of a detailed description of the basilisk allowed a wide interpretation of what it might be. For centuries, it had been thought of as a snake, but by the Middle Ages it had become a chimera, most often a hybrid of snake and fowl. Pierre de Beauvais wrote in his thirteenth-century bestiary that basilisks were born from an egg laid by a cock on a dung heap. The egg would be hatched by a toad and would give forth an animal that had the head of a rooster and the body of a snake.

From the references in the Bible, the basilisk, along with other serpents, was taken to represent sin, an ever-present danger. The medieval world found sin and poison in many places. In 1264, Pope Urban IV wrote to Charles of Anjou to warn him that assassins had gathered fifty types of poison with which to kill him. The thirteenth-century doctor Peter of Abano lists some ninety poisons and gives detailed instructions on how to make very toxic forms of arsenic by heating and oxidising the metal. Hellebore and aconite were probably the most widely available toxins but arsenic salts were becoming increasingly common as advances in the smelting of metals turned the poison from a rarity imported from the Arab world into something so common it was used as ships' ballast. In 1308, the Bishop of Troyes was accused of killing Jeanne de Navarre, the queen of Philip IV of France, by poisoning her. His trial lasted until 1313 and included many witnesses accusing the bishop of regularly buying poisons from witches and apothecaries.

In her work on Chaucer and poison, the author Margaret Hallissy describes a world in which poison was surrounded by magic and mystery. "In such a state of medical ignorance, no effective cure for poison was available, nor was an accurate diagnosis possible: and thus the idea of poison took on the dimensions of a universal malign force. As such it was readily identifiable with preternatural forces of evil and in particular the devil, the venomous serpent of Eden." Poison was divine retribution, a physical manifestation of evil on earth. Early Christian theologians saw poisonous animals as surrogates for the Devil with venom symbolising sin and harming man's soul in the way poison harmed his body.

Saint Ambrose and Saint Augustine, two of the four "doctors of the Church" who established important doctrinal issues in the fourth and fifth centuries, explained the existence of poisons as a sign of the possibility of sin and thus a reminder for man to turn away from it. For Augustine, snakes were a potent symbol of sin and this extended to others who were regarded as sinful – Jews were often portrayed as associated with snakes in early antisemitic imagery. Ambrose viewed snakes as both evil and wise, a symbol of healing and medicine as well as sin. He noted that the universal antidote theriac was made from the dried flesh of snakes. Serpents are to be imitated as well as feared – imitated in the shedding of skins, in wariness, in vomiting out venom that was like expelling evil thoughts. The very existence of poisons was something of a mystery, part of the larger concern as to why evil existed in God's universe. Ambrose believed poisons existed simply because they were possible and they showed what God was capable of creating. Augustine saw them to be punishment for the sins of Adam. Jerome, the fourth-century cardinal who is the patron saint of theologians, surmised that most aspects of life were a mix of honey and poison. Women, for example, had beauty that masked the sin of lust that lay within.

All around them people saw natural substances transformed but did not understand how these changes happened. Grapes turned into wine, flour and water became bread, and fruit went from being hard and sour to soft and sweet. The basic idea behind alchemy was that metals could also be transformed. People understood that most ores were a mix of different metals, often containing small amounts of gold or silver. There was a hierarchy of metals, from the most mixed ores to the purest gold. Finding the means to purify them was the ultimate aim of alchemy. Islamic alchemists, including some of the greatest minds of the time, Geber and Rhazes, believed metals had the same properties as all matter – heat, coldness, dryness and moisture. Rearranging the ratios of these qualities could change the metals from, for example, cold and dry lead to hot and moist gold. Alchemy was wrapped up in complex philosophical rationales and extended far beyond the transmutation of metals into the search for the basis of life, even for the artificial creation of life. The practical experimentation involved in trying to find the philosopher's stone or the elixir of life threw up some important chemical discoveries. Techniques such as distillation and crystallisation led to the creation of acids and chemicals such as *aqua regia*, the combination of nitric and hydrochloric acids that can dissolve gold – a profound source of excitement for these proto-chemists.

Another offshoot of alchemy was the creation of an array of poisons, ranging from arsenic oxides to acids and quicksilver, always a source of great fascination because liquid mercury was believed to be a halfway point in the transmutation of metals. The toxicity was not well understood but the ill health of alchemists and the occult nature of their trade created an anxious miasma around them. Geoffrey Chaucer found alchemists to be rather nasty, bordering on the toxic themselves. "Men may hem knowe by smel of brymstoon/ For al the world they stynken like a goot," he wrote in *The Yeoman's Tale*. Alchemists were for Chaucer poisoners of the soul as well as the body, leading people into moral degeneration. These noxious associations would linger until alchemy faded into oblivion in the eighteenth century, finally replaced during the Enlightenment by the science of chemistry.

Some deadly poisons were available, mostly from plants, but most medieval poisons were more magic potions than anything else. When people were accused of poisonings, they were often said to have mixed elaborate concoctions of spiders, scorpions and snakes. These were collections of culturally toxic substances such as the blood of lepers and various bitter-tasting herbs, unpleasant but unlikely to be deadly. Given the tendency to let them putrefy, they may well have been highly noxious and possibly contained harmful stews of bacteria, but it is doubtful that they were enormously threatening to a healthy person. They were often distilled, which would have significantly reduced the toxicity they might have gained through putrefaction. Drawing from the world of alchemy, many believed that substances could be transformed through distillation, magic and incantations into more poisonous substances.

Judging the incidence of crime is difficult as we have to work mostly from court documents that record some, but far from all, crimes. In the days before post-mortems (rarely carried out until the nineteenth century and mostly worthless in terms of providing the real cause of death up until the twentieth century) and before the emergence of forensic toxicology, there was almost no way to know whether someone had been poisoned except by catching the poisoner in the act. Court records show most crimes to be sudden and violent, lacking the pre-meditation that is typically part of poisoning. The first law in Britain to tackle poisoning was passed in 1531 and mentioned that the offence "in this realm hitherto our Lord be thanked has been most rare and seldom committed and practised".

Despite this, literary and medical sources from this time suggest a preoccupation with poison. Of greatest concern, as always, were bites from toads, spiders, scorpions and snakes. Anxieties about poisonous snakes or scorpions in England suggest either somewhat fevered imaginings or more likely the incorporation of medical texts from elsewhere with few changes to account for local fauna. From the eighth century through to the re-emergence of classical medicine in the twelfth and thirteenth centuries, leech books were the main source of information on the body and health. These prescribed a variety of potions made from herbs as well as magic charms and spells. Incantations and magic words were important tools of medicine.

Medical ideas were bound up with concerns about sin and the influence of the devil. *Piers Plowman*, an allegorical poem by William Langland written in the fourteenth century, elaborates on the complex links between medicine and Christianity, between sin and disease, salvation and cure. It is also replete with poisoners. Sellers of food poison the people; Lady Meed poisons popes; the peacock – a metaphor for the wealthy man – poisons the earth in which it is buried; Constantine poisons the clergy by endowing the Church with worldly goods; and Satan is forced to drink his own bitter poison. In this view of the world, Christ is the ultimate cure – the most effective treacle or theriac – and the best doctor. Worldly doctors fare less well in the poem. They tend to be seen as devious and deceitful, more likely to kill than to cure. Physicians are more like Satan than Christ. Doctors are not the only groups compared to Satan by the somewhat grindingly pious Langland. Lawyers, minstrels and friars get the same treatment. Langland subscribed to the enduring idea that doctors were motivated by greed and were mostly interested in selling costly medicines while decking themselves out in furs and gold buttons.

Poisonings were considered part of the world of sorcery and magic, one of several "hidden crimes" that were punishable by death. By the thirteenth century, there was no legal distinction between those killings done by stealth and those carried out openly. Magic pervaded the law – property could be protected by amulets and spells, and thieves could be identified by supernatural means. In the earlier legal model, secret offences were those that would be known by God and could not necessarily be known by man. Poisoning, which used secret methods to mimic the snakes which had brought sin into the world, was something that only the Almighty could judge.

What is clear from the legal evidence up to the fifteenth century, is that poisoning was not seen as a crime of ordinary people. There are just a handful of recorded cases involving individuals but there is a considerable body of material on the murder or attempted killing of monarchs and top officials of the state. Necromancy, one of the "dark arts" involving the raising of spirits, and killing through such means as putting pins into a wax effigy of a person are all recorded in court cases. In 1331 a plot was uncovered to kill a royal minister, Robert of Ely. The law that covered treason – a new statute in 1352 – had a handy category for such crimes known as "compassing and imagining the King's death".

Monasteries have often been settings for poisonings, although these are often more fictional than historical. Monks had a mastery of medicines and access to poisons, mostly herbs grown in their gardens. They lived together in claustrophobic, Spartan conditions that must have bred intense rivalries and hatreds. Men of the cloth have never been immune to ambition, desire or the lust for power. The Church explicitly forbade monks and priests from spilling blood or carrying weapons. Monasteries were often the main centres for medicine and many monks would have had the most complete knowledge of their day of herbs and toxic plants.

The many volumes of rules written from the tenth to the fourteenth centuries to govern religious communities contain almost no mention of poisonings. According to the French historian Franck Collard, although statutes in the thirteenth and fourteenth centuries included poisonings in the crimes to be tried by the Church, there is no mention of specific sanctions against monks who poisoned their colleagues. Although some of the guides to religious life mention arguments among monks, none of them say anything about the use of poisons against one another. Hildegard of Bingen, a twelfth-century German abbess and composer who also produced some of the most important medical texts of the time, wrote extensively about poisons, but the idea of nuns using them on one another does not appear in her writing.

If poisoning were a major problem in monasteries, one might expect the books of rules and guides to monastic living to be more explicit on the subject. In many cases of secret crimes of violence or sex, the church rules were vague at best, perhaps being unwilling to highlight problems in the orders. Documents of church trials do not help much. Collard was only able to find twenty-five cases of poisonings and the truth of these could not be verified. In the seven centuries from 500 to 1200, it seems that only around 15 per cent of cases of poisoning that appear in records took place in monasteries or among church men and women. The figure falls to just 6 per cent for the next three centuries. Church records are often vague on the subject – monks were expelled or punished for criminality but little was known about what they did wrong. Likewise, disputes and violence were recorded but not how the men or women harmed each other. When poisonings were recorded, they almost never alluded to what poison might have been used, raising the very real possibility that the alleged victims died of food poisoning or some other cause. There are a handful of recorded incidents in which people were alleged to have been poisoned through the bread and wine taken at Communion. Poisoning may not have been separated out from witchcraft in legal terms, it was only after the end of the mass of witchcraft trials of the seventeenth century that legal authorities became more interested in poisoning.

The lack of records of poisonings does not prove that they did not take place. Many might have gone undetected amid the ravages of plague and early death that afflicted people at a time when life expectancy in Europe was extremely low. But among the recorded violent deaths, poisonings appear to be rare. Even accusations of poisonings were not at all common. Where poison is much more common is in the realm of metaphor and as a stand-in for the great fears of disease and evil that must have been constantly present in the lives of people who understood little about hygiene or prevention of illness. That is not to say that pious men did not kill each other – they just used the same methods as everyone else: knives, clubs and their bare hands.

Histories of science have tended to ignore the medieval period, imagining that somehow Europe went from classical Rome to the Renaissance without passing through a thousand years of history. But this period was one of immense intellectual activity, and provided the basis of what evolved into modern law. Far from being "Dark", it was a time of scientific evolution, and sometimes regression, as well as being a period of intense debate about the nature of poison as a physical substance and as a metaphor. The whole notion of the Dark Ages ignores what was going on in the Middle East, which was living through a period of cultural and scientific advance that would eventually feed into European thought, notably through Spain.

There were important changes in the understanding and perception of the body from the twelfth through to the sixteenth century that led to widespread shifts in medicine and the understanding of poisons.

The devastating plagues of the fourteenth century provoked an array of responses from a rise in antisemitism to a turbulent undermining of religious doctrine. It also provoked a new understanding of the body, which came to be seen as more porous and vulnerable, more likely to be penetrated by poisons and more likely to be poisonous itself in its various emissions.

It was a time of fascination about poisons, in part because of the communication of knowledge from the classical worlds, often through Arab texts, and in part because of the concern that poisons were all around. From the fifth to the fourteenth century there were significant intellectual developments, great changes in society as part of the sweeping transformation of Europe under Christianity and the Middle East under Islam. The study and use of poisons was well developed in the Byzantine empire, where several emperors and heirs to the throne are believed to have succumbed to poisons including carbon monoxide. Assassination was a serious risk in the Byzantine world. Of the 107 men who ruled between 395 and 1453 when the Turks took Constantinople, only thirty-four died a natural death. In these 1,058 years, twenty-three rulers were assassinated; eighteen were mutilated and removed from the throne; twelve died in prison; eight died in war; and twelve abdicated.

Greek texts, kept by Nestorian Christians expelled from Constantinople in 431 CE, were translated into Arabic and would eventually re-emerge as a vital source of knowledge in the Renaissance. Arab writers working from the seventh to the tenth centuries incorporated knowledge from as far away as China and India while maintaining Greek knowledge that had substantially disappeared in Europe outside Islamic Spain.

Ibn Wahshiyya, an alchemist living in Baghdad in the tenth century, was one of the most important writers on poisons to emerge from the Islamic world. He handily provided a bibliography in his work that suggests he drew most of his knowledge from books compiled for Persian kings. Shanaq, a Hindu writer, was also an influence, as were the Greeks Theophrastus and Galen. Ibn Wahshiyya brought together these texts along with local folk knowledge in his *Treatise on Poisons.*

He is one of the first writers to describe in detail the symptoms of poisoning:

> Whenever someone eats or drinks something poisoned, a violent frowning and extreme grimness appears in his face; his colour becomes purple, black, blue or is lost. A serious anxiety and confusion of mind appear at the same time together with a trembling of the heart and a fright without resemblance to anything. There is an abundant flowing of tears from the eyes. A deep redness appears and his eyes light up. He trembles, shakes and emits perspiration which is either very hot or very cold. He turns towards the sides without cause; a laziness, languidness and brokenness of the body overcome him. He laughs excessively without cause. His reason is so confused that he performs nonsensical actions with his hands and fingers, he rubs them often and frequently, claps his hands. He stamps insanely upon the ground, stands up, goes, comes, always stretches his arms.

In contrast to the Christian notions linking poison and sin, Ibn Wahshiyya defines a poison as "something overpowering in its nature... Poison is something that overpowers and destroys that which is called the life force of an animal. When it overcomes this force the functioning of the organs of the body are disturbed."

Even sounds could be toxic. Ibn Wahshiyya warns that a beautiful song performed badly might kill a sensitive man and that the skin of a wolf stretched over a drum could kill sheep. The basilisk makes an appearance. "As for what kills even at a distance, there are the snakes which exist in the valley of Khazluj in the land of the Turks. If a man sees these snakes, he dies; also if the snake looks at him he dies immediately. This was in truth observed at this place."

Ibn Wahshiyya's world was impregnated with poisons although, as is so often the case in these works, the most mysterious lurked in distant lands.

There is a stone which is found in the islands near the land of the Chinese; its colour is like that of marcasite, black and is of iron. If a man sees it, he laughs until he dies even though he has covered his face after having seen it. It does him no good so that his laughing does not cease until his death. The remedy that removes the effect of this stone is a bird which lives in the same islands. It is green feathered and as big as a sparrow. If on seeing the stone, one sees also, by chance, the bird at the same time, then the lethal power of the stone is destroyed.

Most of the poisons he described occurred naturally, but he also described new concoctions developed at the urging of rulers who required undetectable ways to kill.

These substances which were placed in food and drink then became more specialised. The next substance was that which kills by touch, then after this that which kills by smell. Later advances in specialization brought about by those that kill by sound and then by sight. Finally came the circumstance which is lethal to one with a certain young woman. Death occurs immediately without delay. How wonderful the sagacity of the medical men? What a splendid result of reason!

He is careful to point out that he does not give recipes for poisons without also including their antidotes, to ensure they are not used for nefarious purposes:

Another poison whose attribute is that it is fatal with any oil. A beaver is confined in a glass container whose lid is stuck on with clay kneaded with pigeon manure. It is left to be corrupted in donkey manure for seven days. Then the glass container is removed and its lid opened. On it is poured the urine of a boy who has not yet obtained puberty. It is again buried in the donkey manure for seven days and removed. Then some narcissus bulbs are dried and pulverised well. It then has a very fetid odour. Indeed as soon as it is mixed with it, it annuls the stench immediately so that it becomes nil and has no odour at all... Indeed it can be fatal in six hours if it penetrates into the brain.

Ibn Wahshiyya was one of many Islamic scientists whose work significantly advanced the boundaries of knowledge. They also transmitted critical Greek, Hindu and Persian texts down the generations by compiling them into new works. Rhazes (known in the Islamic world as Abu Bakr Muhammad Ibn Zakarīya al-Rāzi) was an extraordinary Persian philosopher, scientist and doctor who is credited with the discovery of sulphuric and nitric acids, as well as being the first person to purify ethanol. He provided the first definitive description of smallpox, giving a vivid account of the symptoms. Although educated in Greek and other forms of medicine, he was very much an empiricist, relying on his own observations. He wrote an extraordinary number of books with titles like *Fruit Before and After Lunch*, *Snow and Thirst* and *Warmth in Clothing*. His knowledge was encyclopaedic, covering subjects well beyond the specialised range of contemporary scientists. His work on poisons stemmed from his interest in alchemy, which he abandoned at the age of thirty after his experiments damaged his eyesight.

Less imaginative, perhaps, but vital for the transmission of knowledge was the work of Ibn Sina, a peripatetic Persian scholar whose name is Latinised as Avicenna. His *Canon Medicinae* squeezed classical Greek knowledge into a system of almost mathematical classification. He correctly described tuberculosis as infectious and detailed the symptoms of illnesses such as diabetes. Even more than the work of Rhazes, Avicenna's compendium of medicine had an extraordinary influence on late medieval Europe. Unfortunately, its wholesale and often uncritical adoption probably hindered the development of science. It was rarely challenged despite its many inaccuracies, something that Avicenna, a peerless debater in his time, would not have wanted.

Avicenna warns against foods that are strong in taste – sharp, salty or sweet – as these might most easily disguise the taste of poisons. He opposed drinking to excess for similar reasons: "indeed it may happen that at some time or another some malignant thing like a lizard, or scorpion, may be dissolved in it." Poisons were said to be commonly dissolved in wine and so antidotes were often prescribed *potu in vinum,* or mixed in wine. Most of his cures involve vomiting or drinking large quantities of olive oil or water, sensible enough advice for anything except caustic chemicals. He believed that antidotes led the poison to the extremities of the body where it could be sweated out. Poisons kill "by change of complexion, putrefaction or by an attack on the members", he surmised. Much of Avicenna's work is devoted to the description of the symptoms of poisoning and to listing toxins, among which he includes cold water drunk on an empty stomach.

A link between ancient and medieval knowledge was the Jewish sage Maimonides. Born in Cordoba in 1135, he was thirteen when the city was taken over by a Muslim sect that demanded that Jews convert or face execution. The family fled to Fez and then on to Jerusalem. He arrived in Egypt in 1166 and lived there until his death in 1204. He was personal physician to the vizier and is said to have been invited to become the doctor for Richard the Lionheart, an offer he wisely rejected as pogroms against Jews were then under way in England.

His works display a deep practical concern for patients and what appears nowadays as a calm and rational approach to a subject that tended to provoke some wild claims. His book on poison was prompted by a request in 1198 from Vizier al-Fadil to provide a manual for the public on how to treat snake bites. The vizier asked him to tell people what they could do on their own to treat poisonings if no doctor was available and to come up with recipes for theriacs that could be used in less urgent cases.

Maimonides modestly claimed that his *The Venoms and Their Antidotes,* also known as the *Book of Al-Fadil,* had little new in it and that he had gathered his information from others. In fact, the book, like all his works, showed an originality of thought that would not surface again until the Renaissance. Far from being completely beholden to classical learning, he drew on his own observations and experiences. His recommendations for treating poisonous bites are not outlandish. He urges a tourniquet above the wound if it is on a limb and gives details on how to suck poison from a snake bite. He includes in his discussion of poison a variety of remedies that were dangerous and should be taken with care. Experimentation played a critical role in his judgements. Bezoar stones were not an effective cure for scorpion bites because he could not observe such an effect. He dismissed a belief that victims of poisoning should only eat unleavened bread, saying there was no basis for this. He even questioned the widespread belief that menstrual blood was poisonous but also remarked that he had not done experiments on this and therefore did not know for sure.

In a remarkably prescient observation for the time, Maimonides distinguished between two types of poison: neurotoxins, like viper venom, which paralyse the respiratory organs, and hemolysins, like adder poison, which cause haemorrhaging, reddening of the skin, rising blood pressure and fever. The work is an extraordinary medical manual, aimed at the average person but a great influence on scientists throughout the Middle Ages. Maimonides was the archetypal Jewish doctor of the time, a man whose work and knowledge led to a strange paradox of medieval life. While Jews were being persecuted and murdered across Europe, Jewish doctors were sought out by monarchs and popes for their knowledge of both medicine and the occult.

For the Church, disease was a sign of divine wrath and therefore not to be tampered with. The leper was cast out not just because of the disease but because they represented a moral contagion. If diseases were supernatural, then cures needed to be as well, hence the popularity of relics and pilgrimages. Jewish doctors such as Maimonides were able to operate outside the system of Christian intellectual and medical constraints, which made them both highly regarded and much feared.

4. POISONING WELLS

If it is the part of a good Christian to detest the
Jews, then we are all good Christians.
Erasmus

The exclusion of Jews from public and political life in Europe goes back to the Roman Empire. Jews were a colonised and troublesome people with a monotheistic faith that was a direct challenge to Roman views of the world. The critical shift in the development of European antisemitism, however, was the conversion of Rome to Christianity. As an emerging religion bound up in Jewish roots, Christianity needed to distinguish itself sharply from what went before. Its early definition was mostly in opposition to Judaism. What started as a series of theological distinctions grew into legal and social controls. Previously Jews had been under the same Roman restrictions as Christians; they were excluded from the legal profession and imperial service and could not sue or testify in court. Set apart from the public, they were both protected by and excluded from the ruling elite, a pattern that would endure for two thousand years.

Romans had political justifications for the exclusion of Jews but Christians had something more powerful: a Biblical rationale that developed a powerful hold over the popular imagination. Jews had killed Christ and for that, they would be punished for all time. This calls for a highly selective reading of parts of the Bible written long after the death of Jesus when tensions between Christians and Jews were worsening. From the earliest days of organised Christianity, the

Church led the way when it came to menacing Jews. As early as 305, in Elvira, a town in Andalusia, a Church council ruled that Christian women could not marry Jews. Saint Ambrose of Milan complained in the fourth century that fines imposed on Christians for burning down a synagogue were too harsh, saying "should not the rigour of the law yield to piety". In 438 the Theodosian Code established Christianity as the only legal religion in the Roman Empire. From then on, the Church aligned itself more often than not with the forces of antisemitism.

In 692, an Ecumenical Council held in Constantinople banned Christians from eating unleavened bread with Jews, receiving medical care from them, bathing with them or having sex with them. By the ninth century the Church in Rome had brought most of Europe under its remit and widened the divide between the faiths. Jews would increasingly be blamed for the many woes of life from disease and natural disasters to the Mongol invasions.

The First Crusade in the eleventh century marks a new chapter in a long and terrible history of pogroms in Europe. Guibert of Nogent, a Benedictine monk writing in the eleventh and twelfth centuries, reported that in the French city of Rouen, crusaders "herded the Jews into a certain place of worship, rounding them up either by force or by guile and without distinction of age or sex put them to the sword". Each of the Crusades, with their ramping up of the ideology of religious purity and their extreme hatred for Jews and Muslims, led to a new round of violence long before any fighters had even reached the Middle East. In England, the history of Richard the Lionheart is often laced with romanticism and heroism but, for Jews, his reign was a period of intensifying terror. On 2 September 1189, as he was crowned in London, at least thirty Jews were burned to death in a celebratory purification of the city. Attacks spread across the country culminating in the massacre and forced mass suicide of Jews in York in March 1190.

Many Jews benefited from the economic expansion of the twelfth and thirteenth centuries. Christian bans on usury, reaffirmed in the Lateran councils and only formally abandoned by the Catholic Church in the twentieth century, and the exclusion of Jews from so many other lines of work, created niches in money lending, banking and trade. These created great wealth for a handful of people and stirred up even greater resentment against the entire community. The important role of Jews in the economy gave them a measure of protection under both Church and civil law, although these were unevenly applied at best. In 1179, Pope Alexander III issued an instruction attached to the decrees of the Third Lateran Council saying that Jews were not to have land or possessions confiscated without trial. They were not to be beaten or mocked when they held their festivals nor were their cemeteries to be desecrated.

However, chroniclers of the time provide a picture of pervasive antisemitism. Invariably Jews were portrayed as rich and greedy, killers of Christ and murderers of Christian children. A belief developed that Jews stank. They were imagined to emit *foetor judaicus*, a reek of sulphur that was suspected of causing epidemics. The decrees of the Fourth Lateran Council in November 1215 ordered Jews to wear distinctive clothes, prohibited them from public office and forbade them from observing Jewish rituals if they converted to Christianity. The council had been led by Innocent III, who was intent on consolidating Church power, and what better way to promote cohesion than finding enemies in Jews and Muslims? The rise in antisemitism was accompanied by the spreading idea that Jews were particularly skilled in sorcery and the black arts. Guibert of Nogent was keen to associate Jews with sexual perversion and murder. In one of his works, a monk masters the black arts by paying a Jew who was skilled in medicine and had communicated with the devil. To seal the deal, the monk drank a draft of semen.

Among the most enduring antisemitic myths is the blood libel. This is the belief that Jews killed Christian children and drained their blood to make *matzo* for the Passover celebrations, to anoint doors during Passover, to make Purim cakes, ease the wounds of circumcision or to use as a medicine. Christian blood was also said to eliminate the *foetor judaicus*, allowing Jews to move around without arousing suspicion. The blood libel probably originated in Greece before the Christian era when Jews were accused of sacrificing children. It re-emerged with a vengeance after Christian attacks on Jewish communities in the Rhineland in the eleventh century. Armies gathering for the First Crusade took it on themselves to force the conversion of Jews, who were offered a choice of Christianity or death at the hands of mobs. Many Jews killed themselves after first killing their children. A preference for suicide over being slaughtered is not hard to understand in these circumstances, but it may also have been linked to a Jewish belief of the time that martyrdom would hasten the arrival of the Messiah and his victory over their enemies.

The alleged ritual killing in 1144 of William of Norwich, later the saintly focus of an antisemitic cult, started a wave of accusations of murder that continued up to the twentieth century and laid much of the groundwork for accusations of mass murder. According to the credulous and inventive account provided by Thomas of Monmouth, a Benedictine monk, William was a young apprentice furrier who was supposedly killed as part of a plot by Jews to avenge themselves against Christians. A belief emerged that unless Jews shed the blood of Christians they would never be able to return to their lands.

While a great many members of the Church were active participants in the development of these antisemitic cults of sainthood – some of which were only disbanded as late as the 1960s – some popes recognised that the charges of child killing were false and tried to stop violence against Jews. Pope Gregory X wrote in a 1272 letter:

> Since it happens occasionally that some Christians lose their Christian children, the Jews are accused by their enemies of secretly carrying off and killing these same Christian children and of making sacrifices of the heart and blood of these very children. It happens, too, that the parents of these children or some other Christian enemies of these Jews, secretly hide these very children in order that they may be able to injure these Jews, and in order that they may be able to extort from them a

certain amount of money by redeeming them from their straits.

Greed, revenge, resentment and envy all combined with beliefs that were supported by enough members of the Church to develop a deep resonance with a fearful population. Not only were there the stories of the murder of Christian children, but there was a revival of the conspiracy theories that Jews had been responsible for the Arab conquest of Spain and the Viking invasions. There were at least 150 trials of Jews for the murder of children in the Middle Ages. Likewise, large numbers of Jews were killed, often by being burned to death, for alleged desecration of the host.

While some did gain some economic advantages during this period, Jewish populations were also put under the tight control of monarchs, essentially becoming royal serfs. It was a position that put them at the centre of competition for power and revenues. Louis the Pious, the son of Charlemagne and king of France in the ninth century, started the process with a law that decreed that only he could hear court cases involving Jews. Across Europe, the control of Jews and all their rights of inheritance and taxation became part of the Royal Prerogative. Being under royal protection was not always an advantage. Local authorities did not collect taxes from Jews and therefore had little incentive to protect them. Likewise, they were a popular target of anti-royalist violence, as attacking Jews became a way of undermining royal power. On the other hand, Jews were able to appeal to the monarch for protection or justice, reminding him that any strike against them was also a blow to his status and finances.

NOT ONLY WERE Jews vulnerable for their supposed knowledge of the occult, their imagined hostility to Christianity and their sensitive economic and political position, but they were even regarded as intrinsically toxic. Women were widely seen as being poisonous themselves, in part due to menstruation. A parallel belief developed that Jewish men menstruated as a mark of the curse put upon them for denying Christ. It symbolically linked Jewish men to women and to Eve's role in the original sin. It also connected them to the betrayal of Christ. Judas hanged himself and his bowel burst open. The heretic Arius, who split the early Christian church with his belief that Jesus was a partly supernatural figure rather than a human being, died of a prolapsed anus in a public toilet in Alexandria.

These deaths, written about with disturbing relish and embellishment by early Christian writers, created the image of anal bleeding being linked to violations of Christ. Passion plays often included a scene in which the actor playing Judas pulled animal entrails and sausages from his costume to re-enact the gruesome death. The spilling of his guts was also often said to have infected the Jewish people with the sin of avarice. According to Bernard of Gordon in his influential 1303 work *Lilium Medicinae*, Jews often suffered from piles because they were lazy and they lived in a constant state of fear and anxiety because of the divine curse mentioned in the Bible. He did not mention the stress of living in a murderously antisemitic age.

Even those who associated with Jews or protected them could also find themselves victims of God's wrath by suffering from this problem. Sheriff John of Chesney tried to protect the Jews of Norwich from a pogrom that followed their accusations that they had crucified the child saint, William. In Thomas of Monmouth's account, "the very moment when, by protecting the Jews, John of Chesney began openly to oppose the Christian law... blood began to flow from his bottom drop by drop." He goes on to say that for three years the Sheriff endured this misery until he died. "Hence let the careful attention of the reader perpend how heavily the vengeance of God strikes the man who is not afraid to pit himself against the Holy Church and Christian laws, as this man did." A policeman's bleeding haemorrhoids recounted by an English monk may have been enough to strike fear into the hearts of anyone who might have protected Jews. Those requiring more proof of Jewish male menstruation had the line in Psalms 77:66: "He smote His enemies in their posteriors. He set them in everlasting shame."

One of the reasons given for why Jews would kill Christian children was that menstruating men needed to replenish their blood.

Jews were pictured harvesting the blood of Christian children, puncturing their bodies with multiple needles. Stories of the blood libel often contained details specifically to mock the crucifixion of Christ; children were crowned with thorns, whipped and nailed to a cross before they were drained of their blood. Ideas of toxicity were bound up with issues of blood, femininity, sin, guile and secrecy, all of which were associated in some way with Jews. From these associations, it did not require a great leap in imagination to connect Jews with mass poisonings. Poison, with its subterfuges, its connection to the occult and its invisibility, is a perfect means of linking an entire people to an unproven crime.

Jews would become known for poisoning wells, although there is only the slimmest historical evidence for this specific accusation before 1321. In the outbreak of antisemitic violence of that year, chroniclers of the time believed the well poisonings were part of a wider conspiracy sweeping across Europe. The Muslim king of Granada was said to have beseeched Jews to poison Christians across France. As they were held in constant suspicion, they refused to poison wells themselves but instead paid lepers to do it. And thus, Muslims, Jews and lepers were neatly bound into a web of crime.

These conspiracies were believed to operate on an extraordinary scale. Philip, Count of Anjou, wrote to Pope John XXII to warn him that Jews in Touraine and Anjou had communicated with Muslim monarchs who wished to convert to Judaism. They would return Jerusalem to the Jews if they were first given Paris and the kingdom of France. In order to unsettle France, the Jews paid lepers to poison wells with bitter herbs and the blood of poisonous reptiles. The proof came in the form of a letter, written in Hebrew on parchment and sealed with an elaborately worked piece of gold in the shape of a crucifix showing a Jew atop a ladder defecating in the face of Jesus. The pope took this as proof of Jewish and Muslim ill intentions and included it in an encyclical calling on the faithful to launch another crusade.

Why was the accusation of poisoning wells levelled at this time? Historians have a range of answers: antisemitic views had caught fire during the Crusades; there was a fear of the growing influence of Jewish doctors; and religious figures found the accusations a useful way to build influence among their constituents. And why wells? It was following a period of rapid urbanisation across Europe after a period of economic and population growth that peaked in about 1290. This created problems around drinking water and management of waste, as well as bringing Jews and Christians into closer proximity. Cities took over the management of water supplies, enacting a raft of new rules to control access. As people moved away from reliance on private wells to public systems, it seems greater anxieties about water-borne sicknesses and even poisonings grew.

Accusations of well poisoning were a means of extorting money and also of attacking those who were politically or socially vulnerable. Other people apart from Jews found themselves targets. Basques, who were often believed at the time to spread leprosy, required letters from the king specifically saying they were not poisoning wells. Genoese, political rivals of the people of Barcelona, were also the targets of accusations in that city. Jews, foreigners, lepers, pilgrims and beggars were all believed to have been involved in the spreading of plague. In Mallorca, a Muslim slave was accused of poisoning Christians by swimming in the sea, filling his mouth with seawater and then spitting it on the doorways of houses. He threatened to kill everyone there unless he was freed. Royal officials refused to allow him to be put on trial unless the local people compensated his owner. They decided it was not worth the money and let the man go.

AFTER THE BLACK Death, there were almost no Jews in the Low Countries but the idea of them as threats to Christendom in literature and art proved durable. Religious imagery, unhinged from any discipline of Church doctrine because of the decline in the number of priests, became extremely violent, with plays and paintings portraying graphically the antisemitism that was rife in Europe. Plays often inserted gratuitous violence against Jews such as brief addendums on the destruction of Jerusalem. In architecture and painting, the

Church was portrayed as the Virgin while the synagogue was a fallen and blinded widow. However, there had been a certain symmetry in the images until the fourteenth century when the widow becomes a hag trying to poison Jesus with her sponge of vinegar. At the end of the century, Italian artists started to identify Jews with scorpions; in Germany the sow was often associated with Jews in paintings and in stone monuments, which sometimes depict pigs either suckling or having sex with Jews.

In the spring of 1298 a butcher called Rindfleisch launched a pogrom against Jews that led to the deaths of several thousand people in Rottingen and nearby towns. This was the start of a particularly bleak century for Jews in Europe. Famine in 1315–17 killed tens of thousands of Jews and the Black Death swept through the population in 1347–49. Urban revolts and massive transformations in the countryside led to immense upheaval across the Continent, turmoil on a scale that was not repeated until the twentieth century. In Normandy in 1320, a teenage shepherd had a vision that the Holy Spirit in the figure of a bird had told him to lead a crusade against the Moors in Spain. A group of peasants joined him and moved south towards Spain, attacking towns along the way. Jews were the chosen enemy and as the group, who would be known as the Pastoureaux, moved south they began massacring communities across France; as many as 140 Jewish communities were wiped out. Authorities finally acted against them when they turned on clerics and threatened the pope in Avignon. When they reached Spain, James II of Aragon forbade them from entering. When they did anyway, he ordered his nobles to protect Jews. Most followed this command but at the fortress of Montclus, more than three hundred Jews were massacred. The *Pastoureaux* were finally broken up by King Philip V who decided they had got out of control.

The massacres unleashed fears that Jews would claim revenge. In the summer of 1321, rumours circulated that Jews had joined together with lepers to poison wells and kill Christians. A drug made from human blood, urine and three secret herbs, along with the consecrated host that was dried and made into a powder, was tied in bags and thrown into wells. A leper had confessed to this when caught in the act. He said he had been paid to do this by a wealthy Jew. Other versions had a poison made from frogs' legs, snake heads and the hair of women. Edward Fenton's translation in 1569 of Pierre Boaistuau's *Certain Secret Wonders of Nature* says the poison in question was "an ointment with a confection of the blood of man's urine, composed with certain venomous herbs, wrapped within a little lined cloth, tying a stone to the same to make it sink to the bottom". Jews allegedly placed this deadly teabag into wells across Europe.

Across France, magistrates were ordered to investigate Jews and large numbers were imprisoned or killed – a beneficial side effect of this for the monarch was the acquisition of their property. Even those Jews judged to be innocent had to pay a fine, raising vast amounts for the treasury. It was part of a cycle in antisemitism. Real violence begat fears of revenge which in turn stirred up yet more violence. Jews were massacred in several places across Europe: Deggendorf in Bavaria, Alsace and Swabia. Only after these massacres were accusations of poisoning raised against them.

TODAY ONLY THE survivors of genocide can understand what it is to have a vast proportion of the population of your land die so quickly. To lose entire families, villages, towns, everyone you have ever known or been connected to in just a matter of months is beyond comprehension to most of us. The survivors of the Black Death experienced loss on a scale that was unfathomable and there was nothing they could do to stop it. Agnolo di Tura wrote in May 1348 in his *Cronaca Senese* some of the most moving descriptions of the plague:

> And in many places in Siena, great pits were dug and piled deep with the multitudes of dead. And they died by the hundreds, both day and night and all were thrown in those ditches and covered with earth. And as soon as those ditches were filled more were dug. And I, Agnolo di Tura, called the Fat, buried my five children with

> my own hands. And there were also those who were
> so sparsely covered with earth that the dogs dragged
> them forth and devoured many bodies throughout the
> city. There was no one who wept for any death, for all
> awaited death. And so many died that all believed it
> was the end of the world.

Di Tura wrote that "it seemed to everyone that one became stupefied by seeing the pain and it is impossible for the human tongue to recount the awful thing". After the plague died down, those left in an almost abandoned city began to live their lives again. "All who survived gave themselves over to pleasure: monks, priests, nuns and lay men and women all enjoyed themselves and none worried about spending or gambling. And everyone thought himself rich because he had escaped and regained the world."

In his account of the plague in Florence written in the 1350s, Giovanni Boccaccio described death as becoming so unremarkable that funerals were no longer held and nobody wept for the dead. In *The Decameron*, the tales told by a group who flee the plague in Florence to a villa outside the city, he wrote: "Nor were the dead honoured with any tears or candles or funeral train; nay, the thing was come to such a pass that folk reckoned no more of men that died than nowadays they would of goats." Instead of grief, there was a desperate attempt to maintain some sort of psychological balance with "laughter, witticisms and general jollification".

Plague became a central part of European life and imagination for several centuries. The French historian Jean-Noel Biraben concluded in his study of the plague that the disease struck somewhere in Europe every year but two from 1347 to 1670. The terror of epidemics and the sense of the unleashing of dark forces on mankind can be seen in Pieter Bruegel's *The Triumph of Death,* painted in the sixteenth century and now hanging in the Prado in Madrid. It portrays a battle between an endless army of skeletons and terrified, defenceless mankind. The land is reduced to a smouldering battlefield, buildings lie in ruins, death plunders the wealth of kings and dogs eat corpses.

It is an image of a time that saw cataclysmic famines, the worst epidemic to hit Europe and the start of a series of conflicts that would later be known as the Hundred Years' War (1337–1453).

The Black Death may have killed somewhere between 25 and 50 per cent of the population of Europe in 1347–48. In some cities, it may have wiped out 80 per cent of the people. It is impossible to do much more than guess the toll. England has some of the best records from those times and documents relating to wills suggests a death toll of around 30 per cent of the population. However, these documents concern a small and generally wealthy group of people who may have been less exposed to the plague, as they would have lived in houses less infested with vermin and were less likely to live with their livestock. Bishops' registers that tracked the appointments of priests suggest a death toll of around half the population. These figures may again be misleading; priests generally enjoyed better living conditions than much of the population and might have had a lower death rate. On the other hand, they would also have administered last rites to many infectious people, raising their risk. Rolls of death taxes paid by peasants to their feudal lords show local death rates of between 40 and 70 per cent. Many countries and cities would not return to the same levels of population for four hundred years. In 1348, around fifty thousand people died in Paris, the largest European city of the day. The city would be hit by a further thirty-six outbreaks of plague between that year and 1480, on average once every four years. Additional epidemics of mumps, smallpox and scarlet fever also took a huge toll.

The plague profoundly changed society, even precipitating a shift from feudalism to capitalism. After considerable growth in the twelfth and thirteenth centuries, economies had slumped. Malnutrition and famine had become common. The whole continent seemed to be caught in a Malthusian bind, unable to sustain itself and yet unable to find ways to change. After the Black Death, feudalism partially broke down, as there were too few people to work the land.

Wages soared and landowners were forced to provide their workers with better food and conditions. Although living standards may have increased in the wake of the Black Death, other forces seemed to create rising resentments. Attempts to re-impose old forms of land tenure led to widely varying treatment of peasants. New poll taxes were egregiously unfair. Sumptuary laws that attempted to regulate what people could wear proved unpopular in cities with the aspiring and newly powerful merchant classes. This culminated in England in the Peasants' Revolt of 1381. The king and feudal lords had acted as if the demographic shock had never happened and precipitated a revolt that profoundly changed political development in England.

Death on this scale makes for massive changes. Priests died and so laymen took on a greater role in religion, undermining the power of the Church and opening doctrinal discussion. The death of the wealthy and their families led to bequests that developed new universities. Cambridge opened three colleges between 1348 and 1362. New universities broke the monopoly on knowledge of older institutions in Paris and Bologna. The decline in people who could speak Latin meant progress in vernacular education, which in turn provoked a backlash and desire to return to Latin's classical purity. Out of death and misery came the beginnings of the Renaissance.

The Black Death also seemed to have altered the mental state of Europe. For the Italian historian Carlo Ginzburg, it was the moment that a growing obsession with conspiracies broke through the surface of European culture. "One gets the impression that in the space of thirty years, in a generation, the obsession with conspiracy had formed a thick sediment in the popular mentality. The outbreak, or more often the imminence, of the plague had brought it to the surface."

Rumours of the involvement of Jews in the Black Death started in Savoy. A Jacob Pascal was said to have distributed poisons to other Jews. Dozens of men were arrested across the region and confessed under torture to their involvement. The stories spread to Switzerland, provoking trials of Jews around Bern, Lake Constance and Zurich. In July 1348 Pope Clement VI issued a papal bull in which he said that Jews should not be persecuted because of the plague, which he said should rightly be blamed on astral causes or divine vengeance. In October, the pope issued another decree that proclaimed the innocence of Jews, saying they died in equal numbers to Christians and that the plague had broken out in places where no Jews lived.

It had little effect in quelling the hysteria. Groups of flagellants travelled across Europe, whipping themselves for thirty-four days to win forgiveness of their sins and immunity to the plague. As they went from town to town, their displays of violent piety often ended in massacres. Jean d'Outremeuse, the Low Countries historian, wrote in his *Chronicle*:

> A general rumour spread and it was commonly said and certainly believed that this epidemic came from the Jews and that the Jews had cast great poisons in the wells and springs throughout the world in order to sow the plague and to poison Christendom; which is why great and small alike had great choler against the Jews who were everywhere taken where they could be held and put to death and burned in all the regions where the Flagellants came and went by the Lords and the magistrates...

The persecution spread across Germany with massacres in Stuttgart in November, and Lindau, Memmingen and Burren in December. In early 1349, the Jews of Basel were burned alive in a building in the Swiss town. In February 1349, before the plague had even reached Strasbourg, two thousand Jews were killed there in a single day. The total death toll in the city probably reached eight thousand as the citizens, led by a viciously antisemitic bishop, massacred those they feared would poison them. In Mainz, Jews rose up and were said to have killed two hundred people while defending themselves, but the Christians outdid themselves in revenge, killing at least twelve thousand. It raged across Europe with even those who opposed violence against the Jews such as Pope Clement VI and Pedro IV of Aragon powerless to control the rampages. The ruling family of Strasbourg was deposed by the mob for trying to stop the killings. Other rulers incited the violence, offering supporters their pick of the best Jewish property after a massacre. The only place that escaped the pogroms was England, not out of any greater level of reason or goodness but because there were so few Jews living there.

Konrad von Megenberg in *Das Buch der Natur,* written in 1350, noted:

> In many wells bags filled with poison were found and a countless number of Jews were massacred in the Rhineland, in Franconia and in all the German countries. In truth I do not know whether certain Jews had done this. Had it been thus, assuredly the evil would have been worse. But I know on the other hand that no German city had so many Jews as Vienna and so many of them succumbed to the plague that they were obliged to enlarge their cemetery greatly and to buy two more buildings. They would have been very stupid to poison themselves.

Other chroniclers, including those writing at the time of the Black Death, also dismissed the idea. "Some say that this was brought about by the Jews," wrote Konrad von Megenberg, "but this is untenable." It was clear to most writers that something so widespread and destructive as the plague must have some sort of universal source.

Gabriel de Mussis traced the origins of the Black Death in Europe to the siege of Caffa, a port on the Black Sea. The Tartars were said to have catapulted plague victims into the town. Genoese sailors fled the city and sailed home, bringing the disease with them. Giovanni della Penna, author of a work on the plague published in Naples in 1348, believed the illness originated in the choleric matter of individuals who became overheated. He accepted that poisonous air might be responsible and that celestial action might make this air toxic but it was the nature of the individual that determined whether they became ill. Most doctors did not necessarily believe it was caused by God but instead preferred the model of astral causes or local physical sources of disease such as the presence of swamps and caves.

Although Jewish doctors wrote about the plague and translated the works of others, they made almost no mention of the accusation of poisonings. One Jewish physician, Isaac ben Todros, did lament that people did not seem to care that Jews died as well: "We have seen the impression that this fever flares up in all our dwelling places and actually among our people first. Marvellously it is within our camp and strikes us a vast, severe and overwhelming blow. And [God] has not stretched out his saving hand to the most noble of the Children of Israel and calamity began with them and after them the Christians." Only long afterwards did Jews write of the association that Christians made between them and the plague. The historian Joseph ha-Kohen, writing two centuries after the Black Death, said Christians believed that Jews "had brought a mortal poison into the world and because of them the great tragedy that afflicts us has come upon us".

There is a debate among historians over whether these eruptions of violence were the result of a slow accumulation of stereotypes or whether accusations of poisoning levelled against Jews, lepers and foreigners were a sudden shift in attitudes prompted by changes in economic and political situations. It is hard to explain why violence erupted in certain areas but not in others. Current views of ethnic conflict suggest that violence against minorities needs to be organised and manipulated by those in power or by those who wish to gain from it, rather than erupting spontaneously out of some inchoate anxiety. Some historians believed that contagious disease was inclined to produce scapegoating, as groups of people under stress needed to redefine themselves and strengthen their community bonds to survive the ordeal of pandemics. In cases of the plague, it was necessary to identify an outsider and exaggerate the differences between the group and the people regarded as polluting and dangerous. Violence served to create solidarity in a group shaken by the ravages of disease.

However, this does not explain why certain groups were selected as scapegoats. Jews may have been chosen because of their ambiguous position in society, as both outsiders who were supposed to be powerless and weak and the reality that many played key roles in trade, finance and civic life. Jean de Venette, a Carmelite friar from Paris, wrote a chronicle of the Black Death about a decade afterwards in which he remarked that while people were used to massive deaths from famines, they were shocked by this new disease that occurred in a time of abundance and seemed to be the result of poisons in the air and water around them. "As a result of this theory of infected water and air as the source of the plague, the Jews were suddenly and violently charged with infecting wells and water and corrupting the air."

De Venette does not seem entirely convinced by the accusations of poisoning, saying that Christians were also caught sprinkling powders into wells. "But in truth such poisonings, granted that they were actually perpetrated, could not have caused so great a plague or infected so many people. There were other causes; for example, the will of God and the corrupt humours and evil inherent in air and earth."

Wind, putrefying corpses, ponds and stagnant water were all believed to be causes, but several writers – the Catalan Jacme d'Agramont and Alfonso de Cordoba – felt that the cause may have been deliberate but did not identify either Jews or lepers as the source. D'Agramont wrote that the large number of deaths in Catalonia, Languedoc and Provence were provoked by "wicked men, sons of the devil, who by means of very false ingenuity and wicked skill corrupt foods with various poisons and medicines". Shortly after he wrote this, pogroms against Jews started across Catalonia.

Alfonso de Cordoba also believed that Montpellier was being afflicted by a rash of poisonings. The plague was not caused by celestial activity but "from the depths of an evil discovered through the most subtle practice of profound iniquity". He went on to describe the way such a poisoning might take place.

> Air can be infected by means of artifice as when a preparation is made in a glass amphora. When this preparation is well fermented, whoever wishes to produce this evil will wait for a strong and steady wind coming from any world region. Then he will walk against the wind and will put his amphora near

a stony place opposite the city or town which he wishes to infect. Going back against the wind, so as not to be infected by the vapour, with the amphora neck covered up, he will throw the amphora with violence against the stones. As soon as the amphora is broken the vapour will spread out and disperse in the air. Whoever is touched by this vapour will die very soon as if he were touched by the pestilential air.

The imagery is strikingly similar to contemporary novels and movies in which a biological attack starts with a small vial of a nerve agent, smallpox or the Ebola virus being crushed underfoot in Times Square or some other public place. The glass vial breaking and releasing an unstoppable danger into the world has been a staple of the "biohazard" genre.

DURING THE FIRST class of the day on 21 March 1983 at the girls' middle school in Arrabeh, a village just outside Jenin on the West Bank, a seventeen-year-old felt her throat become sore and itchy. Saying she was having problems breathing, she went to a window to get some fresh air. Soon she complained of nausea and dizziness and was sent home by a teacher. A few hours later, more girls said they felt unwell and could smell something unpleasant like rotten eggs or old fish. Teachers let the girls out of their classes and started to search for the source of the problem. By mid-morning some seventeen girls were affected and were taken to hospital as the health authorities searched the area.

The next day the Palestinian director of health services visited the school and said he too could detect a sharp smell. Some thirty-four students complained of blurred vision but no toxins could be found in the school area when a specialised group of investigators returned with the heath director. On Sunday, 27 March, the Israeli Ministry of Health sent a mobile lab to the school. Again, no chemicals or poisons could be detected. On the following Saturday, fifty-six girls at the Zahra

Middle School in Jenin said they were ill and were taken to a clinic for treatment. By the end of the week, 367 students were ill, two-thirds of them girls from schools around Jenin. By the first week of April, 310 girls were hospitalised in Hebron with what was already being called "Jenin Syndrome". Already anxiety was spreading about what was causing the illnesses. The Israeli newspaper *Ha'aretz* reported: "According to military circles' estimates, poisonous substances had been spread on the window or sprinkled on the curtains of the classrooms. Laboratory tests by health officials, by chemical warfare experts of the military and by police forensic specialists, have not been completed yet but there are indications that nerve gas was used in the affair."

As more people, including several Israelis, fell ill, investigators were baffled by what was causing the sickness. There seemed to be no link to toxins in the food or water and there were no signs of contamination with pesticides or other chemicals. Soon the epidemic had spread across the West Bank and more than nine hundred people had presented themselves at hospitals for treatment, with some suffering serious relapses after being released. Reports appeared that a toxic yellow dust had been found spread in schools. Palestinian doctors in the West Bank announced that girls who had fallen ill showed signs of damage to their reproductive systems. Suggestions were made that the poisonings were part of some deliberate strategy and that girls who were coming up to marriageable age were being targeted. A story spread that Israel was anxious about the rising population of Palestinians. Fearing being swamped, the Israelis were trying to render young Palestinian women sterile.

Four days after the first girl fell ill, the Palestine Liberation Organisation (PLO) denounced what it called "a war of mass poisoning unprecedented in the world". Despite the lack of a precedent, the PLO soon found one, saying in a statement: "It is a new Nazi war against the Palestinian people." A PLO spokesman in Tunis, Ahmed abd al-Rahman, continued the theme: "The United States is responsible for the continuance of the genocide and repression that have been conducted for the past two weeks by Israel against the Palestinians in the occupied territories."

Coming soon after the massacre by Lebanese Christians of Palestinians in the Sabra and Shatila refugee camps in Beirut under the averted eyes of the Israelis, what appeared to be an attempt by the Israeli government to poison Palestinian teenagers caused an uproar across the Arab world. Yasser Arafat, recently forced to flee to Tunisia after the Israeli invasion of Lebanon, announced that it "has been established that the purpose of this poisonous substance given to the young students is to stop procreation... This is genocide after attempts to slaughter our people have failed elsewhere."

By the beginning of April, the charges were escalating. Hamad al-Aydi, a senior PLO official in Kuwait, accused the Israelis of using nerve gases and radioactive materials. "Such barbaric behaviour is an unprecedented wicked means of mass killing that the world has never seen before," he said in a report on the Kuwaiti news agency. Panic surrounding the sickness of the girls may have started to abate in the West Bank, but around the Arab world there was a near hysterical condemnation of Israeli crimes. On 4 April, the Security Council met at the United Nations in New York to discuss what it called "cases of mass poisoning in the occupied Arab territory of the West Bank". In one of those ironic moments that the United Nations often throws up, the issue was brought before the Security Council by the government of Saddam Hussein. The Council, presided over that month by the US Ambassador to the UN Jeane Kirkpatrick, called for a full investigation. In just two weeks, a few cases of illness among some schoolgirls had become an international incident attracting the attention of the world's powers. A sickness that had affected some 967, but had killed no one, had been labelled an act of genocide.

On 29 April, the Centre for Disease Control and Prevention (CDC) in Atlanta published a report on what it called the "Arjenyattah epidemic" that described the outbreak of sickness as "of psychogenic origin and was induced by stress". The CDC concluded that the rotten egg smell from low levels of hydrogen sulphide around the school may have triggered the event. The gas may have come out of the latrines, although subsequent testing did not show any release of the gas. No toxins were found in the school: "Various objects, suspected by residents in affected villages as having possibly caused illness, were subjected to toxicological study. A yellow powder from the schoolyard at Arrabeh was identified as pollen. Powder from a tin at the Yattah schoolyard was found to be calcium carbonate. Residue in a cola can from the school at Yattah was identified as cola. No toxins were detected in those or in other samples." In addition, the CDC performed gas chromatographic, mass spectroscopic and emission spectrographic analyses on nearly fifty blood and urine samples and on twenty-one control samples. Although low concentrations of chlorinated hydrocarbon pesticides were tentatively identified in several patients, no consistent patterns of any toxins were evident, and no consistent differences were found between cases and controls.

By the beginning of April, the Israeli authorities and others, including the International Committee of the Red Cross, had concluded that the outbreak was psychosomatic. Around 70 per cent of the cases were among girls aged between twelve and seventeen, ages at which girls have succumbed to cases of mass hysteria elsewhere, including well researched incidents in Alabama, Massachusetts and Ontario, in which schoolgirls had become sick in large numbers for no apparent reason. Small communities under stress seem most vulnerable to outbreaks of hysterical behaviour. The Canadian psychiatrist Francois Sirois identified seventy such outbreaks between 1876 and 1972, most of which affected young women and girls. Often there is a trigger event such as the outbreak of an illness or a natural disaster. Most took place in some sort of close environment such as a school, a convent or a hospital in which rumours spread quickly and countervailing information did not easily penetrate.

Even though various investigations all showed that there was no proof of the use of any chemical agent, the accusations did not abate. A supposed investigation published by a London-based weekly *The Middle East*, and later reprinted in the *Journal of Palestinian Studies,* could not bring itself to make the most ludicrous charges but instead left it open as to whether the incident was "poison, panic or plot?" The paper accused the Israelis of hindering the investigation, of hiding evidence and of sacking Palestinian health workers who tried to expose the alleged poisoning. After saying that the "real cause will never be known with any certainty", it blames this on the Israeli government and indirectly attributes a motive to them by quoting from a 1940 British War Office manual on chemical warfare that lists the objective of using such weapons, including: "To lower the morale of the civil population and induce a will to compromise or surrender by causing widespread disablement, anxiety and discomfort."

The accusations had a powerful resonance because they tapped into a deep vein of antisemitic belief that goes back two millennia. The libels against Jews as poisoners of wells and killers of children were once part of mainstream Christian belief. Now they have been taken up with a vengeance in the Islamic world. The belief that Jews made *matzo* bread from the blood of gentile children is still peddled in the Arab world today. That Jews are responsible for poisonings – from wells to poisoned food – are still commonplace across the Middle East.

THERE IS AN important coda to the story of Jews and the plague. Waldemar Mordecai Haffkine, a Russian-Jewish émigré and descendent of those Jews who fled eastwards in the fourteenth century to escape the plague pogroms, developed the first vaccine against the disease in the 1880s, cutting mortality by 90 per cent. His work on cholera and plague vaccines saved the lives of millions. He had his own brush with antisemitic accusations of poisoning when in 1902 villagers in the Punjab died of tetanus after being inoculated with one of Haffkine's vaccines. He was accused by a commission of inquiry of being at fault although he was later cleared of any wrongdoing; it turned out an assistant had sealed the vaccine with a non-sterile top. Another Eastern European Jew, Selman Waksman, who, like Haffkine, had been educated in Odessa, developed the antibiotic streptomycin that would cure bubonic plague. These two men saved the world from the disease.

5. PLANTS DIE, SHRUBS WITHER, DOGS RUN MAD

This woman is every man's nightmare.
Justice Michael Hyam sentencing
Dena Thompson (aka The Black Widow)

Cardinal Ferdinando Ponzetti wrote in his treatise on poisons published in 1521 that in Libya, there were women with double pupils in each eye who, like the basilisk, could poison with a glance. The Islamic scientist Ibn Wahshiyya believed that sounds could have an effect that might be described as toxic or at least narcotic: "When a man loves and desires one and when he hears her voice, his colour changes. Often he trembles and he becomes languid when exposed to that." The fact that women were able to alter the physical feelings of men merely with their voices was a step towards imagining an array of powers that were encapsulated in femininity. Given the understanding of the physical world at that time, it was a small jump from men feeling the physical effects of women to starting to regard them as potentially toxic and dangerous. There was much dispute, however, about the intrinsically poisonous qualities of women. Debates raged

over whether women could convert doses of poison
into nutrition and therefore remain non-toxic and
whether they were immune from their own poison
in the manner of snakes. If they were like snakes,
some writers questioned whether they might also
be an antidote to poison, in the way that the flesh
of vipers was often used to counteract toxins.

Dena Thompson was no casting director's idea of a black widow.
Short, plain and pasty-faced, she looked like the dowdy bank clerk
she had been. Her bland appearance concealed a much darker life, one
driven by greed, deception and violence. Thompson is now serving
a life sentence for the murder of her husband and the British police
are investigating the disappearance of a man she lived with in the
1970s. Thompson's crimes were "utterly ruthless and without any
pity", according to the judge who presided at her trial. In an unusually
direct rebuke of the convicted, the Justice Michael Hyam told her,
"Nothing can excuse you for the wickedness of what you have done."
Thompson was convicted at the Old Bailey in December 2003 of
poisoning her husband Julian Webb by mixing a lethal dose of aspirin
and the anti-depressant drug dothiepin into the curry she served to
celebrate his thirty-first birthday.

Thompson nearly got away with murder. If it had not been for
an incident with her third husband involving a bondage session that
escalated into an accusation of attempted murder, she probably never
would have faced trial for killing Webb. Thompson was cleared of
trying to kill this husband when the jury accepted her S&M defence,
although she was convicted of defrauding him of twelve thousand
pounds. The police, sensing a pattern in her behaviour, decided to
re-examine Webb's death in 1994, which had previously been ruled
a suicide.

Webb's body was exhumed and Thompson put on trial for murder.
The motive was her fear that he would discover that she was already
married, that she was awaiting trial on fraud charges and that their

plans to move to the United States to make a new life were part of a scam to grab his savings. As he lay dying after eating the poisoned curry, she refused to call a doctor and stopped family and friends from visiting him. Inevitably, the press dubbed Thompson "the black widow". The police announced they were looking for Stoin Costov, a Bulgarian man that Thompson lived with in the 1970s who had disappeared. The suggestion was put around that there might have been other victims. After the trial, a police investigator told journalists that "men can sleep safe tonight knowing that she has been taken off the streets".

Doubtless there are a handful of women currently plotting to poison some innocent man but not nearly enough to justify the language of fear that surrounded Thompson's trial. Women who kill still shock and women who poison arouse an anxiety that extends back to when a man first wondered what exactly his wife might have slipped into the cooking pot. It has been an enduring fear, tied up with sexual anxieties and fears of disease. Women were less able than men to wield swords or spears and so were said to favour less physically demanding and more indirect methods. Poison was the chosen means of murder in the domestic arena. Its ancient history began with love potions and cures produced by women. It was later associated with midwives, abortionists, manufacturers of cosmetics and witches. Poison was linked to perversion, to sexual diseases and to prostitution. Throughout history and in most cultures the associations of poison are resoundingly feminine.

Since the earliest Roman laws on poisonings, those who sold poisons were liable to as harsh a punishment as those who administered them. The Neapolitan poisoner Giulia Tofana might well have claimed that she never sold poisons, at least not with the intent to kill. Her main product, probably a solution of arsenic and belladonna, was also used as a skin tonic. A sip of a weak solution of the metal oxide would have a vasodilatory effect, reddening the cheeks in a more natural manner than rouge. Swabbed onto the skin, arsenic would lighten the skin, a desirable characteristic as a pale peaches-and-cream skin was the ideal of beauty, even in sunny southern Italy. What other uses the women found for this cosmetic was entirely up to them, according to the vendors.

This was not a defence that would have served her well. Since their origins, as far back as 4000 BCE in Egypt, cosmetics have had an array of moral values attached to them. The Greeks were often critics, as was the Roman writer Ovid. In the early Christian Church, it became established that cosmetics tempted women towards vanity and undermined the beauty that God had given them. From the sixteenth century onwards, ideas of sin attached themselves firmly to make-up. Painting faces or dying hair was a sign of deception. It was taking something made by God and hiding it behind a layer of falsehood. Hamlet, in one of his cruellest comments to Ophelia as he drives her to suicide, says: "I have heard of your paintings well enough. God hath given you one face and you make yourselves another."

The puritanical mind has always been aggrieved by this, being obsessed with control of women's bodies. Some complained of women praying while wearing make-up. "Howe can they begge pardon when their sinne cleaves on to their faces," wrote Thomas Tuke, author of *A Treatise Against Painting and Tincturing of Men and Women*. It became a popular motif in dramatic works to have someone fall victim to toxic make-up. In Barnabe Barnes's play *The Devil's Charter*, first performed in 1607, which traces the rise and fall of the Borgias, Lucrezia dies from poison secreted in her make-up. "My brains intoxicate, my face is scalded," she screams. "Rank poison is ministered to bring me to my death. I feel the venom boiling in my veins."

Some cosmetics were indeed fiercely toxic. A work titled *Tracte Containing the Artes of Curious Paintinge, Carvinge and Buildinge* contains a section on the dangers of make-up in the sixteenth century. One of the most toxic was sublimate of mercury, used to whiten the skin:

> Sublimate is called deadly fire; because of its malignant, and biting nature. The composition whereof is of salt, quicksilver and vitriol, distilled together in a glass vessel. This the surgeons call a corrosive...Wherefore

> such women as use it about their face, have always
> black teeth, standing far out of their gums like a Spanish
> mule; an offensive breath with a face half-scorched
> and an unclean complexion... So that simple women
> thinking to grow more beautiful, become disfigured,
> hastening old age before the time and giving occasion
> to their husbands to seek strangers instead of their wives
> with divers other inconveniences...

Make-up may have made you look like a donkey and caused your husband to stray, but none of this stopped women using cosmetics to achieve an enduring ideal of pale skin and blondness, even among those with dark hair and complexions. Caterina Sforza, daughter of the Duke of Milan, foe of the Borgias and an extraordinary political figure in fifteenth-century Italy, even published a book called *Experimenti* that gave recipes for whitening creams and rouges. It was the first celebrity beauty book. Considerable effort went into lightening hair as dark hair was considered a major social handicap. It was said that the first sign of the saintliness of Godelive of Bruges was her uncomplaining endurance of her ugly dark hair and thick black monobrow.

By the early eighteenth century, women across Europe were applying a thick layer of white paint across their whole face and neck to disguise blemishes or darkening of the skin. The make-up, known as *fard,* was often made of bismuth and vinegar or more dangerously of *ceruse,* a white pigment made of toxic lead carbonate. Lead also made an appearance in rouge, often made from minium (lead oxide) or cinnabar (mercury sulfide) applied to the cheeks in a variety of ways to signal social origins and aspirations. In eighteenth-century France, prostitutes wore darker rouge, the bourgeoisie lighter, pinker tones. Court women wore it in great swathes across their faces, the provincial aristocracy used small circles of it at the centre of their cheeks.

By that century some of the more extreme moralism against cosmetics had died down. Syphilis and smallpox had created a world in which many people, even in the upper ranks, had embarrassingly bad skin. When aristocrats and priests decided that they would cover up their own scars, the authors of tracts on the evils of face painting backed off. Products that were once considered luxuries had become commodities and gradually cosmetics that had mostly been used by aristocrats filtered down through most of the population. Syphilis ushered in a new period of sexual puritanism. Bath houses, places not just for washing but also for sexual liasions of all kinds, were mostly closed. This led to a rapid decline in hygiene for many people and thus led to the emergence of a whole new industry of cosmetics and perfumes. Covering up the results of this lack of hygiene became an important cultural force in fashion and clothing. Wigs were said to have become popular as a way of covering up the hair loss caused by syphilis.

With the Enlightenment, there was a brief shift to an emphasis on natural beauty. Concerns focused on the artifice of make-up and its deceptive powers. Men were lured into seeing a false youth and beauty that women might not possess under the thick layers of paint. There was a plan for a law in England in 1770, never enacted by Parliament, that would have punished such trickery:

> All women of whatever age, rank profession or degree, whether virgins or widows, that shall from and after such Act, impose upon, seduce and betray into matrimony any of His Majesty's subjects, by the scents, paints, cosmetic washes, artificial teeth, false hair, Spanish wool, iron stays, hoops, high-heeled shoes or bolstered hips, shall incur the penalty of the law in force against witchcraft and like misdemeanors and that the marriage, upon conviction, shall be null and void.

The link was clear between make-up and deception. It was the same fear that men had of being poisoned, of women taking something in their power, be it beauty, sex, domesticity or food, and turning it against men. Make-up led to vanity and from there it was a rapid careening down towards all sorts of moral dissipation. It was an easy journey from selling make-up to selling poison.

THE MANDRAKE BY Niccolo Machiavelli was one of the most successful dramas of its time. Machiavelli's renown as the writer of comic trifles for the stage is overshadowed by his political reputation, but in sixteenth-century Italy he was highly regarded as an entertainer. One performance in 1522 had to be abandoned because of the crush of an over-enthusiastic audience. The play is about the mandragora root, a plant surrounded by an array of legends going back to ancient Greece. It was believed that anyone who pulled the root from the ground would die; to harvest them, the plants were tied to the tails of dogs who were then lured forward with a scrap of food so as to pull the root from the earth. The dogs died but the harvester survived. Mandrake roots were also believed to emit a terrible scream when they were pulled out and the sound alone could cause madness. The roots, which vaguely resemble people, with two legs and a body, were a sure way of getting a woman pregnant. However, there was a catch. The first man to sleep with a woman who had eaten the root would die as a result of the poisons in her body. Mandrake made a woman deadly and the only way to draw out the poison was for another man to sleep with her.

Machiavelli's comedy has it all: infertility, toxic women, husbands pimping out their wives to strangers, mothers conniving in the infidelities of their daughters, husbands being cheated on by young wives and certain death associated with sex. In these themes, the Florentine author found the ingredients for a hilarious Renaissance sex romp. Nicia, an elderly and intellectually limited lawyer in Florence, is married to Lucrezia, a beautiful young woman who has not yet produced any heirs. A doctor prescribes mandrake root to Nicia, warning him that he must find someone else to sleep with his wife before he does in order to draw out the poison. In fact, the whole thing has been dreamt up as a way for Lucrezia's would-be lover to sleep with her. Nicia sets himself up as to be cuckolded through a clever and cruel trick. Renaissance audiences, perhaps inured to the nastiness of the play, adored it all.

Women do not just poison; they are poisonous themselves. They have poisoned with a look, a touch, a kiss, the sound of their voice or their breath, indeed their very nature has been tainted. The idea of the toxic woman held a fascination for men in the Middle Ages and Renaissance as Machiavelli's play shows. Margaret Hallissy in her work *Venomous Woman* uncovers this literary theme, showing how poisonous women emerge in the thirteenth century, reappear in the late Renaissance of the sixteenth century and then come back again in the nineteenth century. The timing is no coincidence. These were periods in which the status of women underwent rapid change, not always for the better, in the early twelfth and thirteenth centuries when the more misogynistic ideas of the Church were developing and ideas about witchcraft were starting to brew. In sixteenth-century Europe, women were being forced out of their traditional roles as healers as more formalised systems of medicine developed. Societies were also dealing with new diseases and cures from the New World and a shift towards medicine being connected with business and the trade in new drugs. In the nineteenth century, women's lives were changing at the same furious pace as science and industry. Shifts in medical science, changes in religious attitudes and changing dynamics of travel, trade and knowledge all contributed to emerging obsessions with poison.

The publishing sensation of the thirteenth century was the *Secretum Secretorum*, a collection of the supposed correspondence between Aristotle and Alexander the Great. The origins of the work are murky but it is thought to be derived from an Arabic text known as *The Book of the Science of Government* that dated back to the tenth century, which in turn may have come from the Indian work *The Arthashastra*. *Secretum Secretorum* was one of the most widely read books of this period, translated into an array of languages across Europe and widely cited in sermons and by other writers. Aristotle was one of the great ancient thinkers then becoming popular again and Alexander's history was surrounded by mystery and romance, as it still is today. Even as a fake – *Secretum Secretorum* was almost certainly first written in Arabic, not Greek – it had an immense impact on the re-emergence of classical thinking during the Renaissance.

One of the more popular stories in *Secretum Secretorum* was about an attempt to kill Alexander. An Indian queen feeds one of her daughters small amounts of poison every day until she becomes toxic. She is sent to Alexander as a gift with the aim that she would poison him with a kiss. Aristotle suspects a plot is at hand and saves Alexander's life by pouring an antidote around the woman "with the nature of a snake". Later variants of the story developed in which Alexander is sent a whole troupe of poisonous dancing girls. In another the girl is planted in a snake's egg and then raised by a snake who endows her with the capacity to poison. The means of killing the poisonous woman varies. A variant of oregano known as dittany, lime juice, garlic and the saliva of fasting men are all employed as antidotes to toxic women.

The story has its origins with the works of Kautilya, the Indian chancellor to King Chandragupta Maurya (350–275 BCE), who ruled an empire stretching across the Indian subcontinent from a city twice as large as Rome. Kautilya was the genius behind the throne, advising Chandragupta with a skill unsurpassed by political advisers since. Women were a critical part of Kautilya's war plans. They were necessary to maintain the morale of soldiers, he wrote in his manual *Arthasastra*, and therefore encampments of courtesans were to be established along highways. Women were powerfully addictive, according to Kautilya, and therefore it was possible to use them to sow discord among the enemy. To undermine a group of leaders, a ruler should "make chiefs of the ruling council infatuated with women possessed of great beauty and youth. When passion is roused in them, they should start quarrels by creating the belief (of love) in one and by giving (love) to another." In times of war, women should be sent to poison the generals of the opposing army. Although Hindu laws of war said poison was not to be used, Kautilya was a great believer in the effectiveness of assassinating a king, often by sending women or children to administer the toxin.

The earliest known mention of the poison damsel comes from the *Kathasaritsagara* of Somadeva. It includes a passage that also might be the earliest mention of chemical warfare: "He tainted by means of poison and deleterious substances, the trees, flowering creepers, water and grass all along the line of march. And he sent poison damsels as dancing girls among the enemy's host and he also dispatched nocturnal assassins into their midst."

Indian mythology and stories also include poisonous men. In a story dating from the sixteenth century, Mahmud Shah, ruler of Gujarat, was said to have been raised on poison by his father in order to frustrate any attempts by others to poison him. It was a popular myth taken up by several travel writers. They describe Mahmud eating fruits called chofole, certain herbs and leaves of such plants as a sugar orange called tamboli and then oyster shells. He then spits this onto someone he wanted killed and the person dies of poisoning within half an hour. He had a court of many thousands, including a vast harem, because any woman he slept with was found dead in the morning. His clothes were only worn once and were so toxic they could not be touched. It was alluded to in the poem *Hudibras*, the mock epic written in the seventeenth century by Samuel Butler:

> *The Prince of Cambray's daily food*
> *Is asp and basilisk and toad,*
> *Which makes him have so strong a breath*
> *Each night he stinks a queen to death.*

Some of the descriptions by European writers of Indians with toxic red saliva may be distorted descriptions of betel nut. The nut is mixed with leaves and lime before it is chewed, producing a deep red mix of juice and saliva. Spitting betel nut juice into someone's face was a great insult in India, the symbolic equivalent of a poisoning.

IN THE MIDDLE Ages, the story developed a Christian overtone. The envenomed maiden represented luxury and gluttony. She tempted men and poisoned their souls. The story was a favourite for sermons and evolved into an enduring literary theme, being revived at times when anxieties about science and the role of women peaked. It also appeared to be a common fate that befell aristocrats, killed by women they desired. King Wenceslas III of Bohemia was said by the poet Ottacker to have been killed in 1306 by his mistress Agnes, who was paid to poison him. "How could you mix poison with the fathomless sweetness which you carry in your delicate body," Wenceslas asked his toxic lover. King Ladislaus of Naples was also rumoured to have been felled by a poisonous woman in 1414. He had invaded Florence, although his armies were stopped by an outbreak of plague. To avoid another attack, Florentines bribed a doctor from Perugia, whose daughter was the king's mistress. Her father gave her a cream containing aconite that she was to smear over her body before sleeping with the king. The ointment killed them both.

The idea of the poisonous woman was also bound up with misogynistic views of sin. These ideas had hardened during the Lateran councils in the early part of the last millennium. The Church imposed celibacy on priests, mostly to limit offspring that might take over Church lands, and many of its attitudes towards women and marriage became increasingly harsh. Abbot Conrad of the Premonstratensian Community at Marchthal wrote in 1273 that women were so beyond the pale that his monks would live entirely without them. "We and our whole community of canon, recognizing that the wickedness of women is greater than all the other wickedness of the world, and that there is no anger like that of a woman and that the poison of asps and dragons is more curable and less dangerous to men than the familiarity of women."

For Christians, the origins of the poisonous woman and poisoning woman start with Eve. In the first case of domestic betrayal, Eve takes from the serpent and gives Adam what is essentially a poisoned apple. The "Fall from Grace" is precipitated through poisoned food, prompting thousands of years of fears that all women were engaged in similar attempts to drag men down. The snake becomes both a symbol of evil and associated with women in Christian imagery. The lamia – also known as a striga – goes back to the Greek mythology of the messy and complex marriage of Zeus and Hera. Zeus has an affair with Lamia and the jealous Hera transforms her into an animal with the face and breasts of a woman and the body of a snake. She also kills Lamia's children, a mistake in retrospect. In revenge for this, snakes have been toxic and capable of killing humans ever since.

A common folklore and literary theme is a story in which a young man is tempted by a woman who turns out to be a snake. He is rescued by an older, wiser man who teaches him about the risks of lustful women. Women are often associated with poisonous animals – spiders, snakes and basilisks – principally because of fears of female sexuality. Combine concerns about venereal disease, a powerful anxiety about mortal sins fostered by an increasingly conservative Church, add in fears of castration and domination by women, and you get the wildest fantasies of misogyny – the idea that women are responsible for all the mental and physical failings of men.

If women could kill with a look and cast a toxic spell over men, it was inevitable that they should be compared to the basilisk. In *The Parson's Tale*, Chaucer compares the basilisk's toxic impact to the effect lust has on men. In Renaissance poetry and plays, the basilisk commonly represents the dangers of female seduction and beauty. These powers enter men through the eye. In Philip Sidney's *Astrophil and Stella*, beauty is a source of power that verges on the poisonous:

> *Stella, whence doth this new assault arise,*
> *A conquered, yelden, ransacked heart to winne?*
> *Whereto long since, through my battered eyes*
> *Whole armies of thy beauty entered in.*

Throughout the poem, Astrophil is physically undermined by Stella's beauty; as with many Renaissance lovers, he is the victim of the female's basilisk gaze. Female desire, the "blacke beames" of *Astrophil and Stella*, is clearly a dangerous force.

BELIEF IN THE basilisk was fading by the seventeenth century, indeed the English philosopher Sir Thomas Browne wrote in *Pseudodoxia Epidemica* in 1646 that it "had no real shape in nature and is rather a hieroglyphical fansie". But its resonance as an image of feminine danger and the projection of desire endured much longer. "Cockatrice" was slang in Elizabethan England for prostitute, a word that itself derives from the idea of public display. The basilisk was said to have come from Medusa's blood and inherited her capacity to turn men to stone. In Shakespeare's *Richard III*, Anne tries to brush off Richard's attempt at seductive talk of her eyes by wishing she was a basilisk and could kill him.

The idea of the poison damsel was revisited in the nineteenth century by the American writer Nathaniel Hawthorne. "Rappaccini's Daughter" is the story of a natural scientist in Renaissance Italy who raises his daughter in a garden of toxic plants and feeds her a diet of poisons. She is extraordinarily beautiful but as her breath alone can kill butterflies and wilt non-toxic plants, she lives a life of isolation in the garden. A young medical student moves in next door and spots her over the wall. He falls for her but cannot approach her. He is warned of the story of Alexander in the *Secretum Secretorum* but decides to offer her an antidote he has been given by her father. The story ends in tragedy when her poison kills him and his antidote kills her. The story is explicitly about the dangers of science and experimentation, and reflects burgeoning Victorian anxieties. Hawthorne had been fascinated by the rewriting of the story from *Secretum Secretorum* in Browne's *Pseudodoxia Epidemica*: "A story there passeth of an Indian king that sent unto Alexander a fair woman fed with aconite and other poisons with this intent complexionally to destroy him." Hawthorne had also read that in Mexico, people were inoculated against the bites of rattlesnakes by being pricked often with the teeth of a rattlesnake in order to allow a small amount of poison to penetrate their bodies. After a while, not only were they immune to the poison but were also said to be poisonous themselves.

Curiously, the poisonous women of literature are much more likely to be blameless virgins than femme fatales. In Oliver Wendell Holmes's novel *Elsie Venner,* as in "Rappaccini's Daughter", neither Elsie nor Beatrice, both poisonous women, are ever actually deflowered and yet both are just assumed to be poisonous to any man who has sex with them. Beatrice does have a poisonous aura that kills plants and insects around her but it is never actually proven that any man who has sex with her would die. The folklorist N. M. Penzer connects this to the fear of sex with a virgin, fears linked to the idea of vagina dentata and also anxieties about sexually transmitted diseases. The fate of poisonous women is never their fault. They do not set out to become poisonous and yet they suffer, either from the actions of others who make them this way or those who wish to save them. Men cannot distinguish between what women are and what they represent to men.

THE IDEA OF women being impure and in some way physically dangerous within themselves goes back much further than literary references to toxic women and is connected to the near universal taboos that surround menstruation. These do vary in time and across culture and are not universal, but they are widespread and strongly held. The notion of women being repositories of danger goes back as far as the ancient Greeks but was not always connected to menstruation. Hippocrates believed that women had softer flesh than men and absorbed blood to the point that it became painful. They released this blood through menstruation, a process to which he did not attach any notions of impurity or sin. Pliny was less forgiving. "One will not easily find anything more stupefying, more anomalous, than women's periods," he wrote.

The deep fears and taboos around menstruation in Europe owe a debt to the Jewish laws written in the Book of Leviticus, which lays out in elaborate detail when women are regarded as unclean and what they should do to purify themselves after menstruation or birth. In three verses, Leviticus warns of the dangers of having sex during menstruation. Many of these attitudes appear to have declined in the early days of Christianity. Christ's attitude to women shown in the gospels suggests he did not regard them as impure physically. The focus shifted from ideas of pollution to concepts of sin and corresponding response from ritual purification to redemption. Christ's view of women, even women who would be regarded as extremely impure such as Mary Magdalene, was a marked departure from Jewish custom. The New Testament undermines the message of Leviticus with its tolerance for lepers, women and criminals. The doctrines surrounding the Virgin Mary, the Catholic and Orthodox emblem of all that is pure and without sin, also challenge the idea of the intrinsic impurity of women, although this would resurface often.

Judeo-Christian culture associates menstruation with guilt, sin, impurity and poison as do many other cultures, although not all by any means. In Hindu mythology, Indra kills Vishvarupa when he threatens to destroy the universe. Vishvarupa is a Brahmin and thus the guilt of killing him is too great for Indra to bear alone. He shares it among the earth, trees and women, but in exchange gives each a gift. The earth, when dug, will heal again, trees, when cut, will grow again and women, unlike other animals, will be able to enjoy sex at all times. But only women seem to suffer the downside. Indra's guilt takes the form of menstruation for women and the fact that they were believed to be as impure as an untouchable on the first day of their period and as impure as a Brahmin killer on the second.

Some early Christian thinkers were remarkably tolerant, perhaps as the religion consciously set itself apart from both Jewish and Roman attitudes to impurity. Pope Gregory I wrote in 597 that pregnant women could be baptised, a break from earlier custom. Menstruation, in his view, was a mark of sin – it came from the curse of Eve – but it was also a necessary component of fertility, which was God's gift to man. He did not regard menstruation as impure, writing, "If no food is impure to him whose mind is pure, why should that which a pure minded woman endures from natural causes be imputed to her as uncleanness." Saint Augustine also held similar views, saying there was no reason to ban a menstruating woman from entering a church "for this natural overflowing cannot be reckoned a crime". However, in one of the canons of first Nicene Council, written sometime after 325, it states: "For husbands, it is not allowed that they approach their wives during menstruation, so that their bodies and their children will not manifest the effects of

Elephantiasis and leprosy; in fact, that type of blood corrupts both the body of the parents as well as that of their child."

Medieval theologians were far less tolerant when it came to the intrinsic impurity of women. William of Auvergne wrote in the thirteenth century that menstruating women carried such an aura of contagion that their gaze alone would leave a smear of filth on a mirror. Lothario de Segni, who would later become Pope Innocent III, wrote in his *On Contempt for the World* that menstrual blood was so "detestable and unclean that grains that come into contact with it will not germinate, shrubs will wither, plants will die, trees will lose their fruit and if dogs were to eat it they would run mad. Babies conceived (during menstruation) contract the defect of the seed so that lepers and elephantics are born from this corruption". *De Secretis Mulierum*, a work said to be written by Albertus Magnus but generally regarded as by other hands, said menstruating women were toxic to children. At a time when vast numbers of children died in their early years, it was perhaps not as outlandish as it is today to find an explanation in the toxicity of the people closest to them.

The Dutch doctor Levinus Lemnius drew on Aristotle's ideas of the generation of children when he wrote in his *The Secret Miracles of Nature* published in 1559 that men provided the seed and women, in the form of blood, the material that generated the foetus. At the time of menstruation the blood was "corrupt and dirty and unsuitable to receive a beautiful and well-shaped form". Having sex during menstruation would cause the woman to give birth to a monster, a subject of intense curiosity in the mid-sixteenth century when both Lemnius and the French court doctor Ambroise Paré wrote extensively on the subject. In Paré's *Des monstres et prodiges*, he also suggests that the way you have sex could affect the outcome: "It is certain that most often these monstrous and prodigious creatures are a result of the judgment of God, who permits fathers and mothers to produce such abominations through the disorder that they create in copulating like brute beasts." The justification for this came from the Book of Ezra, one of the apocryphal books of the Old Testament purged from the Bible read by most Christians today, that said having sex during menstruation produced monstrous offspring.

The Jewish physician Maimonides wrote that doctors had told him that men were dying of illness shortly after their marriage and that it appeared to be connected to women using their menstrual blood as a poison. "The physicians found out at last that the medium they 'worked' with was menstrual blood, which they added to the food at the start of their period, the smallest bit of which resulting in great pain." He also mentions that this is not included in any of the works on poisons that he had read and that there was little indication of how to treat this type of poisoning apart from the general approach used for many toxic substances such as emetics. At the end of his note on this, he adopts a slightly sceptical tone, saying that he himself had no experience of this but was passing on the knowledge just in case.

Menstrual blood was also a common component of love potions. Burchard of Worms in the eleventh century said that any woman who mixed menstrual blood with her husband's food in order to increase his sex drive should be subject to five years' penance. He added that women who kill a fish, mix it in their afterbirth and then serve it to their husband should be subject to just two years' penance. These were relatively mild punishments. At the height of the Inquisition, the use of menstrual blood in any way was regarded as a serious crime with women charged with such crimes as saving the blood from their daughters' first period in order to feed it to the girls' future husbands.

The notion of women being intrinsically poisonous did throw up some problems for theologians. Among the concerns were questions over whether the Virgin menstruated and whether her menstrual blood had been involved in the production of Jesus in the womb. Theologians were reluctant to remove Mary entirely from the realm of humanity so they agreed that she did menstruate but declared that her blood lacked any impurities. Thomas Aquinas believed that the Holy Spirit had brought entirely pure blood rather than tainted menstrual blood to the Virgin's womb. Milk was said to be purified blood and the iconography of the Virgin focuses much more on her breastfeeding than on conception or birth. As the Virgin was bodily assumed into heaven, there were no remains to hold as relics. Instead of body parts, vials of her breast milk were common objects of adoration. Early collections of stories of Marian miracles include one in which Mary appears and squeezes droplets of milk around the ulcerated mouth of a dying monk to heal him.

Taboos surrounding menstruating women, which often require them to be isolated or to refrain from certain activities, are widespread but not universal and their meaning is varied. In some cases, they clearly exist to protect men from the imaginary toxic aspects of women; in others, they are about protecting fertile women from the dangers that surround them. There has also always been some debate over the toxic nature of menstrual blood. In a reversal of the usual thinking of the time, Hildegard of Bingen believed that men were the toxic ones. In her *Causae et Curae,* she wrote that Adam transgressed God's commands so instead of being pure he now "emits the foam of semen instead of pureness. Had the human stayed in paradise, he would have remained in an immutable and perfect state. But after his transgressions he was turned into something different and bitter". Semen was not just poisonous but had the power to curdle women's nourishing blood in such a way as to form a foetus. "For with Adam's transgression, man's strength in his genital member is turned into a poisonous foam and woman's blood into a contrary effusion."

In the 1920s, a doctor, Bela Schick, best known for developing a test for diphtheria, suggested that menstrual blood contained bacteria that produced "menotoxins" that could produce observable physical responses. It was these toxins that produced effects that had led to the common beliefs that menstruating women could cause plants to wither and animals to die, wine to turn sour and bread to fail to rise. Schick's ideas were never well regarded, but in the 1970s were re-examined with scientists falling on both sides of the argument of whether menstrual blood was toxic. This in itself was not a sufficient explanation for the rituals and exclusions that surround the subject – the poison was much more symbolic than observable fact. Given that "menotoxins" remain completely unproven, there is a sense that even supposedly scientific efforts to prove toxicity are much more about symbolism and culture than anything else.

Lust was most often cited as a reason for poisoning. In Medieval Europe, the common theological view was that women were more highly sexed than men. The Dominican friar Vincent of Beauvais wrote in his thirteenth-century encyclopaedia *The Mirror of Nature* that women were not just more lustful than men but they were more lustful than all female animals except for possibly mares. Vincent was drawing on an array of sources for his book but he mostly learned from the work of William of Conches, who explained the apparent contrast between the reluctance of women, seen as intrinsically warm and wet, to engage in sex and their ferocious libido by explaining that "it is harder to light a fire in wet wood but it nevertheless burns longer and more intensely in it". Lust, the dangerous sexuality of women and evil are brought together in the poisonous woman. It is the deeply libidinous women who poison their husbands to replace them with younger lovers. Women kill out of jealousy and passion whereas men kill for more noble reasons of honour and to prove their courage. Lucrezia Borgia, Marie Antoinette and many others were all portrayed, obviously unfairly, as voracious, insatiable and promiscuous to the extent they consumed men with their sexuality.

There has often been a widespread belief that women were not just poison but antidote – the homeopathic principle was that all things contained within themselves a cure and therefore women could also be the cure. Augustine believed that the Virgin Mary was in some ways the antidote to Eve. "The poison to deceive man was presented him by a woman, through woman let salvation for man's recovery be presented." Statues of Mary often portray her foot on the head of a snake, a Marian avenger for Eve's seduction by the snake as well as a conqueror of sin.

The Spanish physician Diego Alvarez Chanca published his *Tractacus de Fascinatione* in 1499 on the risks of the "evil eye" by which people could transmit illness from themselves in some sort of toxic emanation that would be absorbed by others through their pores, the process of "fascination". The work was quite a break from established views; it had not been discussed as a method of transmission of illness by the greats of ancient medicine and so was not widely accepted. However, it was widely believed that women could cause illness through the "evil eye". The idea was closely connected to the toxicity of menstruation. The poisonous nature of menstrual blood was used by Chanca to show that it was possible for a poison to exist within a healthy body. Women were more likely to fascinate than men because they were more unstable, they ate more corrupted food, and during menstruation they gave off toxic vapours through their eyes. The problem with making a link between fascination and menstruation is that even those women who did not menstruate or were no longer menstruating were able to poison from a distance. The physician Lopez de Corella asked the question, "Why can old women fascinate children" in his *Trescientas Preguntas de Coasa Naturals*. A lot of doctors recommended that old women not be allowed in the delivery room because of their potentially dangerous impact on the newborn child. Hate or envy in an old woman would produce a stronger boiling of the venomous blood and therefore greater amounts of vapours to be given off. Chanca even warned old women against looking children in the eye lest they inadvertently harm them.

For a time, the idea of a poisonous woman was the subject of serious scientific debate. Andrea Bacci's *De venenis et antidotis* (1586) questioned whether it was possible to create a poisonous woman by giving her small doses of poison. The toxin would be turned into nutrition, he argued, and thus it was impossible that Alexander had been killed by a poison damsel. Sir Thomas Browne, author of *Pseudodoxia Epidemica* and a great debunker of beliefs in the seventeenth century, wrote that "a poison is not without a poison unto itself". Ingested poisons, he wrote, would be "refracted and subdued" so as to become antidotes to what they originally were. Animals that could digest these poisons would act as an antidote to them.

The poisonous woman did not die with the basilisk or even with the scientific knowledge that women could not kill with their voices, eyes, breath or blood. In *The Blood of Fu Manchu*, a 1968 film based on the novels by Sax Rohmer, Fu Machu immunises a group of enslaved women by injecting them with the venom of a black cobra.

Fu Manchu, played by Christopher Lee in yellow face make-up, then sends them out to assassinate various world leaders through the "kiss of death." Poison Ivy in the Batman comics and movie is a botanist who can also deliver poison with a kiss. She is not intrinsically poisonous, but skilled nevertheless in the creation of toxic potions and hallucinogens. In Gabriel Garcia Marquez's novel *One Hundred Years of Solitude*, Remedios the Beauty has a toxic effect on foreigners who appear in Macondo. She wanders around naked, her head shaved but her beauty undiminished. However, she ends up causing the death of an American who spies on her in the shower. He falls from the roof and his skull cracks open to reveal an oil that contains her unique scent. In the X-Men comic book and movie series about persecuted "mutant" humans with strange powers, the character Rogue can kill with a kiss, draining energy and consciousness from her unfortunate boyfriend. Her mutant power is to absorb the memories and life force of others, a modern echo of Circe.

6. MORE WINE, MORE EXCITEMENT

*'Fill up, fill up to the brim,' cried Alexander Borgia
in his happiest voice from the head of a table that
stood in the centre of the magnificent dining hall.
'Let pleasure reign in every heart and the sparkling
wine go around.'*

Anonymous, nineteenth-century pamphlet

The basilisk was many different animals at
different times. Some were lethal, others clearly
not. Its features roam across the spectrum of
wildlife from chicken to fish to snake to deer. It
was linked to the salamander, dragons and the
hydra. The resemblances helped convince people
of the possibility of its existence. It was described
by its similarities to other animals; it was part this
and part that. But when natural sciences moved
from noting similarities to marking out differences
among animals through systems of classification,
the basilisk could not survive. Natural history also
shifted from the description of all animals that
might ever have been, particularly those described
by classical authors, to the identification of those
that could be observed or collected.

In late Medieval Europe, the re-emergence
of classical knowledge brought with it enormous

understanding of the world, but it also revived ideas like the basilisk and set them firmly in the minds of believers. If medical luminaries from the ancient world such as Galen and Dioscorides wrote that they existed, then they surely existed. If contemporary evidence was lacking, it was because people had not looked hard enough or that animals such as the basilisk existed only in remote lands. The study of poisons had a difficult time breaking away from ancient Greece and Rome. The fragments of the writings that survived created an enduring set of ideas about poisons, poisonous animals and antidotes. The subject was also bound up with the occult and witchcraft. Poisons were fraught with danger and a clear and present threat to life. In this environment, it was hardly surprising that even Leonardo da Vinci, a great observer of nature and a man relatively unburdened by dogmatic ancient knowledge, should have believed in their existence.

It was at dinner with Adriano Castellesi, also known as Cardinal Corneto, that Pope Alexander VI was said to have fallen victim to his own plot to kill the newly installed cardinal. A servant, who had been ordered to pour Castellesi a poisoned cup of wine, instead pours it into the goblets of the Pope, his son Cesare Borgia and Cornelius Jansen, whose presence is somewhat mysterious as the Flemish theologian was born in 1585, more than eighty years after Alexander's death.

Lucrezia, the Pope's daughter, who is also at the dinner despite being known to have been at home in Ferrara at the time, cries out that they have been poisoned. Alexander responds: "Beautiful devil. She has spoken the truth. I am indeed poisoned, I can feel it in every vein. But the antidote... the antidote..."

"Is here," cried Donna Lucrezia as she drew a vial from her bosom and held it up to the light. "Here it is but it is not for thee. There is only enough for one." She gives it to Father Jansen, who survives while her father dies. "Dead?" gasped Father Jansen, inquiringly, as he turned over on his side, and gazed towards his late enemy with a look that seemed to have lost none of its philosophical calmness.

The author of *One Link in the Chain of Apostolic Succession; or, the Crimes of Alexander Borgia*, an anonymous nineteenth-century tract published in Boston, made no secret of why he was highlighting the life of such a notorious Pope. "I regard the church of Rome as a thousand-headed hydra, each head nurtured to its present fullness by a thousand crimes. I look upon it as a festering hell on earth, beneath whose seeming piety and sanctimoniousness are enacted scenes of blood and cruelty more revolting than any ever witnessed in the original Pandemonium." The Church, he warned, is a "hideous reptile", a "terrible guillotine" and a threat to "the rights of all Protestant Americans".

Three and a half centuries after Alexander VI died in Rome, the Borgia pope was an enduring club with which to beat the Catholic Church. It was his family's association with poisoners that helped cement in place a deep-seated idea that Italians were master poisoners and that this dark art reached a peak of sophistication during the Renaissance. The Borgias, a family that included two popes and was central to European history and culture across three generations, are remembered more for their murderous behaviour than any achievements. Immediately after Alexander's death, ever darker threads started to be woven into stories about the family, including a sexual licentiousness that knew no bounds. His daughter Lucrezia was accused of incest and poisoning; Alexander was said to have been a Jew who won the papacy with the help of the Devil.

The family had no shortage of enemies. On top of being powerful, greedy, arrogant and often violent, they were also from Valencia.

For the Roman elite, the Borgias were too foreign, too Spanish and most certainly not part of the Italian aristocracy that shared out the papacy among various great families. Political and religious life was violent and the Borgias, particularly Alexander, had participated with gusto in the corruption and viciousness that characterised Church and State.

But this does not entirely explain the stain that has permeated the Borgia name for so long. Other popes had children. Alexander's predecessor Innocent VIII was the father of at least two children that he acknowledged, although a local wag remarked when he was chosen as pope that "justly could Rome call this man 'father'". Alexander's successor, Pius III, aged and senile by the time of his month-long reign, was the father of a happy brood of twelve. Julius II had three daughters. Sixtus IV made no fewer than six of his nephews cardinals.

Julius II, who commissioned St Peter's Basilica and the ceiling of the Sistine Chapel, was certainly more extravagant than Alexander, who was known in his time for being somewhat austere in his personal tastes. Julius liked wars: he often styled himself as a warrior pope and dressed up in lavish military costumes. He also liked expensive art and was a patron of Michelangelo and Raphael. During his reign, work began on the new St Peter's, an extravagance that was one of many factors that led to the Reformation. He was widely reviled at the time. The philosopher Erasmus wrote a scathing description of Rome under Julius II in his *In Praise of Folly*. The Borgias, however, captured the imagination of later centuries in such a way as to sustain their image as the epitome of evil. It is hard to see that they would be so regarded if they were less associated with poison or if Lucrezia Borgia did not represent a critical mass of attributes associated with poisoning in the minds of later generations – a powerful, beautiful, Catholic, Italian woman with allegedly deviant sexual tastes and multiple husbands. There is also the matter of Alexander's death, which was immediately declared a poisoning by chroniclers of the time and attributed not to some other malevolent force but to the Pope himself. There was a poetic justice in the idea of Alexander poisoning himself and his son, but it is far from clear what actually happened on that August night.

Rome is not a healthy place in the summer. The fetid months from June to September were times of epidemics and the city's history from before the Christian era to the twentieth century is marked with terrible outbreaks of disease. A miasma of illness was believed to hang over Rome, fed by the surrounding marshlands. In 1503, the weather was hot and fevers were ravaging the population, killing not just those in the slums but aristocrats who were no less vulnerable, mostly to malaria. Francesco Fortucci, a diplomat in the Florentine Embassy in Rome, wrote to his superiors on 22 July 1503, begging to be allowed to leave: "People are dying in great numbers." His ambassador, Alessandro Bracci, was already dead and his replacement was sick. The letters got increasingly desperate as July passed into August. "I thank the Signoria for leave of absence because I myself am uneasy and almost out of my mind with fright; for so many people are dying of fever and there is also something like the pest."

The Pope was not at ease. Early in August he had been upset by the death of his nephew, Cardinal de Monreale. Alexander was seventy-two. He was tired and suffered from fainting spells that some doctors later speculated might have been a form of epilepsy. Early in August, he was said to have remarked that "this is a bad month for heavy people". He may also have been mindful that it was not a good time for popes. Alexander's five predecessors had all died in July and August – Callixtus III on 6 August 1458, Pius II on 15 August 1464, Paul II on 26 July 1471, Sixtus IV on 12 August 1484 and Innocent VIII on 25 July 1492.

On 9 August, another of his nephews, then Captain of the Vatican Guard, also died suddenly. Money was a worry, as it always was, given the papal networks of patronage and the displays of status that had to be maintained. Alexander had amassed a fortune as a cardinal but had to buy the votes in the conclave that elected him pope. More was needed for a campaign in Tuscany that would cement papal power more firmly. In May, the Pope had held a consistory, a ceremony in which new cardinals were inducted into the curia as princes of Rome.

Each paid a substantial amount for the position. Giustinian, the Venetian ambassador in Rome and a notable and gossipy chronicler of the times, estimated that each paid 120,000 ducats (a ducat would be worth around forty pounds today). It was still not enough, so Alexander and his son Cesare decided that they would take over the estates of Adriano, Cardinal di Corneto.

Corneto was a skilled diplomat who had been Papal Nuncio in Scotland and Bishop of Bath and Wells. He now had command of the Vatican finances and had amassed a considerable fortune. Having just completed a palace in Rome (later the residence of the English ambassador to the Vatican and now known as the Palazzo Giraud-Torlonia), he requested the honour of entertaining the pope for dinner. Tables were set up in a garden and the papal party arrived for the evening. Later accounts of the dinner would have it that Cesare had conspired with his father to kill nine cardinals to get their money. They were to be poisoned with *la cantarella*, the legendary Borgia poison. There has been much speculation over the ingredients of this, but it is likely that it was arsenic. Although arsenic trioxide is itself an effective poison, *la cantarella* was processed in a way that was believed to have made it more toxic. A hog, or in some versions a bear, was poisoned with a massive quantity of arsenic and then slit open. As it putrefied, the liquid that formed was drained off and distilled until a white powder was left. This process would certainly have contributed to the mystery of the poison, although it would have done nothing to make it more effective.

Alexander liked a good party, although this was more to do with a fondness for display than personal greed. His coronation as pope was exceptionally elaborate. Statues of bulls, the Borgia emblem, were set up around Rome. From the forehead of one flowed a crowd-pleasing river of wine. Alexander himself was inclined to eat modestly to the point that Roman grandees rather resented dining at the Vatican because of the sparse fare. But he was sexually athletic, even into his old age. He had been notorious as a young man for his wild ways. After a rowdy party in Siena in 1460, Pope Pius II wrote to the young cardinal to rebuke him. "We have heard that the most licentious dances were indulged in, none of the allurements of love were lacking and you conducted yourself in a wholly worldly manner. Shame forbids mention of all that took place, not only the acts themselves but their very names are unworthy of your position. All Siena is talking about this orgy."

Appointed a cardinal at the age of twenty-six by his uncle, Callixtus III, Rodrigo Borgia rapidly ascended to the position of vice-chancellor, a post that gave him immense power within the Vatican and access to a fortune. His elder brother, Pedro, received many of the temporal gifts of the pope, titles and positions such as Prefect of Rome. His greed and his antagonism with the powerful families in the city made him so unpopular that when his uncle died, Pedro was driven into exile and died shortly afterwards. Rodrigo took his place as the patriarch of the Borgia dynasty. He had the good fortune to cast the deciding vote for Pius II, who despite his letter of rebuke, regarded Rodrigo with some warmth.

Despite spending much of the considerable fortune earned from his bishoprics of Valencia, Cartagena and Portus and from the sale of indulgences, Rodrigo did not emerge from the 1484 conclave as pope. Innocent VIII won, mostly by signing whatever promises the cardinals asked of him. Johann Burchard, the master of ceremonies and chronicler of the Vatican, described the new pope signing documents without even reading them. He would soon renege on them all. Rodrigo was eventually successful on Innocent's death, winning the conclave because the cardinals were divided between two other candidates. He displayed none of the false modesty expected of popes even then and instead scrambled rather too eagerly into the lavish robes set out for his first appearance. Unlike his predecessor, the new Alexander VI honoured the promises with which he had purchased the papacy. His fellow cardinals were happy to finally have a man known for his honesty and intelligence on the papal throne.

THERE IS LITTLE likelihood that the pope was poisoned at the dinner given by Corneto. We know that Alexander died on 18 August and that the banquet was held on 6 August. The pontiff fell ill on 12 August, according to multiple dispatches from ambassadors who avidly followed his health. It is feasible that he contracted an illness at the banquet but it is unlikely that he could have been poisoned but not shown any symptoms for six days. Despite descriptions of poisons that could cause death after a delay, there is no evidence that such a toxin existed in sixteenth-century Italy.

On 14 August, the pope was still vomiting and sick, according to Beltrando Costabili, the ambassador of Ferrara, who gave a detailed account of the illness, knowing that Lucrezia, who was then married to the prince of that state, would be extremely concerned. What sets this account apart from those of the more lurid Venetian ambassador Giustinian, is a sentence that showed how little real surprise there was about the pope's illness. "It is not astonishing at all that His Holiness and His Excellency should be ill; for all the courtiers, especially those who are in the palace, are in the same state by reason of the unwholesome conditions of the air, which, there, they breathe." Even the pope's doctor, the Bishop of Venosa, was sick at home and had to be summoned from his bed to treat the pontiff.

Up to the nineteenth century, many historians claimed that Corneto's dinner had been held on the twelfth and that Alexander died the next day, although his death was kept hidden for five days to allow his son Cesare to consolidate his position. The death of the pope tended to unleash mayhem in Rome. Those in the Vatican who relied on the dead pope for their positions made off with whatever valuables were around, and the palaces of the papal family and favourites were ransacked by mobs. Franceschetto Cybo, the son of Innocent VIII, Alexander's predecessor, was so anxious about what might happen on his father's death that he ransacked the Vatican himself when he thought his father was dying. His actions were premature. His father recovered and he had to return the treasures, presumably a little ashamed. Certainly Cesare had ample reason for concern about his father's health. He must also have known that his own power would evaporate on the death of his father. But Burchard does not mention Cesare taking any such action until the eighteenth.

On that day Cesare "sent Michelotto with a host of people who shut all the doors of the pope's apartments and one of them drew a dagger and threatened Cardinal Casanova that if he did not give up the pope's keys and moneys he would stab him and throw him out of the window". That sort of behaviour was much more Cesare's style than a secretive poisoning.

Burchard, no friend of the Borgias, says nothing about the state of Alexander's corpse on the eighteenth; although if he had been dead for five days, he almost certainly would have. The next day when the body lay in state, it had already decomposed considerably in the heat of a Roman summer. Most of the accounts of a poisoning tend to be based not on the accounts of those who spent that torpid August in the Vatican, such as Burchard and Giustinian, but on writings by historians who were not present and were writing accounts much later.

The account by the Vatican master of ceremonies is quite straight-forward. The pope fell ill on the twelfth and three days later "tertian fever" set in after he had been bled by his doctor. He died on the eighteenth in the presence of the Bishop of Carinola and other attendants. Burchard dressed the body but says nothing about any state of advanced decay, although he does note that his papal ring was missing, presumably taken by those Cesare sent to claim his possessions. As the Vatican descended into increasing chaos, Burchard had the body placed in a chapel out of the way, in part due to fears that a Roman mob might disfigure the corpse. By late in the afternoon, the heat of the summer had taken care of that.

> When I saw the corpse again, its face had changed to the colour of mulberry or the blackest cloth and it was covered in blue black spots. The nose was swollen, the mouth distended where the tongue was doubled over and the lips seemed to fill everything. The appearance of the face was far more horrifying than anything that had ever been seen or reported before.

The corpse was finally taken to be put in a coffin by six labourers who told "blasphemous jokes about the pope or in contempt of his corpse". In a final insult, they made the coffin too short and narrow. His mitre had to be removed and his body rolled in a carpet and pummelled into the coffin.

The Florentine historian and diplomat Francesco Guicciardini dwells on the rumours in his own account of Alexander's death, but he was just twenty years old at the time and not in the Vatican or even in Rome. His account repeats the story of the treacherous servant who served the wine intended to kill Cardinal Adriano to both Alexander and Cesare. Alexander died two weeks after the fateful dinner at a time when the city was stricken with typhoid that had already killed thousands. Almost everyone who had attended the banquet in early August had fallen ill, according to letters from the Venetian ambassador Giustinian. The ambassador of Ferrara wrote accounts of the illness, at first suggesting that malaria might be the cause but later suggesting that many people suspected poison. It is unlikely that up to a dozen people might have fallen sick from malaria at the same meal; some sort of gastrointestinal problem or typhoid is more likely.

It might have been arsenic. Historical accounts do suggest some of the symptoms: gastrointestinal pain, burning in the oesophagus, violent cramps and skin problems, which in the case of Cardinal Adriano were so severe that he was actually said to have shed his entire skin. (Although he attended the conclave for the election of the pope a month after the death of Alexander and there are no comments on his poor state of health or peculiar complexion.) Arsenic poisoning can take some time to kill someone. It is not unfeasible that they might have died two weeks after ingesting the toxin.

What tipped the balance and provided what for many was enduring evidence of poisoning was the rapid blackening and decomposition of the body. Francesco Gonzaga, the Marquis of Mantua, was with the French army a few miles from Rome when the Pope died. On 22 September 1503, he wrote to his wife about Alexander's end:

Those who know the secret say that in the conclave following the death of Innocent he made a compact with the devil and purchased the papacy from him at the price of his soul. Among the other provisions of the agreement was one which said that he should be allowed to occupy the Holy See twelve years and this he did with the addition of four days. There are some who affirm that at the moment he gave up his spirit, seven devils were seen in his chamber. As soon as he was dead, his body began to putrefy and his mouth to foam like a kettle over a fire which continued as long as it was on earth. The body swelled up so that it lost all human form. It was nearly as broad as it was long. It was carried to the grave with little ceremony; a porter dragged it from the bed by means of a cord fastened to the foot to the place where it was buried as all refused to touch it. It was given a wretched interment, in comparison with which that of the cripple's dwarf wife in Mantua was ceremonious. Scandalous epigrams are published every day about him.

ROMANS WERE A tough crowd. Several writers recorded the disdain the city felt for the Pope: "Is it any wonder if the Borgia after death poured so much blood from his mouth? / It was all that he had drunk and not been able to digest." Burchard quotes another: "Violence, fraudulence, madness, ire, lust and the sponge-like thirst for blood and gold. / Alexander VI here I lie, Rome rejoices since my death is her life." Even Machiavelli, a supposed supporter of Cesare, wrote in the first book of the Decennali: "Valencia was taken ill and to find rest / Alexander's glorious spirit was taken up among the blessed / Close on his heels trod his three favourite minions / Wantonness, simony and cruelty."

Ercole, the duke of Ferrara and Lucrezia's father-in-law, was barely moved by the death, remarking peevishly in a letter to his ambassador in Milan that "there was never a Pope from whom we received fewer favours than this one" and that "his death caused us little grief".

Writers of the time commented on the hideousness of Alexander's corpse, with one describing it as "more monstrous than words can tell and without human form". But others also mention Alexander as lying in state in the normal way. Likewise, the ambassador from Orvieto described kissing the feet of the dead Pope as he lay in state four days after his death but failed to mention anything distressing about the body. Alexander's bones were reburied in the Spanish Church of Santa Maria in Monserrato and thrown in with those of the other Borgia pope, Callixtus III. "In this box are the remains of two Popes, Callixtus and Alexander. They were Spaniards," says the terse inscription. Lucrezia was nowhere near Rome when her father died.

Under suspicion for poisoning the pope, Corneto fled Rome, returning in 1513 with the election of Giovanni de' Medici as Leo X. In 1517 he was implicated in a plot of disaffected cardinals to poison Leo. The pope suffered from a chronic anal fistula and the plan was to get the doctor involved in tending this painful condition to apply a poisoned ointment to it. The plot was uncovered and Corneto fled to Venice to escape the fate of his co-conspirator, Alfonso, Cardinal Petrucci, who was executed. Corneto was stripped of his positions. He was murdered by a servant while travelling to Rome following the death of Leo in 1521.

The most enduring stories of Alexander's poisoning come from Guicciardini and Paolo Giovio (1483–1552), the physician, historian and bishop. The latter was the first to describe *la cantarella* or "the poison of the Borgias". He said the poison did not work immediately but penetrated through the veins and sapped its victims of their lives. Giovio and Guicciardini were both historians for hire, conjuring up works to please their political masters and so their rewriting of the death of Alexander, no friend of their patrons, is hardly surprising. The descriptions of Alexander's death are a rare moment of purple prose for the usually cautious and controlled Guicciardini. The pope's poisoned body is described as *"nero, enfiato e brutissimo"* ("black, bloated and hideous"). The historian revelled in Roman enthusiasm at the death of the Spanish pope:

> All Rome thronged with incredible rejoicing to see the dead body of Alexander in St Peter's, unable to satiate their eyes enough with seeing spent, that serpent who in his boundless ambition and pestiferous perfidy and with all his examples of horrible cruelty and unheard-of avarice, selling without distinction sacred and profane things, had envenomed the entire world. And nevertheless he had been exalted by the most unusual and almost perpetual good fortune from early youth up until the last days of his life, always desiring the greatest things and always obtaining more than he desired.

In Guicciardini's eyes, Alexander was all the more monstrous for being another example of immorality rewarded and evil unpunished.

STORIES OF POISONINGS, mostly by Alexander, tend to over-shadow a broader understanding of the Vatican at this time. It is often seen as a period of decline for the church, when it was more involved in politics than the saving of souls. At the start of the Renaissance, the Catholic Church was deeply split, Rome was dilapidated and the Vatican was in a state of physical and spiritual decay. Classical Rome was more than just neglected. Stones were pulled from the ruins to build elsewhere; marble facings and statuary were heated in kilns to make cement. By the end of the Counter-Reformation, Rome had been rebuilt much as we know it today. The Vatican Library had become one of the greatest collections in the world and the city's churches has been decorated with some of the most superb art ever created. Alexander is often pilloried for his careful positioning of his family in church posts or in marriages to powerful figures, but he was only doing what every pope did. By the time of Alexander's election in 1492, he was the only non-Italian cardinal and those around him were almost all from important families such as the Medicis, Farneses, Estes or Gonzagas. The only thing that distinguishes Alexander from other cardinals and popes of that time was his foreign birth.

In the eighteenth century, Voltaire wrote of Alexander's death:

> All the enemies of the Holy See love this horrible anecdote. I myself do not believe it at all and my chief reason lies in its extreme improbability. It is evident that the envenoming of a dozen cardinals at supper would have caused the father and son to become so execrable that nothing would have saved them from the fury of the Roman people and of the whole of Italy. Such a crime could never have been concealed.

Voltaire blamed Guicciardini for the story, saying, "Europe has been deceived by you and you have been deceived by your feelings." Guicciardini, for all the elegance of his writing, relied on hearsay and Roman gossip. But his influence was enormous, shaping many of the myths of the Borgias that endure to this day.

Alexander's supposed pact with the Devil, mentioned in Francesco Gonzaga's letter, first appear in histories nearly fifty years after his death. Francesco Negri, an Italian Protestant, wrote that the devil had appeared to Alexander. This was taken up by other Protestant writers, most notably John Bale, a former Carmelite monk who had left the Catholic Church and was in exile from England during the reign of Queen Mary. Bale described Alexander as "a very riotous tyrant" who had become pope by making a deal with the Devil. The source for Bale's writing is Raffaele Maffei, an Italian writer who dedicated his *Commentaria* to Alexander's hostile successor, Julius II. Despite this, Maffei is remarkably fair in his assessment of the Borgias, praising Cesare for his skilled governance and describing the pope dying peacefully in bed with his children gathered around him. Bale ignores almost all of this, embroidering critical stories of Alexander with relish.

Johann Wolf, the Lutheran author of an illustrated history of the sixteenth century called *Lectionum memorabilium et reconditorum*, also claimed that Alexander had sworn obedience to Satan in order to win the papacy. He also repeats the idea that Alexander had been promised eighteen years on the throne but was cut short by the Devil. Wolf's book contains an extraordinary woodcut of an animal that was said to have washed up on the banks of the Tiber in 1496, during Alexander's reign. It is an ass with a long, flowing mane, large and distinctly human breasts and limbs covered in scales. It is missing one of its hands. One foot is a cloven hoof, the other the claw of a bird. On its hip is a bearded human face and from another head emerges a fire-breathing animal. This grotesque animal "represented the form of the Roman papacy so perfectly and splendidly that it could not be the work of any human being but must be accepted as God's own representation of the papal abomination", Wolf wrote.

A century after his death, the mythology surrounding Alexander was well established and much embellished. The German writer Georg Widman incorporated him into the Faust legend in a work published in 1599. Faust praises Alexander as "a very paragon" among evil popes and describes him first invoking the devil by gazing into a crystal bowl while studying in Bologna. Widman also includes a description of Alexander's incest with his daughter, whom he strangely names Helena rather than Lucrezia. But Widman also highlighted another aspect used against Alexander that links him to a long history of poisonings. He accuses Alexander of being "a baptized Jew", which explained his capacity for evil. Accusations that the pope had been Jewish dated back to the late fifteenth century when he was described as a "Marrano", or one of the Iberian Jews who were baptised as Christians but practised their own faith secretly. The accusation had been levelled by several Italians, including Savonarola, the Dominican friar who had led a puritanical religious insurgency in Florence. The pope's Spanish origins and his decision to allow Jews expelled from Spain in 1492 to live in Italy, all raised suspicions that he might secretly be Jewish.

A Roman poet later praised Alexander's successor for freeing the city from enslavement to "the children of a Marrano shepherd", a line that brings together some of the layers of prejudice against the Borgias. In the 1530s, these ideas surfaced in Luther's *Tischreden* ("Table Talk"). The pope was, according to Luther, "a Marrano, that is, a baptized Jew, who believed in nothing". Stories of incest and poisoning appear, with Luther writing: "this I heard as certain in Rome." Marlowe mentions the death in *The Jew of Malta*, written around 1589:

> *As fatal be it to her as the draught*
> *Of which great Alexander drank and died:*
> *And with her let it work like Borgia's wine*
> *Whereof his sire, the Pope, was poison'd;*
> *In few, the blood of hydra, Lerna's bane*
> *The juice of hebon, and Cocytus breath,*
> *And all the poisons of the Stygian pool.*

In 1607, Barnabe Barnes's play *The Devil's Charter* was published in London, shortly after it had been performed for James I. Guicciardini was a central character – the chorus figure like the ghost of Machiavelli in *The Jew of Malta* – who provides necessary background. The play starts off with the story of Alexander winning the papacy by making a deal with the Devil. In an early scene Lucrezia stabs her husband to death, having steeled herself with the lines:

> *Let none of Borgia's race in policies*
> *Exceed thee, Lucrece. Now prove Caesar's sister,*
> *So deep in bloody stratagems as he.*
> *All sins have found examples in all times;*
> *If womanly thou melt, then call to mind*
> *Impatient Medea's wrathful fury.*

The play continues with the familiar scene of the poisoned wine, sealed with Cesare's own ring but poured for the wrong guests as the Devil swaps the bottles. They rush back to the Vatican to find an antidote that the Pope desperately tries to administer to his son as

Cesare complains: "I feel Vesuvius raging in my guts."

> *Here, Caesar, taste some of this precious water,*
> *Against all plague, poison and pestilence*
> *A present help, I bought it of a Jew*
> *Born and brought up in Galilee.*

They die, the Devil having decided that Alexander has served his time as Pope and can be taken off to hell. The play was written shortly after the attempted assassination of James I in the Gunpowder Plot. There had also been an attempt to poison the monarch and the high-profile murder of Sir Thomas Overbury in the Tower of London that led to the trial of the Earl and Countess of Somerset. The anti-Catholic theme and the elaborate poisoning must have been enjoyed at court, although the play is somewhat ham-fisted.

The Borgias were to make regular appearances in English drama, when anti-Catholic feeling was running high. Nathaniel Lee's *Caesar Borgia* was published in 1680 when concerns were aroused that Charles II might be succeeded by a Catholic monarch. The poet John Dryden wrote an introductory verse to his friend's play, which ends:

> *You [the English] know no poison but plain Rats-bane here*
> *Death's more refined and better bred elsewhere.*
> *They have a civil way in Italy...*
> *Murder's a Trade so known and practis'd there,*
> *That 'tis infallible as is the Chair*
> *But mark their feasts, you shall behold such Pranks*
> *The Pope says Grace, but 'tis the Devil give thanks.*

Certainly in England, fizzing with anti-Catholic sentiment for much of the sixteenth and seventeenth centuries, there was a belief in the intrinsically evil qualities of all popes, not just Alexander. An anonymous pamphlet published in 1679 called *A True History of the Popes of Rome* lists all possible sins committed by those wearing the triple tiara:

Paul III empoisoned his mother and niece that the whole inheritance of the Farneisham family might accrew to himself. His other sister also whom he carnally knew, upon fancying her freeness to others, he slew by poison. Alexander VI by poison made away Gemen (Djem), the brother of Bajazet (Bayazet), for the hire of two hundred thousand crowns and by the like medicine he purged away senators and cardinals; but preparing the like sauce at a supper for others, mistaking of the bottle, he drank of the prepared cup and died like a miserable wretch... Innocent the IV, being corrupted by rewards, offered a bottle of the liquor to the Emperor Conrade; this the said emperor verified by his public letters. Gregory VII by his companion Brasatus poisoned Nicholas XI and by the service of this villain, he impoisoned six or eight cardinals, obstacles in his way to the Papacy. Paul III did his best not only to raise the Christian princes against Henry VIII of England but also plotted against him to destroy him. Gregory XIII loaded with indulgences Parre and others to have laid violent hands upon Queen Elizabeth.

The image of Catholicism being poisonous in itself was recurrent, as was the need for an antidote in the form of good Protestant practice. Poisoning was associated with communion. "The Italyans above all other nations, most practise revenge by treasons and espetially are skilled in making and giving poysons," Fynes Moryson, author of *An Itinerary*, wrote in the early seventeenth century.

For which reasons the Italyans are so wary, espetially during a quarrell as they will not go abroade nor yet open their doors to any knocking at night or so much as put their heads out of a window to speak with him that knocks. For poysons the Italyans skill in making and putting them to use hath been long since tried to the perishing of kings and emperors by those deadly potions given to them in the chalice mixed with the very precious blood of our Redeemer. In our times it seems the art of poisoning is reputed worthy of princes practise.

Frederick Rolfe, the fantasist and ardent Catholic convert known as Baron Corvo, dismissed accounts of the Borgia poison in his eccentric history of the family published in 1901 but the mythology has remained remarkably durable. Even the nineteenth-century Swiss historian Jacob Burckhardt, in his *The Civilisation of the Renaissance in Italy*, brushed off the Borgias as somehow un-Italian and also linked them to poison. He wrote, without offering any evidence, that they had a professional poisoner on staff. However, Burckhardt is keen to present their vices as Spanish vices and not Italian. Writing of Alexander's reign as pope, he said, "Strictly speaking, as we are now discussing phases of Italian civilization, this pontificate might be passed over, since the Borgia are no more Italian than the House of Naples." He goes on to mention that Lucrezia wore Spanish costume when she entered Ferrara, that the family spoke Spanish at home, were entertained by Spanish buffoons and that Cesare even had a Spanish poisoner, Sebastian Pinzon. A Sebastian Pinzoni did confess during the reign of Julius II to poisoning the Cardinal of Modena for whom he worked as a secretary and gentleman-of-the-bedchamber. Pinzoni certainly had some connections through the Church to Alexander but there is no evidence of him ever working for Cesare.

Cesare had as much influence in the shaping of the Borgia legends as his father. Paolo Giovio established much of the mythology about Cesare that was later embellished by others. Alexander's son was as bad as any of the tyrants of the Roman world, according to Giovio, who also suggested, without any evidence, that Alexander had connived in Cesare's murder of his own brother. Tomaso Tomasi's *Vita del Duca Valentino* was not published until 1655 and was a huge success, becoming the major source of information for many later works on the Borgias. Tomasi transformed the Vatican into a version of the court of Louis XIV with Alexander's mistress and the mother of his children, Vannozza dei Cattanei, playing a central role. There is no historical evidence for this but it marked the emergence of a critical role for the Borgia women that would be taken up with a vengeance in later depictions of the family.

Even the admirers of the Borgias did little to help their reputation. Machiavelli based *The Prince* on Cesare Borgia and praised him for his skills in managing the Papal States. But Machiavelli himself developed a reputation for almost demonic levels of evil and his works were eventually put on the papal index of banned books. Protestants may have maligned the Borgias but Catholics rarely came to their defence. Even the Spanish held them in some disdain, although their histories tended to avoid the more outlandish charges.

Machiavelli's approval of Cesare's tough rule over the area of Romagna and his "extraordinary *virtu*", extensively expressed in *The Prince*, did little to help the family with subsequent historians. Most people were less interested in Cesare's skills in government than what Machiavelli termed his "well used cruelties". Suspicion of Machiavelli was intense in England, where he was held up as an icon of a particularly wicked – and very Italian – form of government. Machiavelli was sometimes described as Cesare's inspiration. William Struther, a Scottish preacher, wrote of Cesare in a work for Charles I: "What could the world look for of him, who was the son of such a father as Alexander and the pupil of such a tutor as Machiavelli who took Cesare as the object and centre of all his wicked devises and set him out to the world as the most perfect exemplar to be followed." Struther went to on remark of Cesare's own poisoning: "Borgia was never so rightly placed as when he was put in the belly of a mule so as to draw the poison out of his body. He was never more suitably clad than with such a carcass and that belly was never worse filled than with such a monster." For Struther, the fall of the House of Borgia was a sign of divine justice. "This was an example of God's just judgment to all tyrants who will conquer and rule a state in contempt of God."

The legend of the toxic Borgias was greatly expanded by Alexander Dumas. With his novelist's eye, he took the works of

Giovio and embellished furiously. The alleged Borgia poison *la cantarella*, Dumas wrote, was produced by catching a wild bear, giving it a large dose of arsenic and stringing it up by a hind leg. In its death throes, the bear would foam at the mouth, producing a steady stream of a toxic liquid that was collected in a silver bowl and then sealed in vials. Dumas's influence in creating the nineteenth-century villainess Lucrezia was exceeded only by Victor Hugo, whose play created the enduring image of her as a relentless poisoner.

> Who, actually, is Lucrezia Borgia? Take the most hideous, the most repulsive, the most complete moral deformity; place it where it fits best – in the heart of a woman whose physical beauty and royal grandeur will make the crime stand out all the more strikingly; then add to all that moral deformity the purest feeling a woman can have, that of a mother... Inside our monster put a mother and the monster will interest us and make us weep. And this creature that filled us with fear will inspire pity; that deformed soul will be almost beautiful in our eyes...

So wrote Hugo in the preface to his play which was later adapted into an opera by Gaetano Donizetti and was influential in shaping popular ideas about her. The play and the opera have few foundations in history. Lucrezia did not have an unknown son and she did not poison him and his friends in what is one of opera's most dramatic mass murders. The idea of her wickedness was eagerly taken up by novelists and filmmakers in the twentieth century. Between 1926 and 1952, six films were made about Lucrezia. She does not come out well in any of them.

Hugo's preface to his play illustrates what makes her such a subject of fascination. It is the combination of evil and beauty, low-mindedness and privilege that fuels her dramatic potential. She is a perfect fictional character. She was a symbol of womanhood tainted – the allegations of incest, the multiple marriages and the poisonings.

She was an emblem of the taint of Catholicism – she represented all that was wrong with Italy and the Church in the eyes of outsiders. Of course, to Romans and Italians, she was a foreigner and an interloper. Almost every serious biography of Lucrezia shows her to have been a dedicated wife and mother. She was a pawn in much dynastic intrigue, as was often the case for the daughters of powerful men, but almost certainly innocent of most of the accusations made against her.

Lucrezia was vital to the popular nineteenth-century invention of the Renaissance, a period of intriguing combinations of high culture and low morals. More than seventy-five books have been published on her life since the nineteenth century. She is probably more written about than any other woman of that time, despite playing a relatively minor role in the Borgia story. Her contemporary image came out of these works, not anything based on what her life was like in the sixteenth century. The Pre-Raphaelite artist Dante Gabriel Rossetti painted her washing her hands in a basin having just poisoned a husband. She looks no more troubled than any of Rossetti's square-jawed women – based on his mistress Jane Burden – as if wearily answering a dull husband while applying hand cream. In the background on the table is a sinister, stoppered vial and her dying second husband, Duke Alfonso Bisceglie, is seen reflected in a curved mirror being supported by Lucrezia's obviously complicit father.

Alexander's death also fits a wider cultural understanding that poisoning was something that befell popes. In the lore surrounding the papacy, stories of poisonings are common. John XIV was imposed on the Vatican in 983 by Emperor Otto II. Romans did not take to him and when Otto died, the pope found himself isolated and vulnerable. He was said to have been poisoned on the orders of Boniface VII, his predecessor who had been expelled from the Holy See and sent into exile in Constantinople. Pope Clement II, who ruled briefly from 1046 to 1047, was alleged to have been poisoned and there were suggestions that he died of lead poisoning although it could have been accidental. Pope Benedict XI, an unworldly Dominican, died in Perugia in 1304 after supposedly eating poisoned figs brought to him by a man dressed as a nun from the Saint Petronella convent. Leo X was thought to have been poisoned by Francis I of France after having survived several early attempts on his life. Hadrian VI, a dour and virtuous Flemish pope who tried to curtail some of the extravagances of the Church, was said to have been poisoned within twenty months of ascending the throne in 1522. His doctor had his house garlanded with flowers in praise of his actions and the historian Ludwig Pastor described his death as being met with "frantic joy". Hadrian was sickly and sixty-three when elected and nobody had expected him to last long. The Church did not make the mistake of electing a non-Italian pope for another four and half centuries until it chose Karol Jozef Wojtyla, the Polish cardinal who became John Paul II in 1978.

Guilio Medici became Clement VII when he took the papal throne and would reign during the years of the Sack of Rome in 1527 and the loss of England to Protestantism. His death in 1534 after a failed and bitter papacy was inevitably attributed to poison. Clement XIV, who dissolved the Jesuit order, then unpopular among European rulers, was said to have fallen victim to their skills as poisoners when he died in 1774. By the time of his death, he had become so paranoid about being poisoned that he refused to kiss the feet of statues in the Vatican lest they had been painted with some toxin.

More recently rumours have circulated about the death of Pope Pius XI in 1939. He was said to be about to offer a belated condemnation of fascism when he died. The evidence is skimpy and centres mostly on the fact that his doctor was related to the Italian foreign minister. The death of John Paul I, who reigned for just a month in 1989 before succumbing to what was officially described as a heart condition, has also raised the spectre of deliberate poisoning. He was both the youngest pope and had the shortest reign in four centuries. Conspiracy theorists have linked his death to various Vatican and wider Italian corruption scandals. John Paul was the first pope to die unattended since 1600 and the refusal of Vatican officials to allow an autopsy fuelled suspicions. The book by the British journalist David Yallop alleging that John Paul was murdered because he was going to excommunicate those involved in the P2 Masonic Lodge sold more than six million copies and still provides grist for conspiracy theorists.

Just as the deaths of earlier popes inspired dramatists, Francis Ford Coppola included these suspicions in his movie *The Godfather III,* in which the pope is served a poisoned cup of tea.

SOMETIMES MERELY BEING Italian was enough to generate an accusation of poisoning for those close to the powerful. In August 1536, the Dauphin of France was playing tennis when he started to feel unwell and was unable to catch his breath. He asked his secretary, Sebastiano de Montecuccoli, to fetch him a glass of water. Shortly after drinking it he collapsed and was taken to his bed. He died just over a week later. Montecuccoli immediately fell under suspicion. He was Italian, he had formerly worked for the Holy Roman Emperor, enemy of Francis II of France, and he had in his possession a work on poisons. Briefly tortured, he confessed that he had been paid to kill the Dauphin. He was executed by the gruesome method of *ecartelage* in the Place de la Grenette in Lyon. Each of his limbs was tied to a horse which was then driven in different directions. The dauphin, who had a long history of poor health, had probably died of pleurisy.

Part of the perception of poisoning as an Italian vice came from the fact that so much knowledge of poisons and the lore surrounding them came from Italy, the key location of scientific learning and publishing. Italy was also where classical works were being rediscovered, translated, published and commented on. It was also a critical arena for emerging modern science based on observation and experimentation, a process that began in Europe around the fourteenth century.

Although many of the works reiterated knowledge from ancient Greece, Rome and the Arab world, there were also many additions that advanced knowledge of poisons. Sante Arduino wrote his work *De venenis* (1414–1426), drawing on experiments and knowledge he learned from fishermen in Venice. Far from being the usual classical compiler of the times, he even sought out poisonous species, getting bitten by sea scorpions in the process. Arduino's work was still grounded in the medieval occult view of toxicology. Poison was a manifestation of evil in the world. It was demonic and those animals that were poisonous were part of the Devil's work. Poison could act through any of the senses. It was understood to act in the hidden manner of magic and could therefore be countered in the same ways.

Antonius Guaynerius's tract on poisons, published in 1481 but probably written forty years earlier, is a compendium of earlier knowledge, principally from Galen, Peter of Abano and Avicenna. But it also contains more contemporary observations, drawing both from medical practice and popular lore. He rejected the idea that some poisons acted through heat or cold, disrupting the balance of the four humours. The physical mechanism by which poisons entered the body was through toxic vapours. These were emitted by the basilisk, for example. Antonius Guaynerius was dismissive of two popular views of the time. Poisons could not be made to work at some set time in the future, nor could one person poison another through the action of the mind. He said many people had lost faith in gems but repeated explanations put forward by Albertus Magnus on how they work. He wrote that an antidote to airborne poisons and the plague was a stone engraved with the figure of a man "girded by a serpent whose head was held in the figure's right hand and tail in the left". Gems have such power that they could alter the mind. He wrote of a prince who feared the aggression of a neighbouring king and sent him a saddle adorned with jewels. The gems altered the king's disposition, making him more peaceful.

From the publication of Guaynerius's work in 1481, almost nothing was written on poisons in Italy for fifty years. The historian of magic Lynn Thorndike wrote: "In view of the orgy of poisons attributed to the Borgias and other Italians of the Renaissance period, it may surprise some that no work on poisons seems to have been printed until the third decade of the sixteenth century. If poisoning was such a significant problem, why is there so little writing about it at the height of the Renaissance, at a time when princes and popes were supposedly often victims?" Although the few works that were published do address some of the issues of protections against poison, they mostly dwell on threats such as the basilisk, which was hardly a commonplace method of assassination. Scientific writings were not widely available and probably rarely penetrated popular folk wisdom, but they were available to those in courts who often commissioned them.

There were probably only about twenty plants and minerals that were then known to be poisonous, and doubtless they were used on occasion. But there was also much scepticism about the effectiveness of poisoning, which must have been quite difficult to achieve. A document in the Venetian archives mentioned experiments on pigs with poisons that had been sent from Vincenza and were described as "not very good". There is little doubt that given the terrible standards of hygiene, medicine and surgery, people were often inadvertently poisoned, but there is not much evidence for the vast array of deliberate and secret poisonings that were said to be going on in Italy.

At least in public, many of those accused of poisonings abhorred the method. A letter dated 1548 from Cosimo de' Medici, held in the Medici Archives, only points to an ancient source of advice on poisons but has no current poisoner to turn to:

> In regard to that man who offered to poison Piero Strozzi's water or wine, impelling you to send a courier here [to Florence] to obtain a recipe from us, we inform you that we have never looked into such matters nor authorised others to do so. This is not our custom and even if we should wish such a thing, we would have no idea to whom we might turn in this city. And should such an occurrence [Piero Strozzi's poisoning] come to pass, we think it best that neither you nor I be the source. However, if that person you mention really is inclined to carry out his plan... he might find that Apollonius [of Citium] gives some recipes. In any case, we do not know where to obtain such things, nor have we any interest in doing so, since we find such matters excessively horrid...

Cosimo de' Medici would have been happy to have seen Piero Strozzi dead. As a commentary on the letter says: "Since Cosimo's accession to the throne in 1537, the greatest threat to his security was posed by the underground network of Florentine republicans living in exile, largely in Sienese territory and in Rome."

It is notable how little Machiavelli has to say on the subject of poisoning in his political writings, given the knowledge of it he showed in his plays such as *The Mandrake*. In *Discourses*, he writes at length about conspiracies, focusing on the attempt in 1478 to kill Lorenzo and Giuliano de' Medici, which later became known as the Pazzi conspiracy. Towards the end of the section, there is a somewhat throwaway dismissal of the uses of poison.

> This is all that needs to be said about conspiracies and if I have taken account of those in which the sword and not poison has been used, it is because they are all of one and the same pattern. It is true that the use of poison is more dangerous owing to its being more uncertain, for not everyone has the commodity and those that have it must needs be consulted and the necessity of consulting others means dangers to yourself. Again for a variety of reasons a poisoned drink may not prove fatal, as those discovered who were to kill Commodus, for, on his throwing up the poison they had given him, they were forced to strangle him if they wanted him to die.

Machiavelli expresses no great moral outrage about the idea of poisoning; he describes it as being no different to killing with a sword.

The evidence cited for the rampant use of poisons in Italy is scant. A record from the Council of Ten, the secretive Venetian group of officials who maintained security in the Republic, dated 13 September 1419, says that a Biagio Catena, archbishop of Trebizond on the southern coast of the Black Sea, had been contracted to kill "Marsilio de Carrara by means of Francesco Pierlamberti of Lucca, and wishes to travel in person with the said Francesco, that he may assure himself of the actual execution of the deed". The council voted to pay him for a horse, accommodation and expenses for the poison.

A Franciscan monk, John of Ragusa, also offered his services as a poisoner to the Republic. The notation, from 14 December 1543, has two members of the council requesting a pension of 1,500 ducats a year for the monk whose price list for murders was said to start at sixty ducats for the Duke of Milan, a hundred for the pope, a hundred fifty for the King of Spain and five hundred for the Sultan. The entry has the ring of a scam or an accounting fraud about it. There is no mention that the Council of Ten, an often sinister and repressive organisation given to bouts of paranoia and corruption, ever took John of Ragusa up on his offer. There were no suspicious deaths at the time. Pope Paul III died in 1549 aged eighty-one; Charles V of Spain died in 1556, suffering from severe gout; he was also Duke of Milan at the time. Suleiman the Magnificent died in 1566. There is nothing to suggest John of Ragusa had anything to do with any of their deaths, despite his price list.

7. GAZING AT PISS

*Alle Ding' sind Gift und nichts ohn' Gift; allein die
Dosis macht, das ein Ding kein Gift ist.*
        Paracelsus

Basilisks were also known as cockatrices, a word that is surrounded by confusion from its emergence in the twelfth century until the animal died its mythical death about four hundred years later. The origins come from an anecdote in Pliny's natural history about an animal that would dart into the open mouth of a crocodile to clean its teeth and eventually gnaw its way into the gut of the animal, killing it.

Medieval writers such as Brunetto Latini, the Florentine philosopher, teacher of Dante Alighieri and author of the early Renaissance encyclopaedia Il Tesoretto in 1262, confused the animal described by Pliny – a type of mongoose known as a Pharaoh's rat – with water snakes and came up with the idea that the cockatrice was a fish that enters the body of animals and tears them apart. Cockatrice appears in Wyclif's Bible of 1382 as a translation of the word "basilisk" or "regulus". The word "cockatrice" ended up meaning a variety of possible animals from weasels through to basilisks. By the twelfth century the basilisk had taken on two meanings – the ancient serpent that could kill with its sight and the medieval chimera that was part reptile and part cock.

In Alexander Neckham's *De Naturis Rerum*, cockatrices are created when toads hatch chicken eggs to form an animal that was often portrayed as a rooster with a snake's tail. Cockatrices soon entered heraldry in twelfth-century England and were portrayed on coats of arms alongside dragons and other fabulous animals. The heraldic version often had a second head at the end of its tail and it was this head that was said to be lethal. The cockatrice took on an important symbolic role in alchemy, alongside the salamander, an animal that the sixteenth-century doctor and alchemist Paracelsus believed to be formed of only one element, fire, rather than the more usual mix of four elements. Its ashes could freeze mercury into silver.

In 1574, four Venetian noblemen visiting London were invited to dinner by some members of the Privy Council, among them Lord Burghley, the Elizabethan statesman who ran the government. The small group of twelve conversed in Italian. Later the Venetians were granted an audience with the queen, who also spoke with them in their own language. Only a handful of the Englishmen had ever set foot in Italy and yet many well-educated nobles of the day spoke Italian. Italy was "the nation which seems to flourish in civilization most of all other at this day", according to the first English account of the country, William Thomas's *History of Italy*. Italy was the seat of the Renaissance, the centre of knowledge, manners and culture. Baldassare Castiglione's *The Book of the Courtier* guided the behaviour of gentlemen at court and all who aspired to join them there.

This was the peak of sixteenth-century English admiration for the Italians. It was to sink remarkably in the next few decades until Italians were emblems of evil, dangerous and even poisonous. Unchaperoned travel to Italy came to be regarded as foolhardy, something that would only lead to the importation of bad habits to England. While classical culture was to be admired, contemporary Italian culture would come to be frowned on as degenerate and unsuitable for aristocratic Englishmen. Roger Ascham, tutor to Princess Elizabeth and author of *The Schoolmaster*, a book that had many echoes of Castiglione's guide to manners, warned that "Italy now is not that Italy it is wont to be and therefore now not so fit a place as some do count it for young men to fetch either wisdom or honesty from thence".

The connections between England and Italy were complex. Italian manners and culture had been much admired. It was undeniably the heart of the civilised world in the minds of European Renaissance men. But England was racked with divisions between Catholics and Protestants who remained intensely suspicious of the popes and their temporal and spiritual ambitions. The English were engaged in a complex process of defining their identity and developing a state, which involved considerable anxieties about foreigners and their influences.

Italy became the setting for English dramas that showed the worst depravities imaginable. William Shakespeare, Christopher Marlowe and John Webster all described Italians as adept murderers, skilled and shameless in their use of poisons. The culture became synonymous with deceit and crime in a way that it would never entirely shake off in the English mind. Italy went from being seen as a cultured ally to becoming a much more sinister part of the imaginary landscape. In 1592 Thomas Nashe described it as "the academy of manslaughter, the sporting place of murder, the apothecary shop of poison for all nations". Nashe, an Elizabethan satirist who had a long and varied career as a writer of everything from anti-Puritan propaganda to pornography, had never been to Italy and was prone to embellishment, but his words mark a profound shift in attitudes towards the country in just a few decades.

The word "Italianated" took on a sinister tone. The phrase *"Inglese Italianato e diabolo incarnato"* ("The Italianated Englishmen is the devil incarnate") came into use. Italianated first meant a tendency to greed and fraud. Ascham complained of the "fancy that some have to go beyond the sea and namely to lead a long life in Italy". Education in Italy could lead to an excess of sophistication and contempt for religion. Finally, and most worryingly for the English, it led to collusion with papists and sedition against the state. "I was in Italy my self but I thank God, my abode there was but nine days," Ascham sniffed. "And yet I saw in that little time, in one city, more liberty to sin that I hard tell of in our noble City of London in nine years."

Mentions of poisonings by Italians in the works of English travellers and writers were rare in the first half of the sixteenth century. Andrew Boorde, a doctor and author of *First Book of the Introduction to Knowledge*, wrote that Romans were known to poison the stirrups or saddles of travellers and that the poison would be activated by the heat of the body. As political intrigues of Elizabethan times stirred up greater anxieties about Italy and Catholicism, the vision changed and became much darker. Catholic nobles revolted in the North of England in 1569. The Duke of Norfolk was arrested for his role in the rebellion and executed in 1572. Although Italians had played next to no role in this conspiracy, except for the Pope excommunicating Elizabeth I, the word "Italianated", was soon attached to anyone involved. Lord Burghley was warned not "to accept physic from anyone Italianated" as a Doctor Gifford had supposedly been sent from Rome to poison him. Mathew Parker, the archbishop of Canterbury, in 1572 published a biography of his predecessors, including Reginald, Cardinal Pole, who had urged various rulers to wage war against Henry VIII and then returned to England when Queen Mary re-imposed Catholic rule. Parker blamed the influence of Italians for Pole's treachery:

> When he had remained there some months in safety in the very lap and bosom of the Pope himself, he emerged infatuated and changed, as if he had drink the cup of Circe, from an Englishman to an Italian, from

a Christian to a papist. I omit here the word which
the Italians use to describe our Englishmen who are
changed into their likeness, which they consider a great
and monstrous metamorphosis contrary to both human
and divine nature.

The rise of Protestant radicalism heightened the suspicions of
foreigners who were seen as taking English jobs, hoarding resources
and engaging in a variety of mostly imagined crimes. This xenophobia
combined with an interest in Italian literature and Italy as a setting
for drama. Italian fiction became a means for readers and writers
to address a great many of the tensions and issues at home. Ten of
Shakespeare's thirty-four plays are primarily derived from Italian
material and four others depend heavily on it. From 1560 to 1600,
more than four hundred titles from more than two hundred authors
were translated from Italian into English, much to the concern of the
more puritanical like Ascham who complained that such translations
were now sold in every shop in London "the sooner to corrupt honest
manners". Much of the interest in Italian literature stemmed from a
backlash that was under way against the increasingly state-enforced
religion that was imposed under Elizabeth I. Italian literature provided
an engagement with the human experience that was lacking in the
Protestant morality being stamped on the country.

Shakespeare's Italian plays are rife with poisonings, some
accidental, some deliberate. *Romeo and Juliet* has the most specific
links to Italian poisonings and medicine, but *Hamlet* is the most
toxic, its deaths most disturbing and enduring in the imagination. The
play within the play in *Hamlet*, "The Murder of Gonzaga", is drawn
from Italian sources, according to Hamlet himself. It was said to be
based on the death of Francesco Maria della Rovere, Duke of Urbino,
a soldier, patron and political figure in the early sixteenth century.
The duke died in 1538 after a long illness. Although Hamlet says of
the play, "The story is extant and written in very choice Italian," it
is hard to find much more than a fleeting reference to the poisoning
of Urbino, and mostly by those historians who had a well-developed

tendency to attribute the deaths of all important figures to a deadly powder or potion. The historian Giovio declared that della Rovere had died "not by natural destiny but through the malice of certain men". Although there is no direct account of this being a poisoning, it was soon accepted as such by historians. James Dennistoun wrote: "Whoever may have been the author of the foul deed, it is agreed that the perpetrator was the Duke's Mantuan barber, who is generally said to have dropped a poisoned potion into his ear." Della Rovere was said to have died in the bucolic surroundings of his villa in Pesaro. Like the Duke in the play, he was quite old and had been married to a much-loved wife for a considerable time. Shakespeare blended ideas from this story with imagery and a new understanding of both the body and medicine that was developing when he was writing the play.

Claudius kills the King, Hamlet's father, with "the juice of cursed hebona" poured into his ear. This was Shakespeare's invention and reflects both a dramatic and unusual way of killing and the anxieties of the time about the ear being vulnerable to real and metaphorical poisons. None of the source materials that Shakespeare used for *Hamlet* – Saxo's *Historiae Danae* and Belleforest's *Histoires Tragiques* – mention the king dying in this way. However, this rare method of administering poison had entered the public consciousness during the sixteenth century. Not only was there the murder of della Rovere, but Paré, the eminent French physician and author of many writings on poison, was accused in 1560 of killing Francois II of France by blowing a poisonous powder into his ear.

Marlowe used the idea in *Edward II*, in which a character talks of "blowing a little powder" into the ear to poison someone. In 1560, Bartolomeo Eustachio had established that there was a link between the outer ear and the throat, something that aroused much interest by confirming the physical permeability of the body through the ear. The idea of objects and poisons getting into the ear sent a collective shudder through people. It was a time of frequent and painful earaches for many people due to polluted water and few effective remedies. Paré, despite his reputation as an "ear murderer", had written of the defences in the ear, the "crooked windings" as he put it, in which "little creeping things and other extraneous bodys as fleas and the like" would get stopped from penetrating the body further. He devoted an entire chapter to "Of the Stopping of the passage of the eares and the falling of things thereinto". Clearly there was some concern about what people were putting in their ears, which Paré notes were sometimes blocked by such things as "fragments of stones, gold, silver, iron and the like mettals, pearles, cherry-stones or kernels, pease and other such like pulse".

Concerns about ears tie in with more general anxieties about pollution and penetration of the body by poisons, notably the poisons of words and unapproved religion. As the Shakespeare scholar Tanya Pollard notes, poison was seen to be almost ubiquitous and extremely hard to avoid. William Crashaw wrote in *The Parable of Poyson*:

> There be poisons for our Meate, for Drinke, for Apparell, for Arrowes, Saddles, Seats, Stirrups, for Candles, Torches. Nothing that comes about a man, nothing that he toucheth, or that toucheth him but Mans wickednesss hath fitted and prepared poison for it... Nothing a man takes into him or puts upon him, nothing that toucheth him or comes neer him, that can be safe from bodily poison, if Gods providence preuent it not.

Murder by poison was all around. Sir Edward Coke, the chief justice in the early seventeenth century, believed that poisoning was the worst of all crimes because "of all the others can be least prevented, either by manhood or providence". Sir Francis Bacon wrote that poisoning:

> is easily committed and easily conceal'd and on the other side, hardly prevented and hardly discovered. For murder by violence Prince have guards and Private Men have Houses Attendants and Arms. Neither can such a murder be committed but Cum Sonitu with some overt and apparent acts that may discover and trace the Offenders but by poyson the cup, itself of

> princes will scarcely serve in regard of many Poysons
> that neither discolour nor distaste; It comes up a man
> when he is careless and every day a man is within the
> Gates of Death.

Just as physical poison could enter anywhere, so could moral poison.
A greater concern for the spoken word than the visual image emerged
with Protestantism. The poet John Donne believed that ears were
central to spiritual and even physical fertility.

> The eares are the Aqueducts of the water of life; and
> if we cut off those, that is intermit our ordinary course
> of hearing, this is a castration of the soul, the soul
> becomes a eunuch and we grow to rust, to a mosse, to
> a barrenesse, without fruit, without propagation. They
> say there is a way of castration in cutting off the eares:
> there are certain veines behinde the eares, which, if they
> be cut, disable a man from generation.

Shakespeare's own ambivalence about the power of words and
drama is expressed throughout *Hamlet*. The Ghost tells Hamlet in
considerable detail of the King's death, describing the transformations
of the body vividly, including the "vile and loathesome crust" that covers
his body, the curdling of blood caused by the "leprous distilment".
The Ghost even warns that his own story through its "lightest words"
could "harrow up thy soul, freeze thy young blood". The Ghost's
appearance is even presented by Barnardo with a warning that the
play will "assail your ears / that are so fortified against our story".

Later Claudius warns that Laertes "wants not buzzers to infect his
ear / with pestilent speeches of his father's death". Ophelia's songs
of madness are described as almost weapons. Hamlet's words are
presented as having the power of injury: "I will speak daggers to her
but use none," he says before visiting his mother in her bedchamber.
She later says, "These words like daggers enter into my ear." Even
the "whole ear of Denmark" is "rankly abused" by words. Hamlet

is interested in the theatre himself – he orders the staging of a play within a play. Much of the work is to do with the unreliability of images, of deceit and how important words are in conveying poison. In the performance of the play, Lucianus almost chants an incantation as he prepares the "mixture rank, of midnight weeds collected".

Shakespeare makes the most direct connection between language and poison in *Othello* when Iago says, "I'll pour this pestilence into his ear." Some of the anxiety about poison, as compared with other means of violent death such as swords or cannons, was about its deceptive nature. It appeared to be something else – food, drink, medicine, clothing. It could be hidden anywhere. The linkage of sound, words and poison is much more than metaphorical. There has long been a belief in the harmful, even poisonous nature of sound. We now know that sound can cause intense psychological and physical harm. Charms and incantations could be used to cure disease or protect against poison.

Plato had written in *The Republic* and other works that poetry could have a harmful physical and mental effect on people. Language was *pharmakon*; it was a remedy and a poison. Sophists had believed that words had a similar power over the body as did drugs. The idea of words holding such powers was disputed by many who regarded it as a remnant of belief in magic. Words were seen as powerful as they entered deep into the body through the ear and this explained much of the Renaissance concern about the theatre that playwrights such as Shakespeare explicitly acknowledged. The English satirist Stephen Gosson wrote of the deep power of words to penetrate "the entries of the eare and slip downe into the hart and with gunshotte of affection gaule the minde, where reason and vertue should rule the roste". In comparison, cooks and painters had little impact on the soul, Gosson believed. Their works penetrated the tongue or the eye but remained essentially outside the body.

Hamlet dies at the end of the play when as he says "the potent poison quite o'ercrows my spirit". His final words are a plea for Horatio to tell his story and to portray him fairly and honestly. Words are like poison, unstable, penetrating and toxic. Throughout the play, those who listen are drawn into situations that harm them, ears are the route to harm and words contain a considerable peril. Poison represented for Renaissance theatre what the Gods had been in Greek drama – something beyond human power, unavoidable, representative of the weakness and vulnerability of people and "a constant remind of the limits of human strength and intelligence".

OUR FASCINATION WITH the cultural richness of the Renaissance in Europe overshadows an understanding of the deeply miserable quality of life at the time. Death was everywhere. Life expectancy was in the mid-thirties and about a third of all children died before reaching adulthood. If you did make it to the age of forty, you could expect to live another twenty years but probably in a state of chronic ill health. Outbreaks of plague killed tens of thousands each year. Illnesses such as syphilis – rarely deadly but painful, disfiguring and embarrassing – came out of nowhere. Aristocrats ate diets that were heavy in meat and low in fibre, leaving them with constipation, piles and gout. The poor lived in a constant state of malnutrition and vitamin deficiency. Water was often a source of illness so people drank beer and wine in large amounts; the choice was between disease or a persistent hangover.

There was little relief from any of this. Medical treatment was appalling. Doctors were expensive and were almost as likely to kill as to cure. Surgery was monstrously painful without anaesthetics and as often as not, resulted in the death of a patient. Operations were regarded as so barbaric that doctors took oaths never to cut patients. Instead barbers cauterised wounds, lanced suppurating boils and removed gangrenous limbs.

Medical care rarely improved health but that did not stop doctors becoming increasingly important figures. The fifteenth and sixteenth centuries saw the re-emergence of classical medical knowledge. Guarino da Verona rediscovered *De Medicina*, one of the volumes from the encyclopaedia compiled by the Roman Aulus Cornelius Celsus which were republished in 1478. The first collected works of Galen in Greek appeared in 1525 and many new translations followed.

Much of this knowledge only survived because of Arab translations but there was little gratitude for this. Michael Servetus wrote in 1536: "In our happy age Galen, once shamefully misunderstood, is reborn and re-establishes himself to shine in his former luster; so that like one returning home he has delivered the citadel which had been held by the forces of the Arabs and he has cleansed those things which had been bespattered by the sordid corruptions of the barbarians."

Galen was the central figure in the powerful orthodoxy of Renaissance medicine. The French anatomist Jacques Dubois, better known by his Latin cognomen Jacobus Sylvius, wrote, "After Apollo and Aesculapius, Hippocrates and Galen were the greatest powers of medicine, most perfect in every respect and they had never written anything in physiology or other parts of medicine that was not entirely true." With that level of adulation, it was hard to challenge Galen, and few doctors ever did. Galen's view was that the body contained four humours – blood, phlegm, black bile and yellow bile. These corresponded to four tempers – sanguine, phlegmatic, melancholic and choleric. Diseases had their origins in the imbalances of these humours – colds were caused by an excess of phlegm, cholera derives its name from a belief that an excess of yellow bile caused a choleric reaction. Treatments were based on the idea of restoring balance – hot medicines balanced out cold diseases. Blood, phlegm, black bile and yellow bile needed to be kept in a state of equilibrium. This involved prescribing medications or diets that allowed the purging of an element that was in excess in the body. Bloodletting removed excess blood and heat. If disease did not appear in the manner that Galen had described or respond to his treatments, it was not that he was wrong but that the world was imperfect and sometimes did not live up to its own laws. Original galenicals, as drugs were known, were mostly herbal. Later, large numbers of other substances ranging from mushrooms and eggshells to the excrement of rare animals would be introduced into the pharmacopoeia. In some ways, this very limited and mechanical understanding of the body is unsurprising at a time when technology consisted of a few very basic tools – the windmill and the clock were the most advanced technology of the time.

Galenic doctors saw little need to examine their patients. There is a famous story of the artist Albrecht Durer sending his doctor a sketch of where his stomach hurt. Doctors also spent little time concerning themselves with the *materia medica* – they were there for diagnosis and were keen to differentiate themselves from the far less educated herbalists and pharmacists. Diagnosis was based on the study of urine. Doctors believed they could determine the humoral imbalances by detecting differences in the colour, smell, consistency or taste of urine. Urine samples were swirled around in special glass vessels and carefully examined, but patients went untouched by medical hands. Doctors were expensive and reserved mostly for the wealthy. Everyone else depended on folk medicine.

The wealth and corruption of doctors, their ineffectualness and the intellectual stagnation of a profession that relied on knowledge first written more than a millennium earlier was ripe for challenge, as were so many of the religious and intellectual orthodoxies of that time. This came from many quarters but few people had the influence of Paracelsus, a mysterious and contentious figure who opened the door to modern medicine and chemistry. A recent biographer, Philip Ball, describes this itinerant Swiss-German doctor and alchemist as: "protean, mercurial, a mass of contradictions. A humble braggart, a puerile sage, an invincible loser, a courageous coward, a pious heretic, an honest charlatan, fuelled by profound love and by spiteful hate, dining with princes and sleeping in the ditch, both personifying and challenging the madness of his world." He was a difficult man, combative, uncooperative and laceratingly critical of his rivals. He was boundlessly ambitious. His aim was not just to treat illness but to come up with a complete explanation for the workings of the body and the world.

Paracelsus, born Philip Theophrastus Bombast von Hohenheim, lived from 1493 to 1541, a critical moment in European history as it was caught up in a swirl of intellectual and religious dissent. He travelled the length and breadth of the continent as a student, teacher and doctor, studying with some of the greatest minds of his time while also learning from the lowliest of itinerant healers. In the preface to one of his works, Paracelsus laid out a brief history of his education and travels, adding: "He did not talk only to learned doctors but talked to barber surgeons, learned artisans, weavers, witches skilled in the black arts, alchemists, monks, nobles, common men (often vulgar and coarse), clever and simple people."

Paracelsus owed much to his education in Ferrara. It was there he was taught by Niccolo Leonniceno (1428–1514) who specialised in treating syphilis. Leonniceno's translations of Galen and Hippocrates had led him to a belief in the power of observation of doctors and also in the inconsistencies of classical writers. His focus of attack was Pliny, then regarded as an encyclopaedic chronicler of classical knowledge rather than a somewhat indiscriminate collector of folklore as he is seen today. This critical approach to works that had previously been accepted with little questioning had a profound impact on Paracelsus's career.

When the armies of King Francis approached Ferrara, Paracelsus fled the city. He went south to Naples and became a doctor to the army of King Charles of Spain. Paracelsus found himself powerless to treat the syphilis that was spreading rapidly through the population. He took it as a personal defeat: "God has not permitted any disease without providing a remedy." He turned to folk medicine, derided by doctors as a waste of time but widely used by those who could not afford expensive and mostly useless professional attention.

He drew from the sympathetic magical notion of Germanic folklore that like cured like. A poison in the body could be cured with another poison. For Paracelsus, the presence of poisons was a sign of the existence of God – chemistry enabled doctors to turn these poisons into harmless or even beneficial substances. God had made these things available to treat disease.

Alchemy – the transformation of base metals into gold – had a profound influence here. The effort to create wealth out of nothing had a poor record – it was widely seen as an activity for charlatans and magicians even then. But Paracelsus took from it the idea that alchemy separated the essence of materials. This was what the body did when people ate food. "There is no poison in the body but there is poison in what we take in as nourishment," he wrote. God created an alchemist in the stomach that separated out poison from the nourishment, taking what was good in food to create flesh and blood and expelling what was bad. The archeus, as Paracelsus named the alchemist that resided in the human stomach, was not foolproof, just like other alchemists. Sometimes it did not separate the poisons effectively and they accumulated, causing disease.

Nature, according to Paracelsus, itself signalled the possibilities it held for curing disease. Some orchids resembled sexual organs and could therefore be a sexual restorative. Black hellebore, the Christmas flower, revealed its restorative powers by flowering in the dead of winter. Liverwort and kidneywort had leaves in the shapes of the organs they could be used to treat. This was the idea of the microcosm (man) and the macrocosm (the universe) being linked in these patterns. It was up to the doctor, through observation and experimentation, to find the links. Rather than the contrarian views of Galen that believed that hot medicines would cure cold illnesses, this view of the world lent itself to like treating like. It was often believed, for example, that the iron in a sword that made a wound would cure it or that eating the heart of an animal could cure a heart condition. For Paracelsus, it was not quite so mechanical – he rejected the idea of eating kidneys to cure a kidney disease – but he did believe that the cure would be revealed in the macrocosm. This could be done through interpretations of astrology, facial features and other means.

There were, according to Paracelsus, five causes of disease: the influence of the stars; poisoning; the constitution of the patient – humours and hereditary influences; spiritual diseases stemming from mental derangement; and incurable diseases sent by God. Each of these required a different approach, summarised as: *Naturales* – treat contrary with contrary; *Specifi* – empiricists who treat with specific remedies; *Characterales* – magi who knew kabbalist remedies that treated like with like; *Spiritales* – chemical drugs; and finally, prayer to Christ and his apostles. He did not believe that astrology acted directly on men, although he did accept that exhalations from stars were capable of producing plagues and wars. He did, however, favour some notion of heredity – believing that people did inherit important characteristics from their parents.

In his *Labyrinth of the Physicians* (1553), Paracelsus cites no earlier books, instead he told his followers to seek out God through nature and to learn through observation and experience. He urged people to stop trying to make gold and start seeking medicines. He questioned the four elements and the humours of Galen. Rather than an imbalance of these humours, disease was a localised abnormality within the body. Bodily functions all had their parallels in the wider world or macrocosm.

Diseases grew in the body in much the same way that metals were believed to grow in the earth, colds produced mucus in the same way that the earth produced streams and rivers, fevers resulted from a combination in the body of nitrous and sulphurous compounds in the same way they produced thunder and lightning in the atmosphere. Diseases could be explained through chemical reactions and they could be treated with chemicals. Paracelsus drew from the medieval literature of alchemy, although he was looking for medicines and not gold or the philosopher's stone. One belief was that antimony could be heated with gold to remove impurities from the precious metal. Just as gold was the most precious of metals, man was the most precious aspect of nature and therefore it stood to reason that antimony would also remove impurities from man.

He treated kings, princes and philosophers, including Erasmus, but he also developed an extraordinary knowledge of how the poor and uneducated cured illness. Intellectually he was as important as Galileo, Copernicus and other radical thinkers of this age; viewed today, his work is both quite modern and extremely opaque and mystifying. His ideas on disease prefigure how we see illness now and his use of chemicals launched the drug treatments that are the mainstay of modern medicine. His ideas of observation, experimentation and the treatment of patients changed the way people thought about medicine, but his world view was wrapped up in an idiosyncratic philosophical system that is arcane and impenetrable, not least because of his profligate invention of neologisms.

He believed in the microcosm and macrocosm and the influence of the stars over health. He was very much a man of his day – in some ways the medical version of Luther – in that he called for a return to direct study of nature, for the influence of the cure of the common man over that of doctors, the native over the foreign and for a direct connection between man and his health. Like Luther, he was a raging antisemite. He was also rambling and sarcastic; his erratic behaviour and scorching criticisms of others generated extraordinary myths about him. He was said to have communicated with the Devil and found the secret to eternal life, selling his soul in the manner of Doctor Faustus.

"I have not been ashamed to learn from tramps, butchers and barbers," he said. It was this attitude of openness and a refusal to worship uncritically at the feet of the Greeks that made Paracelsus a great innovator. "When I saw that nothing resulted from doctor's practice but killing and laming, I determined to abandon such a miserable art and seek truth elsewhere." He brushed off the Galenic masters, saying: "Their ignorance cannot justify their fantastic theories. All they can do is gaze at piss." Galenic doctors almost certainly did more harm than good, particularly if they chose to bleed weak patients. Most folk medicine probably did little either – in reality almost anyone could become a doctor for almost any reason. Blacksmiths, left-handed people, hangmen or those with crossed eyes were all considered reputable healers.

Paracelsus generated extraordinary changes in the way the world looked at poisons. His famous line – "*Alle Ding' sind Gift und nichts ohn' Gift; allein die Dosis macht, das ein Ding kein Gift ist*" ("All things are poison and nothing [is] without poison; only the dose makes that a thing is not a poison") often abbreviated to "the dose makes the poison" – is to this day the defining notion of toxicology. This idea came out of his detailed study of syphilis. One of the few treatments for this was from guaiac, a dense wood found on the island of Hispaniola (now divided between Haiti and the Dominican Republic). The wood was boiled and the scum that floated to the surface was applied to the horrible sores that syphilis caused to erupt on the skin. The patient then drank the remaining liquid. This was seen as superior to the more common treatment of dosing patients with mercury, causing their teeth to blacken and fall out and leading to excruciating pain and sickness. Paracelsus recognised the terrible toxicity of mercury, saying: "There is in it a concealed winter, coldness and snow." What mattered was not that it was a poison but that it be given in the correct dose. "Is not a mystery of nature concealed in every poison? What has God created that He did not bless with some great gift for the benefit of mankind? Why then should poison be rejected and despised, if we consider not the poison but its curative virtue."

His empiricism challenged the bookishness of Renaissance science and allowed the scientific method to emerge, but Paracelsus himself was consumed with ideas of magic. He believed God had designed a perfect system that obeyed rules. These rules could be worked out by careful observation of nature. This was a markedly different idea from his contemporaries who mostly followed the idea that the Greeks had already revealed the nature of all these systems. Paracelsus was interested in the occult – in the Renaissance sense of the word meaning hidden forces. This could include magnetism, gravity or the nature of matter as much as astrological emanations or the actions of demons.

Paracelsus viewed illnesses as individual entities that needed specific remedies to cure them. These "specifics" as they were known, tended to be distillations of metals that were believed to act as an antidote to a poison within the body. Although Paracelsian medicine has echoes of the ideas of modern scientific medicine in which drugs are targeted at diagnosed illnesses, the reality is that it was closer to homeopathic medicine in the belief that like cures like, while many of the views were closer to the occult than science. _

His medicines were often derived from metals including mercury, arsenic and antimony, rather than the mostly herbal cures of Galenists. This horrified many who saw these treatments as little more than deliberate poisonings. Paracelsians countered the idea by saying that the process of producing the chemical cures through distillation and other methods removed the toxic qualities. They also believed, somewhat contradictorily, that like cured like and thus poisons were needed to counteract poisons.

After his death in 1541, Paracelsus sank into obscurity. He had published quite extensively during his life, mostly works on astrology and almanacs of predictions, but many medical works were unpublished until the 1560s when there was a resurgence of interest in his theories. His works were taken up by disciples, notably the Danish court doctor Petrus Severinus, described by Ball as a "cultivated and sober man who stripped Paracelsus's words of their rant and rhetoric and presented the central ideas in calm and orderly Latin". Severinus's *Idea of Philosophic Medicine* was a lucid take on Paracelsus's ideas that spread them across Europe.

The Counter-Reformation was under way after the Council of Trent in 1564 and the mood was turning against the Neo-Platonism and the Protestantism that was inherent in Paracelsus's work. The most damaging attack came from Swiss doctor and theologian Thomas Liebler, also known as Erastus, who published *Disputations Concerning the New Medicine of Paracelsus* in 1571. He described Paracelsus as a "grunting swine" who engaged in demonic practices and recounts numerous tales of his incompetence and dishonesty. In later works Erastus said Paracelsus was "an evil magician, atheist and pig", adding that his followers were "indubitably wicked". Paracelsus's life was increasingly embroidered with legends, few of them flattering. It was rumoured that he had found the philosopher's stone, that he could raise the dead and was able to commune with spirits. Conrad Gesner, a sixteenth-century Swiss scholar, said that he "certainly was an impious man and a sorcerer. He had intercourse with demons… His disciples practise wicked astrology divination and other forbidden arts. I suspect that they are survivors of the Celtic Druids who received instructions from their demons." The poet John Donne wrote that Paracelsus worked with the Devil's "minerals and fire". He was, Donne believed, a pretender to Lucifer's throne alongside Copernicus and Machiavelli.

The battle lines between Paracelsian and Galenic medicine were never entirely clear. Doctors had used the medicines that Paracelsus recommended before he wrote about them and Galenic medicine was often bent to accommodate these drugs. But there were fundamental differences in the way they viewed the world. These differences would be central to medical debates for the next century, a time in which medical arguments pervaded popular culture and anxieties about poisonings came to the fore. Paracelsian medicine evoked more than a little horror from many writers from the time. In John Webster's play *The White Devil,* the use of human flesh as a drug represents the new medicine at its most extreme: "Your followers / Have Swallowed you up like mummia and being sick / With such unnatural and horrid physic." Webster's plays contain an astounding number of poisonings and references to poisons. Poisonings drive the plots in *The White Devil* and *The Duchess of Malfi*. Many of his plays centre around something evil hidden within something beautiful, particularly in the case of women. Webster captured the fears of the time of medical poisonings, fraud and the dangers that lay hidden in the emerging sciences. As the cardinal says in *The White Devil*: "They are first / sweete meates which rot the eater: In mans nostrils / Poisoned Perfumes. They are coosning Alcumy / Shipwrackes in Calmest weather… Counterfeited coine. You, gentlewoman / Take from all beasts and from all minerals / Their deadly poison."

8. WITHIN HALF A DEGREE OF POISON

It is a precious powder that I bought
Of an Italian, in Ancona, once,
Whose operation is to bind, infect,
And poison deeply, yet not appear
In forty hours after it is ta'en.
Christopher Marlowe, *The Jew of Malta*

By the sixteenth century questions were being asked about the basilisk. The Swiss naturalist Conrad Gesner wrote of the animal but doubted its existence. William Ramesey, doctor to Charles II, was among the first to question the basilisk in writing. On one hand, he was impressed by the number of ancient writers who had described the animal; on the other he wondered why so few people had actually seen one. The Italian natural historian Girolamo Cardano declared categorically in his *De venenis* that such creatures did not exist. The evidence for this was the fact that Galen had never seen one.

Cardano's concern was not that no contemporary person had ever seen one but that a Greek doctor who had lived more than five hundred years earlier had not encountered it. Science was still bound inextricably to the past.

Ramesey had other, more convincing doubts. It was not believable, he wrote, that nature would create an animal that killed all other living things. There was also the problem of how anyone who had seen a basilisk had survived. He hedged his bets by saying that perhaps they did exist but they were probably not as poisonous as suggested or that their deadly toxin might be cured by the herbs prescribed by the Greek doctor Dioscorides.

Just as Renaissance knowledge was challenging the idea of the basilisk, they started to appear everywhere. Not only did printing allow for the reproduction of a vast array of images of such animals but a passion for collecting emerged, notably of extremely rare objects to be housed in cabinets of curiosities. Basilisks were often concocted from the dried or preserved parts of other animals, often from parts of stingrays and other fish. These hybrid monsters were known as Jenny Hanivers, a corruption of a phrase referring to their origins in the docks of Antwerp (also known by its French name Anvers) and they were a popular, and disturbing, collectable.

A story of a basilisk killing two young girls in Warsaw circulated through Europe in the late sixteenth century. A doctor had examined the bodies and announced the girls were victims of a basilisk attack. The city council decided it would hunt the animal in the cellars and a convict who had been sentenced to death offered to kill it in exchange for his life. He was supplied with an apparatus of mirrors so he would not have to look at it directly and entered the building. He found the animal and brought it out, witnessed by a crowd of several thousand. It was the size of

a cock, with a turkey-like neck and frog's eyes.
Similar stories appeared across Europe.

Catholics, women and Italians all represented threats to masculinity, faith, family and country in sixteenth- and seventeenth-century England. Religious turmoil marked these centuries and therefore the connection of Catholicism to poison is not surprising. Italians represented the fear and allure of the Renaissance and Catholicism in relatively provincial and Protestant England. Women had always been a mixed blessing; keepers of the hearth and prison wardens of domesticated men. More curious is the enduring linkage of Jews to poison. Ideas about Jews in England were unmarred by any contact with real people. Jews had been banned from the country in the thirteenth century and were not allowed to return legally until 1655. Those very few that did come as refugees after their expulsion from Spain in 1492 were forced to convert to Christianity or hide their faith.

Despite this invisibility, cultural stereotypes were powerful enough to provide some of the most virulently antisemitic images in European culture, notably in Marlowe's *The Jew of Malta* and Shakespeare's *The Merchant of Venice.* Stereotyping went well beyond hooked noses, red fright wigs, miserliness and accents. People believed Jews had cloven feet, tails and illnesses that drove them to crave human blood. Since the Black Death, these ideas had been kept alive in songs and stories, even in the very architecture of churches, in their stained glass and sculptures. Those Englishmen who encountered Jews abroad were often surprised when they failed to live up to prejudices. At the time the phrase "to behave like a Jew" meant to flail around madly. When Thomas Coryat, a Jacobean writer, met Jews living in Venice he was startled to find that they did not behave in the "phrenticke and lunaticke" way that he had been expecting.

Jews had been expelled from England in 1290 when King Edward I signed a decree on the fast of the ninth of Ab (a date of manifold disasters for Jews including the destruction of the temple in Jerusalem) ordering them all to leave by All Saints Day. Any who remained would be liable to the death penalty. Public proclamations were made that nobody should harm any of the Jews in any way and they were to be given safe passage out of the country. All their land and buildings were to be ceded to the Crown but they could take cash and possessions. Most left for France and stayed there despite efforts by the Pope to expel them again from such cities as Amiens. Quite a few drowned on the crossing. One group was murdered when their ship ran aground on a sandbank in the Thames. Invited by the captain to stretch their legs with him on the sandbank, he abandoned them when the tide came back in, mocking them with the suggestion that they should call on Moses to part the waters again. Some of those expelled ended up in Spain and some in Malta, where Ingles or Inglesi would become common names for Jewish families. Some found refuge as far away as Egypt.

The expulsion was never completed as some Jews, particularly doctors, were allowed to remain. In 1309 Edward II gave an order of safe conduct for the physician Magister Elias to come to England, presumably to work as a doctor. Dick Whittington, the mayor of London now known for his pantomime cat, obtained permission in 1409 for a Jewish physician to enter the country to treat his wife. In 1410 Henry IV summoned Elias Sabot of Bologna, subsequently doctor to several popes, to treat him. The doctor brought with him ten Jewish followers to ensure a *minyan* for worship.

Jews started to come back, although in very small numbers, after their expulsion from Spain in 1492. Half a century later under Elizabeth I, London was becoming a major mercantile centre with a large and cosmopolitan population. The Tudor period was a time of extraordinary change in England. It went from being a Catholic to a Protestant country, returned to Catholicism and then went back again to Protestantism. These lurches were all accompanied by extraordinary levels of religious persecution and fear. Monasteries, the main centres of religious, medical and intellectual life, were closed. It was a time of global exploration and trade that brought new goods, people and ideas into England. English identity was being redefined in a range of ways as the Tudor state consolidated its power and created many of the enduring structures of English life. The imaginary Jew the English conjured up may have been a necessary opposite. People constructed the idea of a Jew and then made out that the English were nothing like that.

Judaism raised fears but in Elizabethan times, they were somewhat remote from reality. After all, the handful of Jews in the country had no significant influence on economic or political life. The real fear in the ruling elite was of Catholicism. Oddly, given their relationship in the rest of Europe, Judaism and Catholicism were often conflated in Protestant England. George Carleton's *Thankful Remembrance of God's Mercy*, written in 1627, accused Catholicism of being the "synagogue" of the Devil and of promoting sorcery and poisoning. On the other hand, some on the fringes of Protestantism identified themselves as being the chosen people and as such "true Jews" and the followers of Moses. Some even had themselves circumcised.

By the sixteenth and seventeenth centuries, the idea of Jews as an invasive and toxic presence had pervaded Europe again. The Turkish siege of Malta was said to have been aided by Jews. This was an antisemitic theme that went back to the Crusades when accusations of siding with Muslims were common. The Jews were believed to have taught Turks about artillery and other armaments, reducing Europe's military advantages. Jewish armies were poised to conquer the Christian world. These forces lay off in distant lands, their existence unverified but somehow plausible to those who feared revenge for centuries of discrimination and violence.

Images of Jews in the sixteenth century are shocking in their viciousness. Jews are shown giving birth to piglets, eating pig shit or drowning in their own filth. The linkages were almost always to excrement. According to the historian Jonathan Gil Harris, literary images of Jews were often of them penetrating bodies or the body politic through the anus. One of the most pernicious and enduring images of Jews was their involvement with poison and disease. Barabas in *The Jew of Malta* mentions, almost in passing, "Sometimes I go about and poison wells," as if it were something expected of Jews.

Not only was there the *foetor judaicus*, the toxic stench that

Jews were said to emit, but there are frequent mention of Jews as intrinsically poisonous. Thomas Nashe's *The Unfortunate Traveller* has a Jewish character Zadoch cut off his own leg to extract a poison from a running sore on it. He also has breath so bad "it is within half a degree of poison". In *Christ's Tears over Jerusalem*, Nashe accused Jews in general of being toxic, saying, "What-soeuer thou art is poyson and none thou breathest on but thou poysonest."

But as with any issue surrounding poisons, there was an ambivalence in the attitudes towards Jews illustrated by the view of doctors. All doctors were suspect, particularly during this period of emerging Paracelsian medicine and the rivalries that existed with Galenic practitioners, but Jewish practitioners came under very close scrutiny for their magical command of cures and their alleged use of poisons. Harris cites the notes written by Gabriel Harvey, a Cambridge don, in the margins of work by George Meier of Wurzburg published in 1579 called *In Iudaeorum Medicastrorum Calumnias et Homicidia,* a document that details alleged murders and other misdeeds by Jewish doctors. Harvey notes on the text that "the extraordinary treachery of the Jews produces the politics of Machiavelli" but also comments that "the *Kabbalah* of the Jews is capable of so many miracles and that extraordinary natural magic they call cosmology". It is a sentiment echoed by Paracelsus himself, whose rantings against the Jews were later used by the Nazis as part of the justification of their antisemitism. Paracelsus was also, however, a great admirer of the *Kabbalah* and the ancient knowledge it represented.

Harvey's notes also contain a biography of Doctor Roderigo Lopez, a Portuguese Jew who became physician to Queen Elizabeth I and was later executed for allegedly attempting to poison her. Harvey disdainfully commented that Lopez was not unusually learned but had risen to the top of his profession through his "kind of Jewish practis". Lopez acquired "much wealth and sum reputation", Harvey wrote, "as well with ye Queen herself and with sum of ye greatest Lordes and Ladyes." But his interest in Meier's work and his own comments that restrictions on Jewish doctors were justified, because "they can easily add poison to drugs". Meier was adamant that Jews should not be allowed to be the doctors of popes or kings and yet it was normal in court to have Jewish physicians. The Castilian court decreed in the fifteenth century that no Jews could be doctors except the king's doctor.

The popularity of Jewish doctors is a strange phenomenon. Leon Poliakov, the historian of European antisemitism, wondered whether monarchs did not share common folk anxieties about Jews. Harris, however, suggests that Jewish doctors were seen as having semi-magical powers that were worth employing if one was sick. It was their very place as toxic elements in the body politic that gave them their medical prowess, their access to obscure and dangerous spirituality and their hold over their patients. At a time when there was an increasing understanding of the notion that poisons might also be cures, there was a view that Jewish doctors might act in a similar way. They could be both toxic and curative. They were *pharmakon*. That ambivalence held great dangers.

Lopez was born in Portugal in 1520 into a Jewish family, although he later converted to Christianity. He fled Portugal and arrived in England in 1559, a year after Queen Elizabeth I came to the throne. He married the daughter of Dunstan Aries, a spice merchant who was Purveyor and Merchant for the Queen's Grocery. Lopez was said to have been a highly skilled doctor, becoming resident physician at St Barts Hospital in London. In 1575 he became a court physician and then in 1586 he became physician to the queen. His position at court was always precarious; many were jealous that this Jew and foreigner had become so close to the queen that she used him to translate state documents. In 1584 he was accused in an anonymous pamphlet of distilling poisons for Robert Dudley, the Earl of Leicester and one of the queen's longstanding favourites.

On 1 January 1594 Lopez was arrested for his involvement in an alleged Spanish plot to poison the queen and was tried in front of a group of judges that included Essex, Sir Robert Cecil and Francis Bacon. The evidence against him was mostly based on the confessions of his two co-accused, extracted under torture. The prosecutor, Sir Edward Coke, told the court: "For the poisoning of Her Highness this miscreant, perjurer, murdering traitor and Jewish doctor hath been provided. The plot and practice were more wicked, dangerous and detestable than all former plots and he – Doctor Lopez – a dearer traitor than Judas himself."

At first Elizabeth I was convinced of Lopez's innocence and refused to sign his death warrant. He was moved from the Tower of London to Southwark Prison where he was allowed to continue to dispense medicines from his cell. He was later sent to the scaffold at Tyburn – where Marble Arch stands today. He met his death as a crowd jeered at him for being a Jew and traitor, the identities increasingly synonymous. The queen continued to believe in his innocence and paid for one of his sons to be educated at Winchester. The crowd mocked his claim just before his execution that he loved the queen as much as he loved Jesus. According to the diarist John Stow, he was "hanged, cut down alive, held down by the strength of men, dismembered, bowelled, headed and quartered, the quarters set on the gates of the city".

In the *Merchant of Venice*, Gratiano says to Shylock:

> *Thy currish spirit*
> *Govern'd a wolf, who, hanged for human slaughter,*
> *Evn from the gallows did his fell soul fleet,*
> *And infused itself in thee.*

The word "wolf" – "*lupus*" in Latin and similar to "Lopez" – connects Shylock to the story of Doctor Lopez. At the time, the literary image of Jews was encapsulated in two figures – Shylock and Barabas, the Jew of Malta. Marlowe's play, a chaotic and energetic drama, was a hit in London in the years around the time of Lopez's execution. It drew on longstanding antisemitic conventions in English drama. Jews were both dangerous and comic figures. Actors in dramas such as *The Conversion of St Jonathas the Jew by the Miracle of the Blessed Sacrament* would don a fake nose and a red fright wig to play the part of the Jew. Barabas in *The Jew of Malta* is a more subtle, but no less evil, figure. He poisons an entire convent, including his daughter

who has fled there. He kills her two suitors to stop her marrying and
betrays Malta to the Ottomans.

WHILE OFTEN BEING viewed with intense suspicion, Jewish
doctors were highly regarded and coveted by the wealthy and
powerful for their connections to Jewish mysticism and ancient
knowledge. There is a famous, probably apocryphal, story of Francis I
requesting a Jewish physician to treat an illness. The doctor turned
out to be a convert to Christianity, a point that enraged the French
king who insisted that another properly Jewish doctor be sent from
Constantinople.

Talmudic scholars approved of the study of medicine in a
way that Christian theologians never did. "The study of medicine
has a very great influence on the acquisition of virtues and of the
knowledge of God as well as on the attainment of true spiritual
happiness," wrote Maimonides in his *Commentary on the Mishnah*
in 1168. The ethos of "my son, the doctor" is bound up in religion as
well as Jewish maternal pride. Jewish medicine developed alongside
Greek medicine and shared two important characteristics – a belief
in free will and a separation of medicine from religion. It drew on
Galenic traditions that learning about the body was a way to show
respect to God. While the Roman Catholic Church often crushed
science under the boot of religious doctrine, Judaism encouraged
a remarkable freedom of thought and experimentation. In 1135,
the Council of Rheims expressly banned priests from practising
medicine. Throughout the Middle Ages, Christianity viewed sickness
as something to be endured rather than cured and from the Lateran
Councils onwards, repeated papal injunctions reiterated the Church's
position that sickness was caused by sin.

Rabbis were not allowed to use their religious position to
make money and so often turned to medicine as a way to support
themselves. Medicine became associated with the best educated,
most respected and most impressive members of the community.
Many Christians and Muslims viewed Jewish doctors as having

mastered special powers of the occult. Jews were cosmopolitan and therefore brought together knowledge from around the known world. Medical knowledge was a useful possession when one was at risk of being driven from one's home – it was portable and it was easy to set up shop as a doctor in a new city in the days before medical regulation. Jewish doctors were often multilingual and had access to books in Arabic and French. Jewish doctors were becoming the vital link between Arab medicine – with its classical roots – and the folk medicines of Europe. The resurgence of Jewish medicine began with Maimonides (1137–1204) and was based on the doctor's widespread knowledge of classical texts.

Jewish doctors were said to contain within them the ambivalence of the *pharmakon*; they were both cure and poison. They were suspected of killing their gentile patients; they were said even to have quotas of the numbers they had to murder. The Vienna faculty of medicine believed it was one in ten, the Spanish one in five. In Webster's *The Devil's Law Case*, Romelio, a character disguised as a Jewish doctor from Italy, remarks that he can "kill my 20 a month / And work but I'th'forenoons". Jews were believed to be toxic to the body politic, undermining ideas of Christian purity and yet strangely, like some poisonous Paracelsian physic, many people regarded them as essential to it, possibly even curative of its other ills. Jews were represented as great dangers to the community, but also clear and articulate critics of the problems of that community and probably necessary to its health.

Alongside knowledge of medical texts, Jewish doctors were also believed to have mastered occult ways. Christian suppression of these beliefs was often halfhearted as many people, priests included, wanted to employ magical cures when they were sick. In the latter years of the Roman Empire, Jewish Magi had been seen as having supernatural powers and this was carried over into Christian thinking. Jewish amulets, medicine and mystical thought remained much in vogue among Christians despite the irritation of many theologians and priests. In fact, Jewish medicine was not particularly based in magical thought. Maimonides described amulets as "these stupid and mad things".

The Council of Albi in 1254 decreed that any Christian who was treated by a Jewish doctor would be excommunicated. As with so many unpopular laws handed down by the Church, observance was rare. At the beginning of the fourteenth century, Arnold of Villanova, a physician with a deep grudge against Jewish doctors, travelled to see Pope Boniface VIII in Rome to demand the upholding of Church laws against receiving medical care from Jews. He complained to the Pope that despite the ban, one invariably only ever saw Jewish doctors entering convents and churches to treat the sick. The pontiff, who trusted his own care to one Rabbi Isaac ben Mordecai, was not much taken by Arnold's complaints.

The Pope did, however, issue another decree but there is no doubt that people paid little attention to it, for the subject came up again at the Council of Valladolid in 1322 when the prohibition was justified with the claim that Jewish doctors "commit treachery with much ardour and kill Christian folk when administering medicine to them". The accusation was more than just folk bigotry. Jewish physicians to Charlemagne's brother Carloman, the Holy Roman Emperor Charles the Bald and the French King Hugh Capet had all been killed after such accusations were made against them. The libel was so common that David De Pomis, in his defence of Jewish doctors, rebuked Christians by saying: "No one has ever witnessed any crime by a Jewish physician and no one has received reliable information of such. It is only because of common prejudice that we are accused and suffer injury. When Christians accept falsehood for truth, they harm themselves more than us, for this is completely contrary to the teachings of Christ." Even at those royal courts that raged against the Jews, exceptions were made for doctors. The marriage of Isabel of Castile and Ferdinand of Aragon in 1479 led to one of the greatest disasters ever for Jews – their expulsion from Spain. Yet these Catholic monarchs continued to enjoy the services of two Jewish doctors, Thomas and Rodrigo da Veiga, who – along with a handful of others – were allowed to stay on when Jews were driven from the peninsula in 1492.

The position of Jewish doctors became less secure with the emergence of increasing restrictions on Jews as the Middle Ages progressed. These restrictions coincided with the increasing emphasis on credentials and university educations. In 1437, the Church's Council of Basel banned Jews from getting university degrees, a position that threatened to exclude them from the practice of medicine. The expulsion of Jews from Spain meant many became Marranos, while others dispersed across Europe. Marranos, Jews who had converted often under duress, remained suspect. The idea emerged for the first time that "racially" Jewish people, even if they were practising Christians, could be excluded from key positions and professions. They were also subject to the terrible persecutions of this period. Garcia da Orta, a Portuguese Marrano doctor who lived in Goa and produced a sixteenth-century scientific masterwork on Indian plants and drugs, was tried by the Inquisition and sentenced to be burned at the stake. He died before the sentence could be carried out. His body was exhumed and burned anyway. From the fifteenth to the eighteenth century some three hundred Marrano doctors were brought before the Inquisition in Spain and many were tortured until they told the Inquisitors what they wanted to hear. A Doctor Garcia Lopes confessed to killing more than a hundred and fifty people, while another Jewish doctor said he had developed a secret poison that he could place under his fingernails so as to contaminate medicines with a single touch.

Amatus Lusitanus, a Portuguese Marrano, was one of the most prominent doctors of the sixteenth century and author of a commentary on the Greek writer Dioscorides. He worked in Holland, Ferrara and Ancona, turning down an offer to become doctor to the King of Poland but treating popes and ambassadors to the Holy See. However, the election of Pope Paul IV in 1544 saw a resumption of the Inquisition. A mob ransacked the library Amatus had been collecting and drove him into exile in Salonica under the protection of the Sultan. His persecutor was a court doctor in Vienna called Mathioli, who had published a rival work on Dioscorides that has been criticised by Amatus. It was easy enough to convert professional jealousy into an accusation of heresy.

The fears of Jewish doctors and the use of poison accusations persisted into the twentieth century. On 13 January 1953, an article appeared in the official Soviet newspaper *Pravda* under a headline: "Vicious Spies and Killers under the Mask of Academic Physicians." The newspaper article began, "Today the TASS news agency reported the arrest of a group of saboteur-doctors. This terrorist group, uncovered some time ago by organs of state security, has as their goal the shortening of the lives of the leaders of the Soviet Union." It turned out that a group of eminent doctors had been arrested or murdered nearly a year earlier. Among them was Boris Shimelovich, a former chief surgeon of the Red Army and head of the prestigious Botkin Hospital, and Miron Vovsi, Stalin's personal physician. At first thirty-seven people were arrested, but the numbers grew vastly as the conspiracy turned out to be an attempt to persecute Jews in Moscow. "The filthy face of this Zionist spy organisation, covering up their vicious actions under the mask of charity, is now completely revealed," proclaimed an article in *Pravda*.

Controversy still surrounds the Doctors' Plot. Some historians have alleged it was the start of a plan by Stalin to round up and deport Jews out of the country's main cities. It had contained all of the age-old prejudices against Jews and Jewish doctors, and it inspired a flurry of copycat conspiracies and a wave of antisemitism in other Soviet Republics. It might have been the start of a new holocaust but for the fact that Stalin died on 5 March 1953. Shortly after Stalin's death, the new leadership admitted that the plot had been a fabrication.

BETWEEN THE AUTUMN of 1939 and the spring of 1945, millions of Jews and other victims of the Nazis died in the worst mass poisoning in the history of man. The Holocaust overshadows all others in scale; it bears no real comparison and all analysis of it seems incomplete. The first deliberate use of poison by the Nazis to kill in large numbers is believed to be the gassing of disabled patients from the Owinska psychiatric hospital in Poznan, Poland, on 15 November 1939. The last people to die in this way were a group of Austrian anti-Fascists at the Mauthausen camp near Linz on 28 April 1945. In between, the Nazis slaughtered millions of Jews and hundreds of thousands of other people, including communists, Soviet POWs, Polish intellectuals and nationalists, homosexuals, Roma, Sinte and religious dissenters. Poison was deemed more efficient than shooting people and it had less of an impact on the soldiers sent to do the killing. In their world, poison was more humane than other forms of killing but the concern was for the killers, not their victims.

The poisoning of people by the Nazis began with the handicapped and people suffering from infectious diseases. Doctors experimented on the most effective mechanisms, finally deciding that carbon monoxide poisoning was the quickest and easiest to handle. The first "shower rooms" – sealed, tiled rooms with fake shower heads and piping into which the gas could be released were developed at hospitals in 1939 to kill sick or handicapped patients. What was described as "euthanasia" soon became mass murder.

When Hitler invaded the Soviet Union in 1941, there were massively more enemies to get rid of. Shootings were difficult psychologically for soldiers and hard to disguise. Morale was undermined by the slaughter of such large numbers of Jews, communist party members and Roma, the first targets of murder. Arthur Nebe, head of Einsatzgruppe B – one of the special SS groups organised to wipe out opponents – tried to kill a group of mentally ill patients by blowing them up in a bunker. The explosion sent body parts flying into trees and was not judged a success. Instead it was decided to use exhaust fumes and an experiment was set up at the asylum at Moghilev near Minsk.

Later, trucks painted with the name Kaiser's Coffee Company on the sides were developed to kill a greater number of people. These vans had tanks of carbon monoxide that could be used to kill sixty people in about twenty minutes. The invasion of Russia in 1941 both raised the number of opponents the regime wished to kill and made it difficult to ship tanks of gas to the east. The external tanks were replaced with piping that connected the vehicle's exhaust to the cargo space. In June 1942 a report sent back to the Reich Security Main Office used a chilling euphemism to report on the satisfaction the SS felt with their new killing machines: "Since December 1941, 97,000 have been processed, using three vans, without any defects showing up in the vehicles."

The same memo, marked "top secret" suggested improvements in the design of the vans.

> Greater protection is needed for the lighting system. The grill should cover the lamps high enough up to make it impossible to break the bulbs. It seems that these lamps are hardly ever turned on, so the users have suggested they be done away with. Experience shows, however, that when the back door is closed and it gets dark inside, the load pushes hard against the door. The reason for this is that when it becomes dark inside the load rushes towards what little light remains. This hampers the locking of the door.

The vans were used extensively across Eastern Europe. "The soul killers", as they were named by Russians, were parked in the courtyard of a building so that people could not hear the screams of those dying inside. Once they fell silent, the vans were driven to a remote location and the bodies buried in trenches or dumped in abandoned mines. From 1941, gas chambers were installed in camps across Germany and Eastern Europe. Some used motors to pump carbon monoxide into shower rooms but this was regarded as being inefficient. Instead they turned to a chemical called Zyklon B – a cyanide compound that was used to clear buildings of insects. It had been used to delouse the clothing of arriving prisoners at Auschwitz and to fumigate cells until the commandant's deputy, Karl Fritzsch, experimented to see its effects on people.

THE HISTORIAN EUGEN Kogon traced the developments at Auschwitz:

> On 3 September 1941 a trial gassing was conducted in block 11 [of Auschwitz I]. Later, one room of the base-camp crematorium was equipped as a gas chamber [Krema I in Auschwitz I]. After these trials, in 1942, two abandoned thatch-roofed cottages in a wood at Birkenau were transformed into gas chambers; they were known as "the bunkers". In the spring of 1943 construction of four modern crematoria [Kremas II-V] was completed on the site of Birkenau itself. Each was divided into three parts: a section for the crematory ovens, a place for prospective victims to undress, and a gas chamber. The bunkers were no longer used except in emergencies.

In the Birkenau gas chambers, prisoners, often newly arrived after gruelling train journeys, were told they were to take a shower and afterwards they would be given tea or coffee. They were crowded into the shower rooms and left for a few minutes to ensure the temperature rose sufficiently to cause the chemical to evaporate. Cans of Zyklon B powder were poured down into metal mesh columns. The chemical hydrocyanic acid is thirty times more toxic than carbon monoxide. It comes in a crystalline form but evaporates at around twenty-five degrees centigrade. Cyanide penetrates the mucus membranes of the mouth and lungs and blocks a chemical reaction through which cells use oxygen, causing death in minutes.

The French pharmacist Jean-Claude Pressac reconstructed the use of the gas chambers in a study of Auschwitz:

> That same night, 1,492 women, children, and old people, selected from a convoy of 2,000 Jews from the Krakow ghetto, were killed in the new crematorium. Six kilos of Zyklon B were poured into the stacks that

opened into the four grillwork columns implanted
between the pillars that supported the ceiling. Within
five minutes, all the victims had succumbed. The
aeration (8,000 cubic metres an hour) and de-aeration
system (same strength) were then started up and, after
fifteen to twenty minutes, the atmosphere, which
had been practically renewed every three to four
minutes, was sufficiently pure so that members of the
Sonderkommando could enter the stiflingly hot gas
chamber. During this first gassing [in the new Krema
II gas chamber], the Sonderkommandos wore gas
masks as a precaution. The bodies were untangled and
dragged to the goods elevator. Hair was clipped, gold
teeth pulled out, wedding rings and jewels removed.

Many of those gassed in Eastern Europe were descendants of Jews
who had fled there after the pogroms that followed the Black Death.
The ideology of hatred that enabled Germans to carry out this genocide
was often phrased in a similar language to those earlier times. Jews
were poisonous, a toxic presence in the German body that had to be
destroyed. They were a source of problems ranging from disease to
communism. Pseudo-scientific doctrines of racial hygiene and social
biology, popular in Germany since the 1890s, created notions of
deviancy and toxicity that had to be eradicated from society. Legal
documents and official statements made increasing use of terms such
as "vermin" and "parasite", suggesting that Germany was under
attack. Hitler often used such language. The historian Eberhard Jackal
compiled a list of terms used in *Mein Kampf*:

The Jew is the maggot in a decaying body, a pestilence
worse than the Black Death; he is a germ carrier
of the worst type, the eternal poisonous mushroom
of mankind, the lazy drone that works its way into
the homes of others, the spider that slowly sucks the
people's blood, a band of rats that fight each other

until they draw blood, the parasite in the body of other peoples, the typical parasite, a sponger, who multiplies like a harmful microbe, the eternal leech, the parasite of people, the vampire of peoples.

Hitler, who had himself been a victim of poison gas during World War I, wrote in *Mein Kampf*:

If at the beginning of war and during the war, twelve thousand or fifteen thousand of these Hebraic corrupters of the nation had been subjected to poison gas as had to be endured in the field by hundreds of thousands of our very best German workers of all classes and professions, then the sacrifice at the front would not have been in vain. On the contrary: twelve thousand scoundrels eliminated would perhaps have saved a million orderly, worthwhile Germans for the future.

The first legal steps against Jews were to remove them from jobs in the civil service, judiciary and education in 1933, when, according to official statistics, there were 437,000 Germans of Jewish faith in the country. Boycotts of Jewish businesses and antisemitic violence came in waves from the early 1930s. By 1935 many towns had signs on their outskirts warning Jews against entering. In that year, the Nuremberg Laws further reduced the rights of Jews, exposing them to yet more discrimination. The laws set out the criteria for being Jewish, sweeping into one arbitrary racial category a range of people including converts to Christianity and secular Jews who had never seen themselves as anything other than German. From 1935, antisemitism became a pervasive force in all areas of public and private life.

Doctors were much more profoundly involved in Nazi atrocities and the Holocaust than one might have expected from a profession that saw itself bound by the Hippocratic Oath. Doctors joined the Nazi Party earlier and in greater numbers than members of other professions. The Nazi Physicians' League, founded in 1929, was an immediate success; its membership and role escalated rapidly from that time. Dotors were also critical in the creation of Nazi ideology and science, mostly that relating to race and disease that was used as a justification for the Holocaust. From 1933 when the Expert Committee on Questions of Population and Racial Policy was set up, doctors played a major part in the development of a system that led Germany on an inevitable path to genocide. This started with an interest in eugenics and the sterilisation of those who were believed to have genetic diseases. It expanded into issues of "racial hygiene", which were greatly influenced by American laws against miscegenation and by eugenics. The development of plans to end those "lives not worth living" led to the notion that the needle that would euthanise the mentally and physically handicapped belonged in the hands of doctors. Killing patients became an ordinary activity for doctors. In 1941 a psychiatric hospital in northern Germany celebrated the cremation of their ten-thousandth patient by giving the staff a bottle of beer each.

Doctors were never ordered or forced to kill people. They did it on their own initiative, developing methods and rationales themselves. Killings were seen as therapeutic to the German people. By wiping out invading poisons from outside, mostly "Jewish poisons", doctors were acting in the tradition of helping people. There was also remarkably little public recognition or criticism of the widespread role of doctors and medical staff in "euthanasia" and then in the mass killings in concentration camps.

At Number 4 Tiergartenstrasse, in the heart of Berlin, an office was set up in a villa confiscated from a Jewish family, to administer the "processing" of psychiatric patients for "wartime economic purposes". This operation, known as T-4 after the office address, began on 9 October 1939 and was suspended in August 1941 by which time more than 75,000 people had been taken from hospitals and homes across Germany and killed.

T-4 would eventually become the model for killing vast numbers of Jews and others in death camps. The original commandants of Belzec,

Sobibor and Treblinka – three of the most terrible extermination camps – had all worked at T-4. Initially T-4 started off killing severely mentally and physically handicapped patients but "euthanasia" was soon being applied to a wider group of patients who were judged to be a hindrance to the war effort. Later doctors would kill patients in psychiatric hospitals to make way for people evacuated from the Allied bombings of major cities such as Hamburg. Among the many future victims of T-4 were people who had been hospitalised with shock following the bombings. Initially the government produced propaganda movies showing the mentally ill as disgusting and incurable. This failed to mobilise public support for the killings so they switched tack.

"Euthanasia" was often carried out in secrecy; the Nazis had recognised that many people, including most doctors, would be willing to go along with it as long as they felt they could maintain a clear conscience. Families of victims were sent letters saying their loved one had died of "influenza, with an abscess on the lung". The letter said that the body had been cremated to prevent the spread of infection. Dates and places on the death certificates were faked so that nobody knew how many were dying at the same time and place.

The programme took on the language of science and racial hygiene so that patients were "disinfected" when they were put to death in gas chambers. All Jewish patients were killed in the summer of 1940 based on their racial inferiority, while other Germans were selected initially on the seriousness of their illness. This rapidly descended into an assessment of their social usefulness. People were killed for being drunks, work-shy or anti-social. They were classified under "Guidelines for Evaluating Genetic Health" that included four categories: antisocial, acceptable, average and persons of particular genetic value. Those in the first category could be subject to "euthanasia". Soon "euthanasia" was being applied to anyone who did not conform to Nazi standards.

After the war, German medical associations and many doctors adopted the view that crimes had been carried out by a handful of doctors but that the majority were uninvolved. The trials at Nuremberg of Karl Brandt and twenty-two other defendants in 1946–47 and the Auschwitz trial of twenty-two people in Frankfurt from 1963–65 reinforced the view that just a small number of people were involved in atrocities and that these were mostly concerned with the vicious experimentation on people carried out in concentration camps.

Nazism broke down the barriers that existed to stop doctors killing their patients at will. Medical professionals were active participants in the dismantling of that taboo. The Hippocratic Oath proved to be a very fragile when it came to preventing German doctors crossing over from healing to murder.

9. FESTE, FARINE, FORCHE

I learned in Naples how to poison flowers:
Or, whilst one is asleep, to take a quill,
And blow a little powder in his ear.
　　　　　Christopher Marlowe, *Edward II*

Jacobus Salgado, an itinerant Protestant priest from Spain, was strapped for cash when he authored a pamphlet on the basilisk. Published in London in 1680, it is an appeal for a rich benefactor to buy a rare example of this animal.

"Being lately in Holland, I met with a Doctor in Physick, who having travelled in most parts of the world, brought a basilisk, otherwise known as a Cockatrice, out of Ethiopia; he commiserating my sad condition, did bestow it upon me to the end that by showing this rarity to my honourable benefactors, I might invite their charity towards me. But because I found some who made it a scruple whether there was such a living creature which can kill men by only looking on them; I thought it worth the while to search into the works of many authors. I found that the Romans did call it in Latin Regulus, that is to say Little King. The reason why they did call it so, was partly because he wears a crown on his head, somewhat like a king, and partly because he is a terror to all other living creatures, snatching away their food from them and killing them by the penetrating faculty of his poison."

Salgado included an illustration in his pamphlet of an animal, a cross between a dragon and a cock, perched on a hillside and drawn without any sense of perspective so that it towers over its surroundings. In the foreground three small men are shown, one apparently already dead on the ground, the others in the throes of dying with their hands in the air and agonised expressions on their faces.

"At the time of Alexander the Great, there was one of them which lying hid in a wall killed a great troop of his soldiers by the poisonous glances of his eyes upon them. These rare animals were mentioned in the scriptures, which was enough proof of their existence." However, his final plea for money suggests an air of desperation about his scam. "That which might be reputed a piece of a Mountebank to some, cannot be esteemed shameful for a stranger who being destitute of all earthly comfort doth seek an honest livelihood without doing any injury to no man."

In 1680, a few years before Sir Isaac Newton would publish his *Principia Mathematica* describing gravity and laying out the laws of motion, it was still possible for a man to peddle a basilisk, an animal that nobody had ever seen. Scepticism was rife about the existence of basilisks but mentions of the beast in works from Ancient Greece and Rome as well as its inclusion in the Bible, still confirmed for many that it must roam somewhere. Salgado shaped his tale in just the right way to capture the interest of any potential buyer. His basilisk came from Ethiopia, the quintessentially exotic land of the seventeenth century. Salgado reminded readers of its provenance in classical writings, adding a mention of Alexander. Stories about the Macedonian conqueror had been favourites for centuries and his name was closely linked to the profundity of ancient knowledge and to the mysteries of exotic lands. Salgado was an excellent salesman when it came to evoking all the fear and mystery surrounding the basilisk.

Giulia Tofana was so small that the Spanish soldiers could easily heft her body over the low wall back into the convent from where she had been taken several days earlier. The elderly woman, probably in her seventies, had died under torture, her limbs dislocated on the rack, her skin bruised and broken from beatings. She had confessed to providing poison to hundreds of women. They had come to her, seeking to be free of their husbands or lovers. The murders of these men had finally aroused the suspicions of the Spanish Viceroy of Naples. An investigation was launched and word came back that a woman, originally from Palermo, was selling the potions.

Tofana was a widow who made her living selling make-up. One of her concoctions was a clear solution that when applied to the skin brought a rosy glow to the cheeks. A few drops of the same liquid in food could bring on a lingering sickness and eventual death. The Viceroy ordered her arrested but this was not such a simple issue. Tofana had already fled to a convent. Religious orders, of which there were a multitude in Naples, for centuries a city of priests and lawyers, came under the authority of the Pope and the Church was not inclined to allow the Spanish rulers any latitude. Convents and monasteries were a sanctuary from the authority of the Spanish crown and were a favoured refuge of criminals.

Naples, always a febrile and crowded town, was awash with rumours. Men were increasingly suspicious of their wives. Anxiety pervaded the most intimate sphere. No meal was safe. No wife or mistress was above suspicion. Crowds of the curious visited Tofana in her convent, creating a scandal among the city's elite. In such a grave situation, the diplomatic courtesies between the Spanish and the Church were not likely to survive. The Viceroy ordered her arrest. Soldiers dragged Tofana from the convent and imprisoned her in the Castel dell'Ovo.

Pinioned on a rack, the favoured method of torture in southern Italy since the Inquisition, she named all those who had bought her poison. She called it Acqua di San Nicola da Bari, a reference to the shrine where the remains of Saint Nicholas, the original Santa Claus who was the Bishop of Myra in Asia Minor in the fourth century, was said to exude a fragrant, myrrh-scented balm. This holy relic was used as a cure-all and ampoules engraved with the image of St Nicholas were exported across Europe. This mystical, sacred liquid was the perfect cover for a deadly poison. Sealing her concoction in the same small ampoules as those sold at the saint's tomb, she had given it out to the women of Naples with instructions to put a few drops in their husbands' food each day until illness overtook them. It was a popular way to end a bad marriage. According to her confession, more than six hundred men had died this way.

The Viceroy's battle with the Church was not over. Incensed by the violation of sanctuary, the archbishop of Naples protested, sending a crowd of irate Neapolitans to demonstrate against their high-handed and unpopular Spanish rulers. As one account had it:

> The Archbishop, Cardinal Pignatelli, was highly indignant, and threatened to excommunicate and lay the whole city under interdict. All the inferior clergy, animated by the esprit de corps, took up the question, and so worked upon the superstitious and bigoted people, that they were ready to rise in a mass to storm the palace of the Viceroy and rescue the prisoner.

These were serious difficulties; but the Viceroy was not a man to be daunted. Indeed, he seems to have acted throughout with a rare union of astuteness, coolness, and energy. To avoid the evil consequences of the threatened excommunication, he placed a guard round the palace of the Archbishop, judging that the latter would not be so foolish as to launch out an anathema which would cause the city to be starved, and himself in it. The market people would not have dared to come to the city with provisions, so long as it remained under the ban. There would have been too much inconvenience to himself and his ghostly brethren in such a measure; and, as the Viceroy anticipated, the good Cardinal reserved his thunders for some other occasion.

To turn the crowd against Tofana, it was put around that she had poisoned wells, killing many innocents with her potions. The interest in her diminished when the Church decided it was no longer so eager to defend her. Her broken body was returned to her place of refuge, a brutal offering to both appease and taunt the priests so angered by her seizure. It is not known where she is buried but her name has endured. Supposed poisoning outbreaks in Rome in 1654 were blamed on her potion and her toxic liquid was soon spread around Europe by those who had studied with her in Naples. By the eighteenth century it was said to be available widely.

It was her Acqua Tofana – also known as *aquetta*, or "little water", as well as Acqua di Napoli and Acquetta di Perugia – that was said to have killed popes, laid waste to the aristocracies of Europe and even to have poisoned Mozart. The composer died on 5 December 1791, aged just thirty-five. His death has been surrounded by a swirl of mythology ever since, with a common theme being accusations that he was poisoned, perhaps by a rival jealous of his talents. The stories about poison began with the *Musikalisches Wochenblatt* journal writing from Prague a week after his death that his body was suspiciously swollen.

Mozart's wife, Constanze, fuelled the stories, recounting that shortly before his death, Mozart had said, "I'm sure I have been poisoned. I cannot rid myself of this idea. I know I must die, as someone has given me Acqua Toffana and has calculated the precise time of my death for which they have ordered a requiem. It is for myself that I am writing this." Constanze told this to an early biographer and would repeat the account until her own death in 1842. The accusation of poisoning ensured the survival of Tofana's name, as Mozart's death has created a prodigious industry of researchers who have tried to determine whether he was murdered or was an extraordinary man cut down young by an ordinary illness. Certainly the most obvious theory would be illness, probably rheumatic fever, of which Mozart showed the symptoms. The circumstantial evidence has been widely investigated and dismissed. Constanze did not attend the funeral, not out of guilt at being involved in her husband's death but because women rarely did at that time. Mozart was given a pauper's funeral not because he had been murdered, but because the emperor had imposed a short-lived regulation that restricted expenditure on lavish services. The requiem was commissioned not by Salieri but by Count Walsegg, a local worthy who had hoped to pass the work off as his own.

Peter Shaffer's play *Amadeus*, later made into a film by Milos Forman, contains the basics of the Mozart murder story. A rival composer, Antonio Salieri, sends a mysterious messenger to commission a requiem from Mozart. Mozart writes frantically up to his death, nearly, but not quite, completing the work. Salieri, gripped by a murderous jealousy of Mozart's talents, later attempts suicide and confesses to the crime. The poisoning accusations started off somewhat vaguely in the first biography of Mozart, written by Franz Niemetschek. The accusations took on a life of their own, unmoored from any evidence. They inspired Pushkin to write the first dramatic version of Salieri's efforts to kill his rival and since then the death has been the subject of dozens of books and articles, none of which has come up with any proof of poisoning. The speculation around Mozart's death reflects concern that someone so brilliant, even recognised in his own day as the rarest talent, should die such a sad and ordinary death when his musical powers were still at their height.

There is a curious aspect to the murky stories surrounding Giulia Tofana. She was widely accused of being a poisoner, of selling toxic potions and of being complicit in the murders of many men, but she was not accused of being a witch. For centuries, elderly women, mostly those who sold cures and charms, had been the most common victims of witchcraft accusations but from the mid-seventeenth century onwards there was a change across Europe. The supernatural was no longer attached to these crimes. Poisoning replaced witchcraft as the crime of women.

This reflects a pattern that has emerged in many places around the world. Studies in several African countries suggest that as people become better educated and more urban, anxieties around social relations and potential harms shift from witchcraft to a fear of poisoning. A study of this among an emerging class of urban professionals in Sierra Leone in the 1960s suggested that this fear of poisoning might be related to anxieties about those who might stand in the way of professional advancement. Anyone who might represent such a risk could be legitimately shunned – even if, for example, they were family – because of the potentially toxic threat they represented.

BETWEEN THE THIRTEENTH and nineteenth centuries, more than fifty thousand people were executed or lynched in Europe for being witches. Most were women, most were poor, many were widows and many were elderly. They were killed in all parts of the continent by Catholics and Protestants, in towns and in the countryside. They were tried in religious courts and by civil authorities. Witchcraft touched the entire continent, albeit unevenly, and spread with European settlers to the New World. About half of those killed died in the German-speaking states of the Holy Roman Empire. Even here the patterns of accusation are hard to fathom; accusations were made by Catholics and Lutherans. Some cities were almost immune, in others accusations continued for years. Areas ruled by prince-bishops who combined spiritual and temporal rule appear most inclined to accusations; nine such figures were responsible for more than six thousand deaths in Germany.

There is clearly no single cause of witchcraft accusations and many factors went into the selection of certain people for targeting. Historians have identified an extraordinary range of possible causes – all of them plausible at different places and different times. Among the most enduring explanations are the emergence of capitalism provoking guilt about those who were left behind, who were subsequently targeted in witch hunts; the psychological impact of events such as the Thirty Years' War that had racked the continent; economic shocks caused by a short ice age that afflicted Europe in the early seventeenth century; a desire by male physicians to repress the business of midwives and vendors of cures; anxieties about fertility in an age in which the population was regulated by controls on marriage and through delayed childbirth; and a rise in mental illness caused by syphilis.

From the late fifteenth century, the focus of witch crazes turned mostly towards women, although in Germany as many as a fifth of all accusations were made against men. The German Inquisitor Heinrich Kramer, also known by his Latin name Henricus Institoris, and his co-author James Sprenger, published *Malleus Maleficarum: The Witch Hammer*, a guide to the evil of witchcraft. Kramer spent the year before he began work on the book in Innsbruck, trying to whip the town into a frenzy against the witches in their midst. He failed dismally. The bishop of Innsbruck was a sceptic on the matter of witches, as were a great many people, and had refused to accept Kramer's view that the town's women were all up to no good. Trials of several women were launched but soon collapsed and Kramer was banished with a warning from the bishop about the potential wrath of the men whose womenfolk he had insulted. He skulked off, undeterred, and wrote *The Malleus Maleficarum*, a work of staggering misogyny and viciousness that was the handbook for the persecution of women across Europe for the next two centuries. As far as Kramer was concerned, witches were exclusively women. Men could be involved in witchcraft but mostly as bystanders and unwilling participants. It was women, by virtue of their weakness of mind and lack of faith that were involved in this evil. Even the word "feminus" was derived from *fe* meaning "faith" and *minus* meaning "less", he wrote, displaying a shaky grasp of etymology.

The Malleus Maleficarum is a Freudian sweetshop, filled with anxieties about sodomy, female sexuality and castration. Kramer devotes an entire section to how witches steal penises, which is worth quoting at some length to give an idea of the author's mindset:

> And, what then is to be thought of those witches who in this way sometimes collect male organs in great numbers, as many as twenty or thirty members together and put them in a bird's nest or shut them up in a box where they move themselves like living members and eat oats and corn as has been seen by many and is a matter of common report? It is to be said that it is all done by the devil's work and illusion for the senses of those who see them are deluded in the way we have said. For a certain man tells that, when he had lost his member, he approached a known witch to ask her to restore it to him. She told the afflicted man to climb a certain tree and that he might take which he liked out of a nest in which there are several members. And when he tried to take a big one, the witch said: "You must not take that one; adding, because it belonged to a parish priest."

His view of witches was by no means universally accepted but it did have a considerable influence; the book was reprinted numerous times after 1485. His focus on women as witches and poisoners came just before a catastrophic outbreak of syphilis across Europe. It raged through Europe in a way we almost cannot imagine today. It may have led to a massive rise in mental illness during this period. This was not the relatively easily cured disease of today. It may have been an unusually virulent form or populations may have developed greater resistance since then but this was an illness that shaped European perceptions of illness, women, sexuality and poisoning for centuries.

One of the best sellers of the sixteenth century was a work by an English author called Philip Barrough that tackled one of the most critical issues of his day – the pox. In his *The Sixth Book Containing the Cure of the Disease Called Morbus Gallicus*, Barrough blamed the disease on the Spanish who brought it back from the New World "instead of their gold" and believed it was spread principally by the armies of Charles VIII of France during his invasion of Italy, as "his armies were much given to venery". The outbreak of the pox among his troops was often portrayed as divine revenge for his cruelty. "The Frenchmen at that siege got the Buttons of Naples (as we term them) which does much annoy them at this day. But the first finding of this grievous sickness was brought into Spain by Columbus at his coming home so that all Christendom may curse the emperor and Columbus."

Barrough suggests, in an echo of the poison damsel myth, that the Spanish in Naples deliberately sent the French troops "courtesans infected with this grief, minding to let them have some of their jewels which they brought out of the Indian country". This has echoes of other attempted poisonings suggested in writings from this time. Gabriele Falloppio suggests that the Spanish poisoned the wells for the French troops, while David de Planis Campi said they mixed the blood of lepers into the wine and caused the disease to spread in this way.

Syphilis rampaged across Europe in the decades after 1490, producing a disease that was embarrassing because it vividly marked sufferers with facial lesions and scarring. The illness was unavoidable, touching high (the pope's son Cesare Borgia and even Charles VIII himself) and low (tens of thousands of soldiers, mercenaries, prostitutes, traders and camp followers who spread it across the continent).

Any disease associated with sex was immediately blamed on women. To be cured it was necessary to abstain from sex. Barroughs wrote:

> Chiefly Venus must be shut out of the door quite,
> especially while this decoction (the wood guaiacum)
> is given. Some by committing this act just once in this
> cure have failed of remedy through the same. There
> be devilish women desirous to be handled and dealt
> withall who will beautify themselves, to inflame men's
> hearts to lust towards them; abandon your company
> and thrust them out of the doors and house.

His work did suggest a very English view of the disease at this time – it was connected to all the things the English did not like much – Catholicism, the French and the Spanish. From its emergence in Europe at the end of the fifteenth century until its effective treatment by antibiotics halfway through the twentieth century, syphilis had a constantly shifting set of cultural associations that linked it with sin in the sixteenth century and with creativity and intellectual vigour in the nineteenth. The first enduring associations were links between the disease and Naples. It is likely that the first widespread outbreaks were among French troops who had attacked the city, leading them to call it the Neapolitan Disease. Other Italians, the next victims of the disease as it was spread by the French troops and other European mercenaries fighting with them, named it *mal francese* or the French disease. The blame passed from country to country. In Japan it was known as the Portuguese disease from the sailors and traders who introduced it there.

Syphilis was associated by many with both foreignness and sin, but as the disease spread through society infecting aristocrats and clergy, the volume of moral outrage was dialled back. Afflicted leaders such as Cesare Borgia were powerful, dangerous and also exactly the people who might commission a book on the subject, and therefore writers were quick to rein in their more extreme judgements. Doctors dealt with the problems by moving rapidly from sermons on the dangers of sin to practical medical guidance about the disease.

In 1530, a Paduan doctor, Girolamo Fracastoro, wrote an epic poem about the disease that included the invented myth of a shepherd named Siphylus. The work, *De Morbo Gallico*, was also aimed at publicising a new treatment for the disease that used highly toxic mercury. Although far from effective and often debilitating in itself, this treatment was used up until the nineteenth century. Fracastoro's poem illustrates some of the difficulties of writing about the subject – it was easier to use either slightly elusive language or to deal with the subject in fiction. Not everyone kept the urge to scold at bay. William Clowes the Elder, writing in his *A Brief and Necessary Treatise Touching on the Cure of the Disease Called Morbus Gallicus*, called syphilis "a notable testimonie of the just wrath of God against the filthie sin".

Gaspar Torella of Valencia was Alexander VI's doctor and also treated Cesare Borgia. In a work on the illness in 1497, he repeatedly uses the phrase "they say" when discussing whether the disease might be God's punishment, seemingly distancing himself from the idea. Addressing his advice at fellow clerics (he was a bishop), he recommends that if they cannot be celibate, they at least find women who are not infected, certainly a pragmatic approach that the Church might frown upon today. Niccolo Massa in his *Il Libro del Mal Francese,* published in Venice in 1566, also mentions other routes of infection – such as wet nurse to infant – that took some of the moral sting out of the disease, as children infected in this way could not be guilty of any sins connected to sex.

The battle lines were being drawn between those who demanded a moral approach to the subject, putting abstinence above everything, and the realists, who wanted to employ what would today be called harm reduction: essentially recognising that people are going to have sex anyway, so they should be taught how to avoid the worst consequences. Gabriele Falloppio, a doctor who taught at Padua and is still known for his description of female reproductive anatomy, was among the first to address ways to avoid the French disease. After sex, men should wash and then cup their genitals in a special cloth pouch he had designed that had been impregnated with medication. Falloppio claimed to have had more than a thousand men follow this practice and not one of them had become infected.

Daniel Sennert, a professor at Wittenberg in Germany, responded with some fury at Falloppio's encouragement of sin and raised the argument that offering people the prospect of prevention meant that they would be wild in their sexual behaviour, an idea prevalent in right-wing Christian circles to this day. Sennert also believed that Falloppio was foolish in his belief that small corpuscles spread the disease and could be wiped away – instead he believed that it was spread by "infected vapours and spirits" that entered through the veins. There was some debate about the morality of coming up with "preservatives" that could be taken in advance to prevent syphilis, with some authors believing it was immoral to offer such products as they would only encourage sexual activity. Others questioned whether they even worked: as if God had intended the disease to focus people's minds away from sex and if abstinence was the best prophylactic, then why would God allow any other prophylactic?

The seventeenth century was marked by a rise in moralism and religious extremism when it came to syphilis. Even masturbation was attacked by one French writer; one might have assumed the solitary pursuit was relatively free of the risks of venereal disease but the author wondered where people's hands had been. By the eighteenth century there was a marked decline in the fire and brimstone moralising of the previous century. The emphasis shifted to understanding the disease and treating it rather than simply describing it as part of the moral downfall of humanity. Many writers still modelled their ideas of disease on humours, but this coexisted with and was being replaced by the notion of infection. James's medical dictionary of 1747 defined the contagion as "an active and penetrating poison which consists of an extremely subtle sulphurous poison". Others believed it to be an acid that had to be passed through contact and could not be transmitted through vapours. These poisons caused a disintegration of the humours in various ways. Others believed that it was a living agent. Antoine Deidier, a professor of chemistry at Montpellier and a galley doctor in Marseilles, believed that syphilis was caused by "tiny living worms which produce eggs by copulating and can multiply readily as do all insects". Deidier died in 1746, long before the *Treponema pallidum* bacteria was identified as the cause of the disease. His description was remarkably prescient.

Mercury emerged as the key treatment for syphilis despite its obvious drawbacks. Iatro-mechanists – those who believed that chemicals had some sort of mechanical impact on the body – believed that mercury travelled in the blood and pulverised the poison that caused syphilis. Iatro-chemists believed that mercury produced a salt or oxide which had an affinity for the poison and absorbed it. There were many ways of administering the mercury. In Italy it was often done in the form of underpants smeared with mercury ointment. Mercury "frictions" were also recommended, although by the eighteenth century caution about administering mercury internally had evaporated. It was used in anti-venereal enemas and through corrosive sublimate or mercuric chloride. Van Swieten's liquor, named for the doctor to Marie-Therese of Austria, was corrosive sublimate dissolved in a solution of alcohol and water.

The advantage over ointment-smeared underpants was considerable. Patients could take doses themselves in private without revealing to everyone that they were afflicted with the illness. Mercury, however, led to skin ulcers and severe neurological effects, all of which were generally attributed to syphilis (one should remember that there was little consensus about whether all venereal diseases were one and the same or what the standard symptoms of syphilis were). There was no apparent understanding of the fact that the sores initially produced by syphilis healed by themselves in a few weeks, while the disease remained in the system and people could still infect others.

The disease also encouraged a multitude of quacks and charlatans. Doctor Bellet's mercurial syrup was one such treatment – even its name suggests something sweet and curative rather than the horribly toxic remedy it would have been. Others offered medicines that were free of mercury, thus avoiding the ever more obvious side effects of this extremely unpleasant treatment. Count de Milly's "safety water", Vergery de Velnos's "new vegetable anti-venereal remedy" and Laffecteur's anti-venereal nectar, the recipe for which remained secret but appeared to contain honey, cumin, marsh reeds and sarsaparilla, were all popular. The latter ended up becoming the highly lucrative official cure used by the French navy.

As with all diseases, the charlatans provoked a backlash from doctors who demanded more regulation and protection for their mostly equally ineffectual cures. Doctors increasingly resented the ever-growing number of people outside the profession who took it on themselves to treat illnesses. One doctor complained that "now even medical students, herbalists, apothecary's boys, monks, blacksmiths, matrons, seamstresses, clothes menders, cobblers etc. etc. make it their business to treat these diseases to the great detriment of the species and society".

The sociologist Stanislav Andreski attributes the early modern spate of outbreaks of witchcraft to syphilis. The idea that witchcraft accusations occurred in response to the stresses of conflict was not supported by a historical record that suggested witch hunts were a mostly peacetime activity and that during wars officials tended to be too busy to pursue them. Likewise they could not have simply been driven by economic factors, as suggested by the anthropologist Alan Macfarlane, because in many cases there had been little real change in economic systems when the accusations were made. Syphilitic shock does fit with some aspects of the witch hunts: the time period from 1500 to 1700; the typical targets of women and midwives; the association with sexual depravity, and the variation in different countries, including a low incidence in England, Russia and Spain where there was also a lower infection rate. It is possible that the disease was one of several factors fuelling the fears of witchcraft and the associated anxieties about poison that were a feature of European life for several centuries.

However Andreski's arguments are problematic – the real surge in witchcraft prosecutions was about a century after the rapid spread and relatively rapid decline of syphilis. Syphilis might have set off a low-level witch craze which later became a raging epidemic. *Malleus Maleficarum: The Witch Hammer* predates the European outbreak of syphilis. Another possible problem is that the ending of the witch hunts did not coincide with any shift in the spread or incidence of the disease. It had even become less virulent nearly a century before the end of the craze. Syphilis may have been a factor, but even Andreski says it does not explain local variations or why certain women were seen as witches. War and general violence may have been a factor, as was the movement of women into the workforce in greater numbers. There is certainly a time lag between the syphilis shock and the greatest number of executions for witchcraft. There is evidence that climatic shock and epidemics of all diseases, not just syphilis, coincide with the greatest activity against witches and the greatest persecution of women for their role in poisonings.

LIKE MANY SOUTHERN Italian women of no education, wealth or position, Giulia Tofana has no place in the formal records. Her story appears today only as a brief historical aside in toxicology textbooks. She supposedly aided the murder of a generation of Neapolitan aristocrats but has gone unnoticed in the histories of the city. It is even unclear if she was ever formally tried. Tofana exists less in history and more in the realm of poison lore and in the cautionary tales about crones dealing in abortifacients and poisons.

But there is more to Giulia Tofana than her mythical status as one of the greatest mass poisoners of all time. Tofana was just one of a great many women, and probably even more men, accused of killing with poison, but her name endured. Tofana features in several non-Italian works that were popular across Europe in the eighteenth and nineteenth centuries. *The History of Inventions, Discoveries and Origins* by Johann Beckmann is a compilation of scientific tales and embellished history that was first published in German in 1777. Beckmann, a polymath who first coined the word "technology", was fascinated by poisoners and was taken with the idea, popular in England and northern Europe, that it was among the Italians and French of the seventeenth century that poisoning reached its apogee. "They were never prepared with more art, at any period, or in any country or employed oftener and with more success than they were in these countries and at that time," he wrote.

Beckmann's main source on Tofana appears to be Doctor Garelli, the Italian physician to Emperor Charles VI who wrote about the Sicilian and her poison, which he believed was composed of "of an arsenical calx, dissolved in Acqua cymbalariae and which I suppose was rendered stronger and more difficult to be detected by a salt that may be readily guessed". Garelli wrote to the German doctor Frederick Hoffman:

> Your elegant dissertation on the errors respecting poisons brought to my recollection a certain slow poison, which that infamous poisoner, still alive in prison in Naples, employed to the destruction of upwards of six hundred persons. It was nothing else than crystallised arsenic dissolved in a large quantity of water by decoction with the addition but for what purpose I know not of the herb cymbalaria. This was communicated to me by his imperial majesty himself to whom the judicial procedure, confirmed by the confession of the criminal, was transmitted. This water in the Neapolitan dialect is called Acqua de Toffina. It is certain death and many have fallen sacrifice to it.

But Beckmann believed that Tofana could not simply have doled out poison to all comers but must have known something about her victims. "At any rate, I am of opinion that the celebrated Tophania, when she engaged to free wives from disagreeable husbands within stated weeks and days must have had certain and very accurate information respecting their constitution and manner of living, or as the physicians say, their idiosyncrasy."

Later writers were to draw on two other sources. Pere Lebat was a French priest whose travelogue about his journeys in Spain and Italy was a best-seller and was translated into several languages. It is from Pere Lebat that most of the details of Tofana's story come. He said he arrived in Naples in 1719 just as the Viceroy had found out about Tofana. For Lebat, the story was mostly one of a clash between the

Spanish rulers and the Catholic Church over their respective privileges in the city where the Spanish ruled over the temporal and the Church over the spiritual. The heavy hand of the Church was a normal source of tension in all the Italian states that suffered from being close to the Pope and yet flexible in the implementation of his wishes. It is Lebat who mentions the Viceroy spreading rumours that Tofana had poisoned the city's wells and granaries as a way of deflecting popular outrage over her seizure in a convent.

The second writer who mentions Tofana and proved a source for later works on her was John George Keysler, an eighteenth-century German religious layman and educator, a fellow of the Royal Society in London and a man whose outlook on the world was one of pursed-lipped disapproval. Reading his works, with their rippling disdain and heavy breathing voyeurism, one can picture him marching through Naples, head held disdainfully high, his face fixed in a moue of disgust.

Like so many travel writers, Keysler was obsessed with the discomforts and dangers of travel. After devoting several pages to the threat of tarantulas and how their bites can be cured through dancing the high-stepping tarantella folk dance, Keysler went on to note that these were not the only dangerous creatures to be found in Naples:

> But according to some a greater evil still remains; and the worst creatures in this delicious country, say they, are the inhabitants themselves, who besides execrable and unnatural lusts are of a vindictive, treacherous, bloody nature. Though national charges generally imply ignorance, narrowness of soul and uncharitableness, it is certain, however, that the history of Naples, almost beyond any other, abounds in sad instances of the excessive depravity of man. Tophana, the noted female poisoner, who first invented Acqua Tophania, is still living in prison here and few foreigners leave Naples without seeing this infernal hag.
>
> She is a little old woman who had entered some

kind of religious sisterhood and on this account, if not on a worse, her life has been hitherto spared. She is said to have poisoned some hundreds of people and was remarkably liberal of her drops, which she gave by way of alms, to wives who from several intimations, she knew would not be inconsolable for the deaths of their husbands. Five or six drops of this liquid it seems answer the purpose and may be lowered and tempered so as to take effect in any determined time... But since lemon juice has been found to be a sort of antidote against it, this vile composition has fallen into some disrepute.

Already, the version put around by Beckmann is contradicted. Visiting the city in 1817, Keysler found Tofana not just alive but a tourist attraction, "an infernal hag" living unmolested in a convent. But even more disappointing is the reality that Acqua Tofana could be neutralised by nothing more than lemon juice. Keysler's telling of the story is very much in the manner of his entire book, simultaneously salacious and prim, sceptical and disapproving. For him, Tofana, grotesque, dangerous and fascinating, was the perfect emblem of Naples.

Keysler described Neapolitans as living in a voluptuous manner, tamed by the lavish provision of *feste, farine, forche* – festivals, flour and the gibbet. The people were "excessively fond of public diversions, clamorous upon the dearth of corn and seditious unless intimidated by severity". He was shocked by the licentiousness of the city. "It must be owned that in no great city in Europe the prostitutes are so numerous and so abandoned; these *donne libere*, as they are called, amounted to eighteen thousand in the city and in one particular part of it is a receptacle for two thousand of them; and yet it is no uncommon thing for ecclesiastics to lodge in those infamous parts of town."

Although Keysler reserves most of his condemnation for the licentiousness of Neapolitans, he also mentions in passing a curious use of Acqua Tofana that suggests her products were employed by the rich and powerful. Acqua Tofana was said to have been the poison used in the murder of Count Gallas, sent by the Spanish king to become viceroy. Neapolitans, dreading his reputation for being authoritarian, killed him before he could cause them any problems. Keysler mentions that the poison could be timed to kill at any time up to three months after it was administered. However, while saying that the Count clearly died from poisoning, he dismisses that some of his servants might have suffered the same fate, saying that it "was no more than the natural consequence of drinking to excess a strong wine and afterwards regaling themselves with ice water when they were extremely hot".

In Keysler's and Lebat's writings, Tofana is something of an archetype: a deceptive, religious crone, linked to cosmetics and to women who murder for the age-old reason – lust – they kill to be free of a husband and to have access to their lovers. She starts to become something of a cipher onto which multiple story lines were projected. But why did Tofana's story, as elusive and vague as it is, have such a hold on the imagination? As is often the case with enduring stories of any kind, it contains within it components of beliefs about poison and mass poisoners that have endured for centuries. Tofana embodied all those groups that attracted poison accusations in Europe at that time: women, Italians or more specifically Neapolitans, Catholics, and the elderly.

A similar case leads us to some sense of why accusations of mass poisoning were made against certain women. The Italian historian Giovanna Fiume examined the extensive records of a trial of Giovanna Bonnano, a seventy-five-year-old woman arrested in Palermo in 1788, nearly a century after Tofana was said to have carried out her crimes. Bonnano, a widow and beggar, according to Fiume, was widely believed to be a witch who associated at night with the *donna di fora* or the "women from beyond". These spirits were seen as controlling the world, being responsible for the fickle nature of fate. But Bonnano was far from an outcast in the slums of Palermo. She was a source of spells and magic potions eagerly taken up by local women, mostly those involved in family disputes. One such woman, a bread seller called Agata Demma, asked for a spell to get rid of her widowed son-in-law who had stopped supporting his four children. She was told to get some unsalted water from a tannery and sprinkle it around the stairs and corridors through which her son-in-law passed. The magic did not work and later Demma lamented to the judges at Bonnano's trial that she had been conned out of a significant sum.

Bonnano fit many of the characteristics of a witch – she was elderly, single and poor. Her priority was survival in a harsh city, and having the credulous pay for spells and potions was one way to live. Although witches did not universally fit this picture, the pattern had been strong. Old women living in poverty and isolation were easily demonised. When witchcraft scares spread uncontrollably through other layers of society, as they did in the German city of Wurzburg in the 1620s, the trials eventually met with resistance from those in the elite who realised that they too might be accused. By the late eighteenth century when Bonnano was arrested, witchcraft trials had almost entirely ceased across Europe. Just as several factors might have produced witchcraft scares – the tensions of the Reformation and Counter-Reformation, poverty, war, economic and environmental stresses, diseases and even possibly ergot poisonings – different forces came together to reduce the number of trials for witchcraft. Blaming misfortune on the occult became more difficult but the ideas that lay behind witchcraft – that someone must be responsible for bad things happening – shifted to a supposedly more scientific and rational basis. Ideas about justice changed across Europe as the Inquisition was wound up and jurists became increasingly sceptical and critical of the idea of witchcraft. The emerging medical profession also increasingly attacked ideas of demonic possession, beginning a process of identifying mental and physical illnesses that caused symptoms – speaking in voices, twitching and motor ailments – that had previously been linked to witchcraft.

Witchcraft was divided up into a range of other "real" offences that were then tried. This process involved the repeal throughout the eighteenth century of witchcraft laws and the abandonment of the idea. Witchcraft had often included the idea of magical harm through poisonous potions, but critics of witch hunting began to press the case that women who harmed their neighbours with potions were actually administering poisons and should be prosecuted for this rather than supernatural crimes. Across Europe a pattern developed. In Wurttemberg in the 1670s, the number of trials for witchcraft dropped from forty-six a decade earlier to sixteen. The number of trials for poisoning rose from six to fifteen. There does not seem to have been a major change in beliefs. Many people still believed in witches and their nefarious powers. In 1911 an old woman was burned to death in a lime-kiln in Perugia after being accused of being a witch. There was, however, a major change in the way courts dealt with these accusations. A growing sense that large numbers of innocents had been caught up in trials led to greater judicial caution. Growing state power from the centre also reduced the latitude for towns to take matters into their own hands.

In Palermo a chance discovery by Bonnano changed her from a witch into a poisoner, moving her into a realm that the new systems of justice could tackle. She heard that a child had nearly died after drinking a potion used to get rid of lice. This delousing liquid, which she described as a "mysterious vinegar liquor", could be sold to people who wanted to get rid of relatives. The "vinegar", almost certainly a solution of arsenic made by a local herbalist Saverio La Monica, had little taste or smell and could be mixed undetected into foods or drinks. It took a week or two to work, allowing time for sick men to go to confession and repent of their sins. Their murderers would not have their victim's eternal damnation to weigh too heavily on them.

Six women used the vinegar on their husbands. One man used it to kill his wife. In each case the death of a spouse offered some prospect of improvement, be it a new marriage or an advanced social position. Giuseppe D'Ancona was a thirty-five-year-old baker who had been married to his wife Rosa for fourteen years. He was a relatively prosperous man in his neighbourhood, but widely disliked because of his aggression and his brutal treatment of his wife. He was in love with a fourteen-year-old girl and had promised her mother that he would be rid of his wife and able to pay a substantial dowry for her. He heard from a local seamstress, Rosa Billotta, that a vinegar could be used to get rid of his tiresome wife. Billotta, who had helped Guiseppe D'Ancona carry out his many affairs, was reluctant to sell him the liquid, given his violent tendencies. When D'Ancona's wife died, there were immediate suspicions, all of which were given ample voice at his trial when his neighbours somewhat eagerly testified against him. He ended the relationship with the fourteen-year-old and went on to marry another woman, providing the court with more disgruntled witnesses.

D'Ancona was rich enough to afford a lawyer, who advised him to deny everything and try to pin the blame on the doctors who had treated his wife during her long, slow death. He claimed there was no proof that he had killed his wife and that the testimony against him was tainted by resentment from the mother of the young woman he had promised to marry. Billotta and Bonnano were "guilty and infamous" and could not be trusted. There was no clear evidence against D'Ancona except for his violence towards his wife and the fact that neighbours testified that he had prepared her medications when she fell sick. The symptoms of arsenic poisoning were indistinguishable from typhoid or cholera, neither of which was understood by medical science at the time.

Bonnano was arrested in October 1788, either after a trap was laid by the police or because she was reported by her neighbours after a dispute with them. She was taken in a cage to a prison where she was tortured briefly even though she had already confessed when examined by judges. But the depositions recorded by the court showed that there was a wide gap between how those involved in the crimes saw them and how the court had decided to treat them. From the start the emphasis was on proving a case of poisoning against Bonnano, while the perception of her among those who had dealt with her was that she was a witch and that the potions she had sold acted through magic. The trial came just six years after the Inquisition, headquartered in the forbidding fourteenth-century Palazzo Steri, had been closed down and the three women who had been held in the secret dungeons there were released. As the Enlightenment took hold in Sicily, there was no room for accusations of witchcraft and devil worship. From the start, the trial of Bonnano and the others focused on the poisonous vinegar. Saverio La Monica was ordered to make up an identical batch before a group of experts. This was tested on two dogs that died a few days after being given the liquid.

Bonnano and the others involved in the trial steadfastly denied witchcraft, as women brought to trial tended to do unless tortured into making fantastic claims of involvement with the devil, but this proved to be her downfall. It is clear from the trial documents that the judges rejected any association with witchcraft, even though this is how many of the witnesses saw the crimes. Bonnano was widely regarded as a witch and her potions were thought to be mostly harmless magic that provided a way for her to earn a living but never actually harmed anyone. For two years the deaths of various men and women passed without rousing much suspicion, even though Bonnano boasted to people about her magic vinegar. The judges, clearly eager to inculcate and display a modern sense of justice, were having none of this. The trial was about the action of poison, not the world of the occult. Fiume compared it to the "maxi-processo" of mafia suspects that were as much staged media events as real efforts to find out the truth of a criminal matter; the trial was about imposing the new values of the state on a sceptical public whose beliefs lagged behind the modernising efforts of the government.

Bonnano was sentenced to death. She was hanged on 30 July 1789, on very high gallows, watched by a large crowd. An account of the execution by Count Villabianca had people sending their carriages on ahead earlier in the day to secure the best spots from which to witness her death. Others in the case received a wide range of sentences from five years in prison to eighteen years of exile. Two men were sentenced to fifteen years rowing on the royal galleys. Fiume wrote:

> Everything about this spectacle seemed to respond to an
> opportune moment politically to offer a grand punitive
> exemplum and show the efficacy of public vengeance,

when, driven by the light of reason, such powers resided solidly in the hands of a centralised authority... yet a disquieting suspicion remains: that the judges, despite their intentions, had unwittingly burned one last witch.

PART TWO

10. EXCELLENT AGAINST ALL VENOME

They poured strychnine in his cup
And shook to see him drink it up:
They shook, they stared as white's their shirt:
Them it was their poison hurt
I tell the tale that I heard told
Mithridates, he died old.

            A.E. Housman, "Terence, this is stupid stuff"

The Indian political manual *Arthasastra* is one of the earliest guides to running a country. It pre-dates Machiavelli by nearly two thousand years and its author, Kautilya, outdoes him in all manner of deviousness. The German sociologist Max Weber described *The Prince* as "harmless" in comparison. *Arthasastra*, which is sometimes translated as just *Economics* and often as *The Science of Governing*, is full of advice for rulers of Hindu kingdoms on how to deploy assassins, poison wells, kill your enemies at a distance by using drums and how to defend oneself from rivals plotting the same moves. In a chapter dealing with the safety of royal harems, the intimate domain that was a place of exceptional vulnerability for a king, Kautilya explained how to ensure no poisons are used against the king or his wives:

"The parrot, the minah and the malabar bird shriek when suspicious of snake poison; the curlew becomes quite tipsy in the neighbourhood of poisons, the pheasant swoons; the amorous cuckoo dies; the eyes of the chakora partridge change their natural colour and become red."

Other Hindu writers of this era, Sushruta and Medhatithi, also described swooning pheasants and birds that would shriek in the presence of snakes or poisons. Birds were an early form of food taster. One writer described how the king's meals were served: "Courtesans prepare the king's dishes, physicians and various incantators gather together; the table-servants present themselves, offerings are made to demons and gods; the cages of the chakoras are speedily carried about." They were believed to be so reliable that one queen, determined to kill her king, presented him with unpoisoned foods that passed inspection by birds but later slipped him a toxic pill once the birds were removed from the room.

The belief that certain birds, notably partridges and peacocks, would cry out in the presence of poison passed to Europe through Arab writers. Birds liked to eat snakes and were thus said to be sensitive to their venom. The chakora partridge lived on a diet of moonbeams, the purest possible food, and this made it immune to poison and able to detect it. Peacocks ruffled their feathers in the presence of poison, an idea that also spread west and is recorded in early works on ornithology. Peacock feathers, smoked in a pipe or reduced to ashes, acted as an antidote or could be worn as a charm to prevent poisoning.

Invisibility is the key to poison's power and the

fear it evokes. Its hidden nature and the difficulty
of evading it mark it out as a particular danger that
requires elaborate and mystical forms of detection.
Much of the history of poisoning is also invisible.
It was a secret crime that could not be proven
with any degree of certainty until the nineteenth
century. There is little prospect of historians ever
proving the use of poisons in earlier deaths. But
the shape of fears surrounding poisonings can be
discerned by looking at the many measures that
people took against it. It is through the detection of
poisons and the antidotes that cultures developed
that we can trace this history.

Going back to the ancient world, there has
always been a belief in the notion that everything
had an antidote. Pliny wrote of the basilisk: "Yet
to a creature so marvelous as this – indeed kings
have often wished to see a specimen when safely
dead – the venom of weasels is fatal: so fixed is
the decree of nature that nothing shall be without
its match."

If poisons could kill, it only made sense that they were powerful
enough to change the physical world around them. Mirrors would
darken, glasses could crack, china shatter and jewels glow. John
of Arrezzo, the eleventh-century Benedictine monk, believed that
snake's horn – a bone-like lump found on the heads of some snakes
– would break into a sweat in the presence of poison. Hildegard of
Bingen wrote that "the heart of a vulture, split in two, dried before
a slow fire and in the sun and worn sewn up in a belt of doeskin,
makes one tremble in the presence of poison". That unseen nature
presented a problem for dramatists who wanted to signal the presence
of something invisible. In cartoons a hazy skull and crossbones might
waft out of an unstoppered bottle or a glass might fizz and bubble

when the contents of a poison ring are emptied into it but on the stage, representing poison was extremely difficult. Writers of Jacobean tragedies, the most poisonous of all dramas, wrestled with the issue. Their plays are filled with their poisoned objects: books, paintings, chalices, even a toxic beaver, all feature as weapons. John Webster provided elaborate signposting for an audience, building up tensions as the inevitable poisoning approached. In Thomas Middleton's *Women Beware Women,* the main character is called Livia, a namesake of one of the most infamous poisoners of the classical world. She speaks repeatedly about deceptions and of carrying out a crime "without the fear of danger or the law". She later says she will "drop plagues into [Hippolito's] bowels without voice". Poisonings never come as a surprise to the audience, as their dramatic power comes from lengthy prefiguring of the crime.

Invisibility is obviously more than a dramatic problem. The question that hangs over the subject is how poisons can be detected, rendered visible and avoided. Until the invention of the mass spectrometer and other analytical equipment, detecting poisons with any certainty was extremely difficult, sometimes impossible. To this day, there are some chemicals that are difficult to identify as poisons, particularly those that are close to substances that are produced naturally by the body. Poisonous chemicals such as arsenic or lead can be precipitated out of a liquid by chemical reactions or sometimes detected through smell, as is the case with cyanide with its rather pungent bitter almond aroma. But some poisons are tasteless, colourless, odourless.

Ancient mechanisms for detecting poison tended to be postmortem. Bodies were said to blacken and decompose more rapidly if the victim had been poisoned. When dissections of bodies were allowed, scarring and ulcers in the digestive tract were also said to show poisoning but these were very unreliable indicators. In the days before refrigeration, bodies decomposed rapidly in all but the coldest climates. Food tasters were widely employed; Ptolemy, the general who would rule Egypt, held the post for Alexander the Great. Later the job tended to go to enslaved people or servants. Today the more paranoid or security conscious still have them. Saddam Hussein's food taster survived all attempts to poison his boss but fell victim to Hussein's murderous son Uday, who was said to have beaten him to death after a trivial argument. Cardinal Richelieu had the slightly less sinister approach of using cats to taste his food. The use of a small mammal that would succumb more rapidly to a dose of poison is a more sensible test for toxins than having a person taste food. Mice were deployed in Thailand during a summit to taste the food for George W. Bush. The great American tradition of politicians being obliged to eat the delicacies of whatever ethnic neighbourhood they are in – knishes, tacos, Polish sausages or pirogi – still causes much concern for the Secret Service. They were forced to warn Bill Clinton against consuming any food thrust into his eager hands. Any food cooked for an American president is monitored by US Navy cooks from the White House Mess to ensure nothing untoward happens.

Anything that was to be eaten by anyone of importance in medieval Europe would have been tested and these tests became part of the elaborate etiquette of banqueting. The procedures, known as assaying, normally consisted of a test with a unicorn horn that would be waved over not just the food but any linens and tapestries used at the table. The salt would be tested by a porter and the pantler, the servant who maintained the cutlery and salt cellars. The food would be placed on a sideboard during its testing. In Italy this became known as a *credenza* because that is where "credence" tests were carried out to ensure food had not been poisoned. Even the napkins used by a prince to dry his hands would have gone through elaborate rituals, being kissed by the pantler and draped over the bare skin of his neck to ensure that no toxins had been infused into it. Those who prepared food for a noble had to be prepared to eat it themselves in public. If the cook was going to poison the soup, he had to die in the process.

The medieval and Renaissance worlds relied mostly on supernatural forms of detection or those that relied on sympathetic associations between poisons and their sources. Snake stones were believed to sweat in the presence of poison. Certain plates were believed to crack if poisoned food was placed on them. Venetian glass was also said to shatter if it came into contact with poison. Bone and horn sometimes had a similar response – snakes and animals such as deer and goats were believed to have antagonistic relations, and deer horn and bone were often used in cures for snake bites. But the most enduring mechanism for detecting poison was the unicorn, an animal bound up with Christian imagery that for several centuries was regarded as the key to preventing death by poisoning.

The first mention of unicorn horn having medicinal effects was made by Ctesias (fourth century BCE), a Greek physician who practised at the court of Darius II of Persia. Ctesias was author of a vast work on India, fragments of which were transcribed about five centuries after it was written and then revised in the ninth century by the Byzantine patriarch Photios I (820–93). The book described unicorns as coming from India, looking like large wild asses and having a horn about a foot and a half in length that sprouted from their foreheads. Their heads were purple, their eyes blue and their bodies white.

> The dust filed from this horn is administered in a potion as a protection against deadly drugs… Those who drink out of these horns, made into drinking vessels, are not subject, they say, to convulsions or to the holy disease. Indeed they are immune to poisons if, either before or after swallowing such, they drink wine water or anything else from these beakers.

Despite the paucity of references to them in classical texts, Ctesias's description of unicorns providing a cure for poison and their exotic origins in India sealed their reputation. His works on India and Persia were the source of much of the travel writing, most of it elaborate fiction, that was enormously influential in the Middle Ages. He described unicorns, pygmies and a wide variety of exotic animals. Other writers drew on the work, only requiring mild embellishments to conjure up the fantastic worlds demanded by Renaissance readers. Edward Topsell, in his history of four-footed beasts written in 1607, described unicorns as being very swift, with jointless legs. They lived mostly in mountain and desert areas. "There is nothing more horrible than the braying of it, for the voice is strained above measure..." They were shy and hostile to other unicorns, he wrote, except when in heat when the males became "tame and loving".

Few people were troubled by the lack of unicorns wandering the European countryside. The animals were mentioned in the Bible and, as with the basilisk, that was a sure sign they existed. There are two references to unicorns in the King James version of Numbers but, as is so often the case, these appear to be mistranslations from earlier texts that refer to a wild ox. It is these references that shaped European thinking for centuries. Bishop Isidore of Seville, writing in AD 600 and borrowing from Arab and classical lore on the animals, described the unicorn as "a right cruel beast" that often fought with elephants. This points to some confusion with the rhino, although these were known by the Romans, who had captured and paraded the animals through their capital. Isidore was also first to write of the enduring link between unicorns and virgins. Unicorns were almost impossible to catch but they could be lulled into submission by a virgin, whose spectral scent was said to be irresistible to the animals. It was said that unicorns would only be attracted to a virgin if she was naked and bound to a tree. As the Virgin Mary gained importance in the church, the use of unicorn symbolism grew. The unicorn's horn was determined by theologians to symbolise the special unity between Christ and God and thus had the power to redeem sin. The horn represented the central post of the cross, the capacity of Jesus to protect people from evil and also the idea that the godless would be gored by the horn as part of their punishment on Judgement Day. The Virgin was shown in paintings cradling the head of a unicorn in her lap, its horn a rather threatening phallic presence nestling close to her untouched loins.

The *Physiologus*, the bestiary that laid down much of the thinking about the animal kingdom in medieval England, said that deer and unicorns attacked snakes and could neutralise their poisons. This was based on the idea of essences and the notion that like would neutralise like. A poison could be countered by something that was like it in essence. Unicorns ate snakes and took in their poison. This poison would gather in the horn, the place where the essence of unicorns resided. The poison was neutralised by the unicorn. Because they only had one horn, the concentration of the antidote to poison was much stronger than if it had been spread between two horns. Therefore, it was very effective at detecting and curing poisons.

The fourteenth-century priest John of Hesse wrote that he had witnessed a unicorn in the Holy Land purifying polluted water so that animals might drink from it. Snakes and other poisonous creatures could taint water but it was rendered clean again with a touch from the unicorn's horn. This led to a spate of attempts to capture unicorns to ensure a pure water supply in European towns constantly ravaged by disease and foul wells. The Hunt of the Unicorn tapestries in the Metropolitan Museum in New York that date from the end of the fifteenth century show a unicorn dipping its horn into a stream surrounded by wild animals, which have come to drink the water that has been purified by the unicorn, or symbolically consecrated by Christ.

A back story was developed, principally by the Dominican friar Albertus Magnus in the early thirteenth century. The unicorn was the first animal named by Adam in Paradise. When Adam and Eve were expelled, the unicorn was offered the choice to remain in Paradise but decided to leave, condemning it to a life of suffering at the hands of men. To Albertus the unicorn was a wild force of nature, symbolising the power of Christ. It was also said to have destroyed Sodom and Gomorrah. The unicorn could only be tamed by the Virgin.

Thus the unicorn raged in heaven and earth until our radiant lady took him in her lap when He penetrated her citadel – that is to say into the lap of her chaste body, so that she could nurse him at her breast and dress Him in humble flesh and where – according to divine purpose – the unseizable animal might be captured by His pursuers, namely by Jews and Gentiles and voluntarily submit to death on the cross.

The unicorn thus becomes Jesus, linking the old testament and the new.

What were thought of as unicorn horns were more often the spiralled tusks of the narwhal, small white Arctic whales that grow a single tooth up to nine feet long. The basilica of Saint Mark in Venice had three such horns, which were used as batons by admirals of the city. Other cathedrals across Europe owned them and they appeared in the inventories of the Medicis, the French monarchs and others. The so-called Windsor Horn was presented to Elizabeth I by the explorer Sir Martin Frobisher; it was later sold for a pittance during the Commonwealth of Oliver Cromwell. The goldsmith Benvenuto Cellini fashioned a setting for a tusk for Pope Clement VIII, creating a full-sized gold unicorn head to hold the horn. One of the most spectacular uses of the horn is the seventeenth-century throne of Christian V of Denmark, now in the Rosenborg Castle. The legs, arms and back of the throne are formed from spiralled columns of narwhal tusk set with gilt tops. Gilt figures sit around the throne, which is topped by a small roof and a broken pediment crowned with an orb.

Unicorn horns were sometimes hung with "snake tongues", actually fossilised shark's teeth, and placed over dining tables to detect poisons. Most often they ended up in the cabinets of curiosities that were a part of Renaissance connoisseurship. As narwhal tusks became less rare with the expansion of trade and whaling in the Arctic Ocean, unicorn horn, most often pulverised, made its way into apothecary shops. Apothecaries put unicorns on their shop signs and sold the horn in packages marked *veneno pello* or "I drive out poison".

The widespread use of ground bones, fossils or teeth instead of "genuine" unicorn horn, led to questions about its worth as a medicine. In 1550, the zoologist Conrad Gessner published *Historia Animalium*, a bestiary based less on the Bible and more on scientific observation. Gessner had great faith in unicorn horn, which he said should be ground with ivory, gold dust, amber and coral and then boiled with redcurrants and cinnamon. But he warned: "In Venice there are various wicked villains and vagabonds who mix pulverised flint, chalk or other stones with soap and make a paste of them which they sell as unicorn horn for when they take a pinch of it and pour it into wine it starts to bubble."

Tests became increasingly elaborate. David de Pomis, an eminent sixteenth-century rabbi and doctor, mentioned one in which scorpions were placed in a jar with a piece of the horn. If the horn did not kill them in a few hours, it was fake. It is not clear whether the experiments allowed any air into the jar. Anxieties about authenticity turned into deeper doubts about its efficacy and even the question of whether the unicorn existed. The Bible said so, as did the ancients, and yet they were not found among the many exotic animals being collected and sent to fashionable new zoological gardens and royal menageries. Nor had any of the increasing number of travellers come back with any convincing evidence. In 1566 Andrea Marini wrote *A Discourse against the False Opinion of the Unicorn* that said the animal had been invented by Arab swindlers. With bracing Venetian scepticism, Marini questioned whether horns would sweat in the presence of poison or whether they could possibly respond to the vast array of poisons that existed. Why had the animals never been seen? he asked, reminding his readers that just because they were described in classical texts did not mean they existed. Harpies and sirens made many appearances in the classics but it was clear they did not present a current threat to shipping.

Owners of expensive unicorn horns must have been perturbed by Marini's assault on their treasures. Andrea Bacci, physician to the pope and staunch believer in the unicorn, came to the defence of the animal. Its existence could not be dismissed just because it was rare, he wrote. The animal was shy and lived deep in forests. When it died, horns ended up in rivers and were washed down to the sea, which explained why they were found there. Believing in the power of unicorns was vital, anyway, he asserted. To dismiss their powers was to aid the wicked; not to believe would only help poisoners commit their foul deeds. "Thus we are duty bound for the sake of the common good to write and persuade the ignorant that what is said of the unicorn is true, because such a belief makes wicked men powerless to cause evil by making them believe that the virtues of this horn will easily discover their failings and bring down ruin on their heads." Bacci's book was dedicated, as was the norm in those days, to his aristocratic patron, Duke Francesco de' Medici. A few years after its publication, Francesco was poisoned along with his wife, Bianca Capella, possibly by his own brother, Ferdinand, Cardinal de' Medici. Recent studies of remains by Italian toxicologists suggest the couple, who died just eleven hours apart, suffered acute arsenic poisoning.

Paré, a royal physician and the leading French doctor of the sixteenth century, was given a unicorn horn that belonged to Catherine de' Medici and after experimenting with it, doubted its worth. Writing in 1582, he said that if unicorns existed, the Romans would have known about them and would have engraved them on their coins. That no two writers could agree on a description of the animal raised his suspicions. Some said them to be wild and only found living among venomous beasts while others said they were tame and easily captured. Paré believed that unicorn horns were elephant bones that had been shaped or teeth taken from fish that were softened and lengthened in some way. "If one grants that the unicorn exists, is its horn efficacious against poisons? I have often made trial thereof, yet I could never find any good success in the use thereof. A great number of physicians of reputation have long since bid it adieu and have thus detracted from its supposed divine and admirable virtues." Paré's opinion was not popular; he was opposed by almost all the medical faculty in Paris.

Laurens Catelan, apothecary to the Duc de Vendome, wrote several works in the 1620s on cures and antidotes, mostly to assure aristocratic owners of expensive jewels and horns that they had not been conned. Catelan hedged his bets a little, saying there were different types of unicorns; one was a bird, another a fish, most probably a swordfish or perhaps a narwhal. Another was a type of snail and finally there were eight varieties of quadruped unicorns including rhinos, onager, elk, reindeer and lycorne, an animal the size of a horse with the mane of a lion, the head of a deer and the feet of an elephant. Catelan lists several objections to its existence and then rejects them all. He was a great believer in the fact that nature kept its most mysterious and precious things secret and they would not be revealed as easily as some enquiring minds demanded. In 1651 James Primrose concluded in his *Popular Errours* that there were many different types of unicorn, including one that was found in Germany and another that was used in India to create drinking vessels that protected against poison. Primrose writes with some disappointment that Paré had ended the practice of touching the king's food with a horn before he ate it.

The philosopher and mathematician Gottfried Wilhelm Leibniz included a chapter on unicorns in his book on the House of Brunswick, a work that in the manner of the seventeenth century, opens with a history of the German principality since the dawn of time. He begins with fossils and shells found buried in the area. Among the fossils was a "monstrous animal dug up at Quedlinberg" to which he devotes an entire chapter and includes an illustration. It is an odd, triangular animal with long forelegs but no back legs. A large horn emerges from its forehead and it has a short tail where one would expect to find back legs. Leibniz acknowledges doubts about unicorns, saying that there were suggestions they came from fish found in the Northern Ocean but says that this animal looks more terrestrial in nature and that reports from Abyssinia and elsewhere suggested that they did exist. Leibniz mentions that the bones were badly damaged when they were dug up and includes an illustration of what scientists recently identified as a fossilised elephant molar. The Quedlinberg unicorn was probably part of an elephant that had been imaginatively reconstructed with a single tusk. Leibniz's ambivalence – at once mentioning that unicorns probably did not exist and yet eagerly finding evidence of them – in part came from his understanding of fossils. Many natural historians regarded them as "games of nature", much like rocks that looked like Moses or images of the Virgin Mary that appeared on eggplants. But Leibniz believed they were the remains of real animals somehow transformed into stone. If there was a fossil of a unicorn, then real unicorns must exist. Although Leibniz was of course right about the nature of fossils, it led him away from the growing view of the time that unicorns did not exist.

By the mid-eighteenth century, the unicorn had almost entirely fallen from favour, its origins dismissed, its medical powers doubted and its Christian symbolism faded almost into irrelevance. The unicorn, the resplendent centre of attention in some of the most glorious tapestries and paintings of the Renaissance, makes its most ignominious appearance in art in the background of the William Hogarth series, *Marriage A-la-Mode*, painted in 1745. A young man visits a quack doctor with his mistress and in the background, alongside other worryingly outdated medical equipment, hangs a unicorn horn.

THE CLASSICAL BELIEF was that all poisons could be matched with an antidote as nature required such balance. At the same time, determining what that antidote might be was a problem. If one were poisoned, it was not always possible to know what had been used. To overcome this, a universal antidote was needed, something that could tackle any poison, even act as a prophylactic against being poisoned.

In 1646, John Evelyn, an English traveller, wrote in his diary that he had just returned from Venice, then, as now, a popular stop on any tourist's visit to Italy. He had witnessed the manufacture of theriac, or Venice Treacle, in a public ceremony he described as being "extremely pompous and worth seeing". For the manufacture of this drug, a courtyard or public square would be draped with tapestries, busts of Galen and Hippocrates set up and large majolica jars of ingredients laid out for public inspection. In front of a crowd of officials and dignitaries, pharmacists would mix the ingredients and pound them in large mortars. The ingredients would have been on display for several days for the public to inspect. A painting of a theriac ceremony that hangs in the Hanau Historical Museum in Germany shows four large tables arrayed on a checkerboard floor. Another table fills the middle of the painting. Each of the tables is covered with dozens of dishes of the more than one hundred ingredients used in the manufacture of theriac. This abundance was both the making and unmaking of theriac.

The first mention of an antidote in Greece is Homer's description of the herb "moly" that Odysseus used to protect himself against Circe's potions. "Therewith the slayer of Argos gave me the plant that he had plucked from the ground, and he showed me the growth thereof. It was black at the root, but the flower was like to milk. Moly the gods call it, but it is hard for mortal men to dig; howbeit with the gods all things are possible." Scientists speculate that this may have been *Galanthus nivali,* or the snowdrop, which contains a chemical known as a cholinesterase inhibitor which can counteract some nerve toxins. The idea of a substance that could fight all poisons goes back to Nicander in the second century before Christ. Nicander probably borrowed much of his material from Numenius, a Greek author born in Heracleia in 250 BC who wrote extensively on diet, cooking, banquets and angling but none of his works survive.

The enduring recipe for theriac came from Mithridates Eupator (132–83 BCE), the king of Pontus in Asia Minor and one of the few serious challengers to the power of the Roman Empire. His armies dominated Asia Minor, had conquered Greece and were preparing an invasion of Italy before he was finally defeated. In a violent era, Mithridates was a brutal king. In 88 BC, he hatched a plot to kill all Romans in Asia Minor and crowds of his people were said to have massacred eighty thousand of the new colonists. He sent chariots armed with whirling scythes through the ranks of Roman soldiers at the battle of Bithnyia. The historian Appian recorded that men were chopped in half so swiftly that they remained breathing for a moment. He captured the Roman legate and had him executed by pouring molten gold down his throat. Accounts of Mithridates are all seen through the lens of Roman historians who were hardly sympathetic to his defiance; he was probably not quite the monster of Roman lore.

Mithridates's contribution to toxicology comes from his determination to find a universal antidote that might lessen the risk of being poisoned, something he feared intensely, perhaps because of his own widespread use of poison. Mithridates, always said to be the first great experimental toxicologist, tested his mixes of antidotes and honey on prisoners who were poisoned or bitten by snakes. Plutarch, writing in celebration of the eventual defeat of Mithridates by Pompey, said the Roman general discovered a cache of secret documents that revealed that Mithridates had poisoned his son Ariarathes and had killed another man because he had beaten the king in a horse race. Each day he took small doses of poison to build up his tolerance and prevent anyone from killing him this way. He was also said to have been accompanied by a team of Scythian shamans known as Agari who were skilled at treating arrow wounds. When he slept, a bull, a horse and a stag stood nearby ready to sound the alarm if anyone approached.

Adrienne Mayor, a historian of ancient warfare, speculates that Mithridates may have known of the Indian sacred code of conduct, the *Laws of Manu*, that includes a line saying: "Let the king mix all his food with medicines that are antidotes against poison." From his experiments, Mithridates devised the Theriac Mithridatium. There is some dispute over what exactly this was. It may have been a mix of rather quotidian ingredients such as figs, nuts and honey but was later embellished by various Romans until Pliny recorded it as having fifty-four ingredients, including blood from Pontic ducks raised on poisonous plants.

Theriac may have had its origin in India, where as early as 600 BCE, the medical writer Sushruta mention two antidotes to poison, one with sixty ingredients and another with eighty-five. Around 300 BCE, the political writer Kautilya urged commanders to dose their troops and animals ahead of battle with a range of substances including small doses of such poisons as aconite and anti-toxins such as charcoal.

The antidotes worked well for Mithridates – perhaps too well, as legend has it that he tried to kill himself with poison in 63 BCE after being overthrown by a son and fleeing to what is now Armenia. It did not work and he had to have one of his Celtic bodyguards run him through with a sword. This news was, according to Plutarch, greeted with great joy among the pursuing Roman army. "Sacrifices were made to the gods; there was feasting and merrymaking. The death of Mithridates seemed to them like the death of ten thousand prisoners."

A century after Mithridates's death, Pliny the Elder was greatly impressed by his inquiring mind and determination to seek a universal antidote. Emperor Nero, also anxious about being poisoned, ordered his doctor Andromachus to investigate the claims. This led to the development of his own version of the drug called Theriaca Andromachi. It may well be the case that Mithradates never produced the drug; the story may have been a Roman invention to make the drug appear more potent. Drugs were often described as coming from elsewhere, often Greece. Setting a pattern that would endure for many centuries, the Romans liked their cures to be exotic and expensive.

Andromachus's key addition to theriac was the flesh of vipers. There was a widespread belief that as vipers did not suffer from any effects of their own poison, they must contain within themselves an antidote to it. They later became such a key ingredient of theriac that they were farmed in "viper gardens" in Venice and in the south of Italy. Galen adopted Theriaca Andromachi as a key part of his pharmacopoeia, ensuring its continued role for the next 1,500 years. It became a driver of global trade, bringing together exotic ingredients to be processed in Europe into one of the costliest substances available. Great fortunes were made through monopolies of drugs, principally to deal with general disease but also because of a deep fear of poison.

Although Galen and others wrote of them during Roman times, there is little literary evidence of their use. Lucretius, Ovid, Virgil and Horace mention many poisonings but not antidotes and not theriac, even though it was supposedly widely available. Tacitus does mention antidotes but only once when Nero's mother, Agrippina the Younger, fearing that she would be poisoned, fortifies herself with remedies. Otherwise they do not appear, despite the long procession of murders by poisoning. This may have been because they were so ineffectual or because the use of antidotes might disrupt the dramatic charge of these poisonings as they were recorded in these histories, often long after the events had supposedly taken place.

Theriac was still produced after the fall of the Roman Empire but did not make a significant comeback until the renewed interest in classical medicine in the Renaissance. Maimonides wrote a short book praising the usefulness of theriac. In 1348, during the Black Death, the medical faculty of the University of Paris recommended that everyone should take a small amount of theriac with their meals. Theriac would not have saved anyone from the plague but that did not dent its growing reputation.

It was the archetypal drug of the Renaissance – derived from the newly rediscovered classics, dependent on expanding webs of trade and exploration spreading across the world, costly, and prepared with a hefty dose of ceremony. Nicander's description of his medicine would have been as baffling to a Renaissance pharmacist as it would have been to a modern one. Was birthwort available? How did one get roots of iris out of season? What is spikenard and does the pellitory have to be dried? Although all this ancient knowledge became available to them, the actual drugs remained out of reach. In Salerno, home to Europe's first medical school, pharmacists in the twelfth century had used another ingredient in place of dried vetch, which provoked bitter criticism. Theriacs from Salerno were denounced as false for the lack of one ingredient.

In Venice, the knowledge and ingredients to make theriac came together in one place, making it the dominant centre for the manufacture of the drug for centuries. The city state had an almost unique set of circumstances that allowed it to become a key location for drug making and the emergence of medical science during the Renaissance. It was at the centre of an intricate web of trading routes and was for centuries the pre-eminent Italian entrepot. Pharmacists were able to procure the vast arrays of plants listed in the classical works. Connections to the eastern Mediterranean, where Venice had many colonies and trading partners, meant access not just to the herbs and minerals used by the ancient Greeks in this area but also connections into the Arab world. The pragmatic Republic, whose leaders tended to be more concerned with commercial opportunities than spiritual limitations set by the Pope, maintained one of the world's best universities at Padua, home to the first chair of *materia medica*, established in 1533. The university frowned on Catholic autos-da-fé and tended to brush off Vatican strictures; its open environment led to some of the most important developments in science and medicine of the Renaissance.

Venice was also a major centre for publishing. In the early sixteenth century, botany re-emerged due to the enthusiasm for classical learning. Theophrastus, Dioscorides, Galen and Pliny were the important writers, purged of the errors that had been introduced by Arab science and medieval translators. Up until the Renaissance, doctors had generally been contemptuous of pharmacy and made little effort to study plants. The emphasis of medical education was on theoretical knowledge and few doctors could even recognise the most commonly used herbs. The re-emergence of botany was principally driven by the needs of medicine but it was also connected to the humanism of the times and the powerful urge to rediscover classical knowledge. The greats of classical science had studied plants and therefore Renaissance man would as well.

There had been scrappy and incomplete translations of Dioscorides, Galen and Pliny in medieval Europe but the works of Theophrastus were unknown until the early fifteenth century, when his *Historia Plantarum* was brought to Italy in a batch of Greek manuscripts. The translation was somewhat loose, the plants very difficult to identify, but it stood as a critical work for several centuries. Pliny's *Naturalis Historia* was a vital part of education for all scholars from the early fifteenth century who would soon start a monumental series of debates over this encyclopaedic work, coming up with vast lists of "corrections" of where Pliny (or his translators) had got things wrong. To buttress their arguments for or against Pliny and others, scholars questioned apothecaries, herbalists and people in the countryside who collected plants for a living.

One of the critical texts in the revival of botany was Dioscorides's *De Materia Medica*. There was some knowledge of this work from Avicenna's *Canon of Medicine* and a work by Serapion that compiled excerpts of Galen and Dioscorides. Peter of Abano had lectured on *De Materia Medica* at Padua in 1300 after seeing the great codex in Constantinople. A number of manuscripts circulated in the fifteenth century, but it was not until 1516 that there emerged a rush of new publications and translations across Europe. A version by the Frenchman Jean Ruel was eventually translated by Pietro Andrea Mattioli, along with a commentary in Latin. The early editions following its publication in Venice in 1554 sold thirty thousand copies, making it an almost unprecedented best-seller.

Venice enforced tough controls over its pharmacists, establishing a reputation for quality that was enhanced by the elaborate public performances of the manufacture of theriac. The costumes and ceremony were there to reassure consumers that they were getting true theriac with all its expensive ingredients. Venice theriac, or Venice Treacle as was it known in England, became one of the first brand-name drugs. A German consul wrote in 1621 that "throughout Germany, Venetian theriacs have always been in customary use as antidotes precious above all others and the most excellent and best that are made anywhere".

Mattioli, a Siena-born doctor, complained in his second edition of a commentary on the works of Dioscorides that botany had fallen into such disrepute since classical times that people could no longer identify the plants used to make drugs. In many cases the ingredients were no longer available or were faked. Part of the problem was the way the information was collected and written about. Plants were described as being similar to other plants – not much help if you did not know the plant to which another was being compared. Few classical texts came with illustrations of plants – Pliny had condemned the practice of painting plants as dishonest as it could not show them in all seasons. Galen refused to describe plants, feeling that descriptions could be too easily misunderstood. All of this hindered the development of botanical classification and certainly made life difficult for those who were trying to follow recipes from classical texts.

Nicander had written a recipe for theriac that showed the immense problems of reconstructing the drug hundreds of years later:

> But that you may with instruction compound a general panacea – it will be very serviceable once you have

mixed all the simples together – let these be birthwort, root of iris, spikenard, of valerian too with dried pellitory, of all curing dried carrot and of black Bryony and with them the spongy roots of a freshly dug peony, sprigs of the black hellebore and mingled with them native sodium carbonate. Pour in too cumin and a sprig of fleabane mixed with the husks of stavascre and great down an equal quantity of bayberry and tree medick and lowly horse moss and gather in some cyclamens. Cast in also the juice of the gleaming poppy and over all the seeds of the agnus-castus balsam too and some cassia and with them cow parsnip and bowl full of salt, mingling them with curd and a crab, but the former should come from a hare and the latter should be from a dweller in pebbly streams.

Venice was the first city in Europe to tackle the problem of ingredients. Not only did it found the first chair in *materia medica,* but a decade later established a botanical garden. It became a collection point for plants found across the eastern Mediterranean and an important centre for teaching. Gardens planted with exotic species, mostly plants with medicinal uses, became another part of aristocratic patronage. Collecting became the mark of an educated man and even a relatively humble scholar could collect plants in the wild or acquire seeds from others.

All of this led to a surge in the number of pharmacies in Venice. Although the population fell from around 170,000 to 140,000 between the mid-sixteenth century and the early seventeenth century, the number of pharmacists soared to more than a hundred and a new rule had to be introduced forbidding the opening of any pharmacy within a hundred paces of any other. Drug prices were fixed each year by the Collegio degli Speziali. Large quantities were produced. Orazio Zattabella at the San Girolamo made between three hundred and five hundred pounds of theriac each year. That was just one pharmacy and the amount used by each patient was quite small – a pill-sized nub would be taken in wine.

Mattioli was a sceptic about theriacs. Many did not contain the correct plants and therefore could not have the impact that was suggested in classical writings. But he did believe that they were getting better as the study of botany improved and more ingredients were rediscovered. However, the list of missing ingredients remained lengthy. Many substances, including cinnamon and balsam, were almost entirely unavailable. Ferrante Imperato used around ten substitute ingredients in his theriac in 1557, but by 1571 he was down to just six. Other debates raged over the quality of ingredients. The Italian naturalist Ulisse Aldrovandi claimed that vipers used in the preparation of some theriacs did not meet the requirements. Some snakes were pregnant, others were captured in Ravenna, near the sea, and would thus be "salty" and therefore inappropriate. Others were not killed in April, when they were at their best. Aldrovandi's comments caused such a raging controversy that the matter had to be settled, eventually in his favour, at an audience with Pope Gregory XIII, at which he brought along some cardinals to argue his case.

Demand to combat the plague had led to frequent adulterations and therefore display and public ceremony became even more important. In Paris, the drug was produced by the Societe de la Theriaque with great fanfare. The vessels used in the preparation of theriac also became more ornate and valuable and were often mounted for public ceremonies on elaborately carved stands. One such piece is now in St John's Cathedral in New York. Decorated with a Virgin and Child as well as a flock of cherubim, the eighteenth-century northern Italian stand has an inscription that reads: "These are the excellent antidotes of Andromachus the Elder which derive from the King of all-producing Pontus. Both have grown in applause; they have been useful to the whole world. Now their fame runs through East and West; thus they rejoice in the favour also of the Blessed Virgin."

The preparation of theriac was lucrative for all involved and by the late Renaissance came in as many variants as there were major European cities. A famed theriac was said to have been invented by Sir Walter Raleigh while he was imprisoned in the Tower of London, perhaps another bit of advertising mystique. Theriacs were often outlandishly expensive and packed full of costly ingredients. *Theriac Celeste* was among the most excessive. It contained pearls, emeralds, rubies and garnets. Over time almost all of the drugs were simplified as it became more difficult to get the ingredients. But the *London Pharmacopoeia* of 1746 gave a formula for Venice Treacle with over seventy ingredients, including dried vipers, roses, liquorice, spikenard, myrrh, horehound, pepper, valerian, gentian, St John's wort, germander and galbanum, all mixed with honey.

There is some evidence that the common mix of ingredients would have had a slightly antiseptic effect but would have done little against any serious illness. As was always the case, enough people recovered from an illness for there to be enough anecdotal evidence that they had recovered after using a drug. Very few illnesses are a hundred per cent deadly – people have survived even blights like the plague or typhoid. Many versions of Venice Treacle contained opium, which might have helped with some digestive complaints and would have provided relief from pain.

In the 1660s, a French physician, Guy Patin, declared that theriac was useless and sales were mostly in the interest of pharmacists. This had little effect on what was still a growing industry. But by the beginning of the eighteenth century the old remedies were disappearing. New approaches against poisoning were emerging, such as the use of emetics and various vomiting instruments such as feathers, leather-gloved fingers, suspended rocking beds to induce motion sickness, leather straps coated with tannin and a stomach brush. Described in 1649, this instrument had a flexible whalebone handle with a horsehair brush. By the 1820s, the stomach pump had come into use.

At the Farmacia Santa Maria della Scala in Rome, it is still possible to buy Theriaca Andromachi. It is recommended for insomnia, bad nerves and digestive disorders. Nowadays it is a mere shadow of itself, containing few of the exotic ingredients that had been ground together over the years. Theriac is a medical footnote today, but it shaped many ideas about drugs and science that endure today. Drugs such as theriacs led to the creation of botanical gardens and to the collecting and cataloguing of exotic plants. Theriacs promoted state control over the manufacture and content of drugs, now one of the most regulated industries in the world. The manufacturers of theriac, always conscious of the public perception of their product, held what were the precursors of the massive marketing efforts put into drugs today. They recognised that drugs were not just about efficacy but about public faith. If people believed, you could charge them whatever you liked.

11. EMERALDS AND HAIRBALLS

A farthing's worth of treacle (even of the common sort)
is much better than a crown or 10 shillings expended
in this dear drug. Or a glass of good wine with hart's
horn in it has fifty times the life.
>   "A Nice Cut for the Demolisher"
>   Anonymous eighteenth-century pamphlet
>   on bezoars, London, 1715

The cures used in Greece and Rome were simpler than the complex theriacs that evolved over the centuries. A treatment for injuries from poisoned arrows was to put iron rust on the injury. A relief sculpture from Herculaneum now in the National Archaeological Museum of Naples, shows Achilles treating the wound he gave to Hercules's son Telephus with his poisoned spear by scraping rust from the weapon into the wound. Pliny suggested powdered verdigris and rust mixed with myrrh as a treatment for poisoned wounds.

Among the other cures used against poison was *terra sigillata* – sealed earth – it was a clay that came from the isle of Lemnos. Its reputation dates back to Herodotus in the fifth century BCE and it is still used today. According to Dioscorides, it was made into tablets with goat's blood and then stamped with an image of the goddess Diana. The clay was sometimes fashioned into cups, or small tablets of the earth were set like gems into metal vessels. Such was the demand for clay from Lemnos that *terra sigillata* was found in a great many places ranging from Sicily to England. Colonial expansion brought with it the desire to find similar earths around the world because of their value. Venice developed an important business in making vessels made of *terra sigillata* that were believed to stick to the lips on contact whereupon they worked their effect. On analysis, *terra sigillata* was found to have oxides of iron, aluminium and magnesia; it would have been absorptive and astringent, possibly having some very limited effect on poisons. Earth from the Holy Land was exported to many parts of Europe for centuries as a curative. God had used it to form Adam and therefore it supposedly had powerful life-preserving properties.

Incantations, charms and gems were all believed effective against poisons. *The Leechbook of Bald*, an important medical work of the tenth century, urged that anyone poisoned by wolf bane stand on his head and let someone "strike him with many scarifications on the shanks then the venom departs through the incisions". Other treatments included hanging by the feet and even gouging out an eye to allow poison to escape through the wound. There was a belief that a poison could cure its own poison – the leaves of a poisonous tree crushed in water and drunk would cure the poisonous action of the tree itself. Arnold of Villanova suggested, without apparent irony, that holding a snake in one's hands could cure snakebites. In some case the word "Faul" was regarded as being effective against adder bites. The writer Bede believed that anything from Ireland, even shavings from an Irish book, particularly a Bible, would counteract poison. Ireland did not have snakes, Saint Patrick having chased them all out, and therefore anything Irish would likely be hostile to the animals and their toxins.

The poor may have been forced to rely on charms and chants to ward off poisons but the aristocracy had an array of special gems that could act against toxins. Turquoise was said to have considerable powers. In Persia, it was said that turquoises could stop horses stumbling and prevent falls and so they were often used to decorate harnesses and saddles. They were also believed to change colour in sympathy with the health of a person or to warn of any possible danger to them. In *The Merchant of Venice*, Shylock's most treasured possession is a turquoise ring. Bloodstones were said to cure snakebites, as were bezoars, stones that never had enough beauty to sustain their value but were once among the costliest gems available.

The persistent fears of disease and poisoning spurred the collection of rare and precious materials to find a cure or antidote. These were bound up in notions of luxury and wonder. The efficacy of gems as a cure had the backing of most of the great medical thinkers. Maimonides wrote: "Emerald acts as a simple antidote against poisoning only when it is juicily green, translucent and reduced to a fine powder. In such a condition, nine grains of it should be taken in cold water or in wine. Ibn Zohr, the Master, said this remedy had been sufficiently proven and admits of no doubt."

Today, choosing the right name for a drug is a vital part of its development. Companies spend millions of dollars on consultants whose aim is to come up with a name that is catchy enough for the public to remember (and ask their doctors for). Not only does much work go into naming drugs, but now there is increasing attention to renaming them – for example Zyban, used for diminishing cravings among smokers, is Wellbutrin, the antidepressant. And thus the brand is extended and more profits roll in. It is not a new idea. Robert Boyle – known for his law on gases and as a founder of modern chemistry – deserves to be named the patron saint of drug company marketing managers for his abilities in spinning a remedy. He offered this as a cure for cataracts: "Take Paracelsus's Zibethum Occidentale of a good colour and consistency. Dry it slowly until is pulverable; then reduce to an impalpable powder which is to be blown once, twice or thrice a day, as occasion should require, into the patient's eye." Zibethum Occidentale sounds like a wondrous ingredient, an exotic, powerful mineral, rare and expensive, perhaps found only on some distant island at the far end of an arduous trade route. Zibethum Occidentale was in fact one of the substances that has never been in short supply. It was human shit.

Few medicines have benefited from this sort of branding as much as bezoars, the nubbly, irregular stones regarded as powerful antidotes to illness and poison for centuries. Among the many names bezoars went under were Pedra de Porc, Lapis de Goa and Malacca Stones. They were never prescribed under their most direct description – goat hairballs. Bezoars were known as far back as ancient Greece and Rome. They were long a part of the Arab pharmacopoeia, appearing in a tenth-century work that began a surge in their popularity as a cure not just for poison but for a vast array of illnesses. From the fifteenth century onwards, the market grew rapidly as they became more commonly available through the new trading routes to Asia and the Americas. In 1574 Nicolas Monardes, a physician from Seville, wrote *Joyfull Newes Out of the Newe Founde World*, a work that was less than snappily sub-titled: *A Booke Which Treateth of Two Medicines Most Excellent Against all Venome*. The book, a widely translated best-seller, was a celebration of everything that had been discovered in the New World, including, with undisguised delight, new sources of bezoar stones. Given that they were worth as much as ten times the value of gold, it was a great prize for the Spanish to control this new source. Monardes was the leading Spanish physician of his day, a wealthy man who tended to the upper classes of Seville. Documents about his business show that he was extensively involved in the trade in medicines, hence his enthusiasm for the new drugs coming in from the New World.

Bezoar stones were part of the emerging patterns of conspicuous consumption of the Renaissance, a time for wonder – "the first of all passions", according to Descartes. Bezoars were often included in *Wunderkammer* or cabinets of curiosities: collections of natural oddities and relics. There are several in the Kunstkammer Wien, including a vast silvery stone mounted in a gilt, emerald-studded stand made for the Habsburgs. The Vatican Museum once had a fifty-six-ounce bezoar in its collection. These cabinets were often stopovers on the Grand Tour and contained strange new objects from the New World or the East.

The origins of bezoars were for a long time shrouded in a suitable amount of mystery. In Ovid's *Metamorphosis*, bezoars were said to be the solidified tears of deer bitten by snakes. In the story, Actaeon is transformed into a stag; while looking at his reflection in a pond, he weeps tears of bezoar. Pliny believed they were formed when animals such as deer or goats ate snakes, an idea picked up by Arab writers who believed them to be a combination of the venom of snakes and healing herbs. The Spanish soldier who wrote to Monardes from Peru about his discovery of bezoars there said they came from eating herbs that were efficacious against poisons. These calcified balls of hair or vegetable fibre were taken to be concentrations of herbal power. Bezoars were said in India and China to hold great powers, beyond their ability to prevent poisonings; the famed Jahad stone of Babur was a bezoar that was believed to stimulate rain and snow.

Bezoars were collected by beating goats on their stomachs until the stones moved into a softer part of the body and could be felt. Cows also produced stones – as do all ruminants – but they were regarded as worthless. Monkeys, deer and various small animals found in Malaysia and Indonesia were all sources of valuable bezoars. Monardes wrote that knowledge of the stones was "hidden from us" for a considerable time, presumably until it was reintroduced to European medicine by the Arabs. Oddly, although European mammals, particularly goats, produce bezoars, they were not much regarded. The items had to be procured at a distance to be worthwhile, which of course opened the door both to considerable fraud and the colonial imperative of finding new, more reliable sources of the stones.

The best ones were to be found in animals that lived in mountains, as they ate the most potent herbs. Monardes dismissed those from Malacca as coming from the wrong type of goat. As a doctor catering to the rich and powerful, Monardes prescribed a considerable quantity of bezoar. He regarded it as something of a cure-all. It worked not just against poisons but also leprosy, melancholia, scurvy, scabs, itching, elephantiasis, against fevers and pestilence, for the effects of bites of venomous worms "it doth extinguish and kill the malice" of pestilential fevers. It even worked against flea bites.

What made bezoars special was their action as an antidote. In a letter to his patron Edward Dyer, the translator of Monardes, John Framptom wrote of "howe common in the worlde the practise of Poyson is and what malyce raignes now among men and how needfull it is that some kinde of persons should feare and provide for the worst and weighing that by our Persian Marchants and by other meanes the Bezaar stone this greate jewell is brought into the Realme and may be compassed in this our time for a little money." Fear always makes someone rich.

The extent of the use of bezoars is hard to gauge, but in his work debunking them published in 1715, Frederick Slare describes a London druggist who was not known to cater to the rich but sold 500 ounces a year of bezoar stone. Slare estimates that it took about ten stones to make up an ounce so this one pharmacist was selling around five thousand a year – in other words, at least five thousand goats or other small ruminants died to provide stones for that one pharmacist. The price was about five pounds an ounce at a time when a clerk would earn less than fifteen pence a day. He also mentions that doctors were earning up to fifty thousand pounds a year, around six million pounds in today's money, by selling powders containing bezoar – in particular gascoign powder. This was actually a powder made from the black tops of crab claws, a type of plant, "but to advance its fame and value it is called a Bezoardic powder", Slare wrote.

Discoveries from the wreckage of the Nuestra Señora de Atocha, a ship that sank off the coast of Florida in the early seventeenth century, illustrate the rarity and value placed on bezoars. There is a gold "poison cup" that has a bezoar mounted on its interior as a permanent antidote to any poison that might lace wine drunk from the goblet. Another egg-sized bezoar is mounted in an engraved and enamelled framework that was designed to be suspended from a chain and dangled over food. Another ten unadorned bezoars were found in a silver canister.

The German naturalist Georg Eberhard Rumpf, also known as Rumphius, describes bezoars as coming from monkeys. Warning of the prevalence of fakes, he writes of several tests in some detail; in Malaysia, for example, they put lime on white linen, which turns yellow when a real bezoar is rubbed on it. Rumphius believed that a better test was to see whether the stone flaked apart when pressed with a hot nail and whether the flakes that came off disappeared completely when burned. Fakes were often made of some sort of resin, often actually mixed with bezoar dust. Chinese merchants faked them by making a paste of bezoar stone and dust and then spreading it over a small stone. Rumphius said that this could be detected by removing the outermost layer of a bezoar to reveal a smooth layer underneath, which did not occur with real bezoars.

Such was the rarity and expense of the medicine that when Charles II collapsed while shaving – most probably of a stroke or heart attack – he was first treated by being bled a pint of blood from his right arm. Eight ounces more of blood were taken from his left shoulder. Then an emetic was administered, followed by two physics and an enema consisting of fifteen ingredients. His head was shaved and a blister raised on the scalp. To purge the brain he was given a sneezing powder followed by cowslip powder. Meanwhile there were more emetics, soothing drinks and more bleeding, and a plaster of pitch and pigeon dung was applied to the royal feet. He was also given melon seeds, manna, slippery elm, black cherry water, extract of lily of the valley, peony, lavender, pearls dissolved in vinegar, gentian root, nutmeg and forty drops of extract of human skull. Finally as a last resort, a bezoar stone was employed. Charles II died shortly afterwards, doubtless sped on his way by his physicians.

Bezoars were part of elaborate recipes that relied on some of the costliest ingredients available. Efficacy was connected to expense and also to the notion that jewels held curative powers and were often used in combination, ground into powders and formed into pills. These pills often included precious metals such as gold and occasionally substances that would have done much more harm than good. Lapis de Goa was one such miracle cure:

> Take Hyacinth, Topaz, Sapphire, Ruby, pearls of each one ounce. Emeralds, half an ounce, Oriental bezoar, white and red coral, 2 ounces, musk, ambergris, of each half an ounce, leaves of gold, no. 40, Make all into a fine powder which bring into a paste with Rose-water and form into oblong balls not much unlike to little

eggs... It is an Antidote against Plague and Poison and cures the bitings of serpents, mad Dogs or any other venomous creature.

The stones were first made in Goa by Gaspar Antonio, a Florentine lay brother from the Pauline monastery. John Fryer visited the monastery, which had been founded by the Jesuits and had a large hospital and pharmacy, in the 1670s. He reported that Antonio was by then very old but was held in the highest esteem because his stones brought in large amounts of money each year – according to one account, more than three thousand pounds annually. Goa stones were brought back by East India Company sailors who were allowed to import a certain number of chests without being taxed. At the end of the seventeenth century in London, Goa stones cost a guinea for a small ball or five guineas for a larger one. Local production was not a great success – pharmacists were unable to reproduce the fine polish of the stones and as John Quincy wrote in the *A Compleat English Pharmacy* in 1718, that people "were more apt to admire what comes a long way".

Medicine was changing in the seventeenth century and these complex compounds were falling out of fashion. Robert Boyle's works, eventually published in a compilation called *The Advantages of the Use of Simple Medicines*, urged people to avoid complicated compounded drugs of dubious worth. Others urged doctors and scientists to rely more on observation. Bezoars and other stones fell from favour but Goa stones lingered on, not as medicines but as a flavouring for punch. As they contained musk and ambergris, they imparted a slight aroma to drinks. They were even used to give a light perfume to linens.

Modern research on bezoars offers a slight glimpse of scientific confirmation of efficacy against poisons. Researchers at the Scripps Research in San Diego found that bezoars were capable of absorbing a small amount of arsenic from a solution, although not to a degree that would have any serious impact on its toxicity.

IN THE METROPOLITAN Museum in New York is a nineteenth-century sculpture of an African woman pulling a thorn from her foot. It stands on a base of filigree gold and is decorated with more gold, jewels and enamel. It was once in the collection of J. Pierpont Morgan, the American banker and collector. It would be a rare and valuable piece if it had been made from marble or wood, but what makes it extraordinary is that it was carved from ambergris. It was made from the stomach contents of a sick sperm whale.

The origins of ambergris, a waxy perfumed substance that was a key ingredient of antidotes, medicines and perfumes for centuries, were a mystery in Europe until the eighteenth century. Arabs trained camels to pick up its scent on the sea shore and believed it flowed from streams into the sea. Sinbad recalled it from his sixth voyage:

> And in that island is a gushing spring of crude ambergris, which floweth like wax over the side of that spring through the violence of the heat of the sun, and spreadeth upon the sea-shore, and the monsters of the deep come up from the sea and swallow it, and descend with it into the sea; but it becometh hot in their stomachs, therefore they eject it from their mouths into the sea, and it congealeth on the surface of the water.

Ambergris was believed to have come from sea birds – because it sometimes contained what looked like beaks – or from hives of bees, because of its waxy consistency. Marco Polo witnessed the hunting of sperm whales in the Indian Ocean and saw that it was found within them but he believed they ate it from the ocean floor. The Chinese called it *lung sien hiang* or "dragon's spittle perfume". The Japanese, aware of ambergris for centuries, had a less poetic name for it, calling it *kunsurano fuu*, or "whale dung". The Japanese name gets closer to the origins of a substance so valued that fortunes were made from chance discoveries of lumps washed up on beaches.

Ambergris is not exactly whale dung but an accretion of a steroid similar to cholesterol, combined with a chemical known as ambreine.

Within the gut of the whale it is an almost black soft substance but on exposure to air it lightens, hardens and, as the ambreine breaks down, develops a distinctive scent, often described as earthy or woody. Rather like truffles or musk, it has a scent that lies perilously close to the disgusting, containing something hormonal and sexual that hooks in the nose. Moderation was the key, as Alexander Pope observed: "Praise is like ambergris; a little whiff of it, by snatches, is very agreeable; but when a man holds a whole lump of it to his nose, it is a stink and strikes you down." As with truffles and musk, ambergris has always been regarded as an aphrodisiac and a pick-me-up. The French gastronome Jean Anthelme Brillat-Savarin raved about the powers of chocolates flavoured with it. Charles II of France ate it shaved onto scrambled eggs.

A detailed account of ambergris, published in London in 1783, revealed the murkier origins of the substance, which is believed to be a response by whales to irritation caused by a build-up of squid beaks in their intestines. Even knowing that it was not a product of magic springs or dragons did not deter users. In the eighteenth century, ambergris sold for a guinea an ounce – equivalent to about £240 an ounce today. It was often adulterated, for at that price it was easy to mix with waxes or other aromatic compounds. The bezoardic powder mentioned in the *London Dispensatory* in the seventeenth century contained sapphire, ruby, jacinth, emerald, pearls, unicorn horn, oriental and American bezoars, musk, ambergris, bone of stag's heart, kermes and sixteen other ingredients. "I am afraid to look upon it," wrote Nicholas Culpeper, the famous herbalist. "Tis a great cordial to revive the body but it will bring the purse into consumption."

THE MOST EXPENSIVE and exotic of all substances that was believed to act against poison was paradoxically also one of the most common substances available. Human flesh, known as mumia, was said to cure poisonings by restoring the life force that toxins drained away from the body. Johann Schröder, the seventeenth-century author of *Pharmacopoeia Medicochymica*, described a preparation known as Aqua Divina. According to the recipe, "the whole corpse (killed of course by a violent death) with bones flesh and intestines is separated into tiny pieces and this done well every part of the body is ground up so that nothing remains unmixed. Then it is distilled with favourable results." Paracelsus was a great believer in the power of the body to heal itself; for him, mumia referred to this life force although he also did, on occasion, prescribe actual preserved body parts. "Flesh possesses an inner balsam which heals and every limb has its own cure in it," he wrote. It was this inner power that gave mumia its medical force; it was a harnessing of the body's own ability to heal.

Blood was also prescribed as a restorative. Blood drawn from adolescents was said to be useful in perking up older people. Frederick Barbarossa, the Holy Roman Emperor, was said to have lived to a great age – he was over seventy when he died on the Third Crusade – by keeping his abdomen warm at night by having small boys sleep on his stomach, an explanation hardly likely to survive the scrutiny of child protective services today. Others slept with puppies, which Francis Bacon described as "creatures of the hottest kind" and were thus also good for keeping the abdomen warm. These little boys and small dogs presumably survived the night to be used again, but most medicinal applications of humans were much more sinister. Marsilio Ficino in his *De Sanitate Tuenda* wrote:

> There is a certain ancient and popular opinion that certain old women, whom we call witches, suck the blood of babies to rejuvenate themselves as much as possible. Why should our elderly who find themselves bereft of any other assistance not likewise suck the blood of a young boy of robust strength I mean who is healthy, cheerful, even tempered and who has perfect blood and by chance in abundance? Then suck it as would a leach or blood sucker from the open vein of the left arm.

The best blood, according to Father Francesco Sirena, a pharmacist at the monastery of Santa Croce in Pavia, was of healthy hot and humid bodies, from people with white or ruddy complexions but not from people with red hair. Epilepsy was often believed to be cured by cranial powder – essentially pulverised skull. The Barigazzi family of barber-surgeons in Capri was known for making a type of poultice of mummy mixed with mother's milk. The plaster, as it was known, was made from the head of a mummified corpse – the family were said to have several of these in their house at any time.

The human body and its various secretions were widely used in medicine. A seventeenth-century work published in Italy by Alessandro Venturini lists all the possible uses of humans as cures – from mother's milk to human fat, dried mummies to excrement, urine and "the dirt to be found around the neck of a man's penis", which was used to cure scorpion stings. Dried testicles were given to women in a powder form after menstruation to make them fertile, according to an eleventh-century recipe from Trotula of Salerno. The blood of freshly slain gladiators in Rome was sold for the treatment of epilepsy, but it had to be taken before it congealed. Executioners in Europe in the seventeenth century boosted their incomes by selling blood from those they had just killed for the treatment of gout and dropsy.

Mentions of human remains as medicines go back at least to the twelfth century, when it was possible to buy "mellified" remains in the Cairo bazaar. The process was described by the Chinese medical writer Li Shizhen in 1597. Old men who knew they might be approaching death but were willing to help others spent the last month of their lives eating nothing but honey. They were finally entombed in a stone coffin filled with honey and left to marinade for a century. After this time the coffin was unsealed and the contents sold as mellified mumia, which was said to cure broken bones among other ailments.

Li's account in Chinese in his *Compendium of Materia Medica* was filtered through long distances and a whole array of possible misunderstandings, but his writings suggest he was not surprised or disgusted. In most cases, bodies that were turned into mumia were not from elderly volunteers. Nicolas le Fevre wrote in *A Compleat Body of Chymistry* that the best mumia came from the bodies of men who had died in sandstorms in the Libyan desert. "This sudden suffocation doth concentrate the spirits in all the parts by reason of the fear and the sudden surprisal which seizes on the travellers."

This description is echoed in *Purchas His Pilgrimage*, the seventeenth-century travel book by Samuel Purchas, in which the author writes:

> They traveled five days and nights through the Sandy Sea which is a great plain champaine full of a white sand like meal, were if, by some disaster, the wind blows from the South, they are all dead men. He supposed mumia was made of such as the sand had surprised and buried quick but the truer mumia is made of embalmed bodies of men as they used to do in Egypt and other places.

He gives more details of the embalming process, which used:

> nitre and cedar, or with compositions of myrrh, cassias and other odours... Some also report that the poorer sort used hereunto the slimy bitumen of the Dead Sea which had preserved an infinite number of carcasses in a dreadful cave (not far from the pyramids) yet to be seen with their flesh and members whole after so many thousands of years and some with their hair and teeth: of these is the true mumia.

Purchas also describes its preparation by Ethiopians who had their own unique method:

> They take a captive Moor of the best complexion and after long dieting and medicining of him, cut off his head in his sleep and gashing his body full of wounds put therein all the best spices and then wrap him in

hay, being before covered with a cloth after which they bury the body in a moist place covering the body with earth. Five days being passed, they take him up again and removing the cloth and hay, hang him up in the sun whereby the body resolves and a substance like pure balm, which liquor is of great price. The fragrant scent is such, while it hangs in the sun, that it may be smelt (says Friar Luys) a league away.

Shröder describes four types of mumia: Arab mummy, a fluid or condensed fluid that exuded from cadavers that was preserved with aloe, myrrh and balsam. Egyptian mumia was a liquid produced by corpses preserved with pissaphaltum (a mix of pitch and bitumen also known as rock asphalt). Another type was made from cadavers roasted in desert sand. There was also a "fresher" form of mummy that was said to come from the flesh of a red-haired man (this may have also meant blond-haired), although no reason is ever given for this. Red-haired people were seen as hotter with a different balance of humours and mumia was a hot medication that balanced out cooling illnesses or poisons. The preference was for a man aged twenty-four who had died of a violent death but did not have blemishes on his body. The corpse was to be exposed to the rays of the moon for a night before the muscular flesh was cut off, sprinkled with myrrh and aloe and then soaked until tender. It was then hung and dried until it resembled what Shröder described as "smoke cured meat that did not smell at all".

As with any expensive medication, people soon started producing fakes. Paré suggested that bodies were being taken from the gibbet in Paris and sold for mumia. Pierre Pomet, the apothecary to Louis XIV and author of *A Compleat History of Druggs*, recommended that buyers look for a product that was a fine shining black, not full of bones or tainted with dirt and that did not smell just of pitch. Pomet was somewhat sceptical about the product, writing that it was most useful for catching fish. Most of the accounts of its manufacture were based on some person other than the author witnessing the events. Given the wide array of substances that were used in medicines, it is not surprising that human flesh would have been included, but the stories of corpses found in the Libyan desert might come from the imaginations of medieval apothecaries eager to play up the exotic, and therefore costly, nature of their cures. The Franciscan friar Andre Thevet described the stealing of ancient mummies from tombs to be sold to traders bound for Venice, from where it was distributed around Europe. His own experiences taking mumia were not positive: he complained afterwards of stomach aches and bad breath. The export of mummies was banned but went on nevertheless. It was surrounded by fears that mummies could curse those that stole them as medicine. Shipping them by sea was said to stir up great storms.

The idea of eating preserved human flesh clearly evoked considerable disgust in the sixteenth century and became a popular literary reference. In Webster's *The White Devil*, it is described as "such unnatural and horrid physic" that can only be vomited up. Shakespeare makes lighter of it when the obese Falstaff says in *The Merry Wives of Windsor*: "Water swells a man and what a thing I should have been when I had been swelled! I should have been a mountain of mummy."

Giovanni da Vigo, a Genoese physician who died in 1525, included mumia in his list of essential medicines that all ship doctors and those practising in villages without apothecaries should have with them at all times. Mumia was mentioned in most medical works and *materia medica*, although there was some debate about whether it was preserved human flesh or the liquid that was released from preserved bodies. The whole spate of medicinal cannibalism across Europe might have been based on mistranslations and misunderstandings of Arab medicine. Rock asphalt was used as a medicine and antidote to poison in the Middle East. It was found mostly in Darabjerd, in what is now Iran, and in the Dead Sea, where large masses of the substance are found floating on the lake. Some are big enough to support the weight of several men. The substance was sold around the Middle East going back to at least 500 BCE. Used as a medicine, it was known as *mumiya,* from the Arabic word for "wax".

Bitumen had been known as a medicine at least since Dioscorides claimed that the substance from the Dead Sea, often known as *bitumen Iudaicum,* was the most effective form. Rhazes and Avicenna used the word "mumia" to describe bitumen and prescribed it for poisonings as well as abscesses, contusions, broken bones and paralysis. When Rhazes's work was translated in the twelfth century by Gerard of Cremona, mumia was described as "the substance found in the land where bodies are buried with aloes by which the liquid of the dead, mixed with the aloes, is transformed and it is similar to marine pitch". Similar mistranslations of Arabic works in which European writers conflated the use of bitumen as a medicine with its use to preserve and mummify corpses started a lengthy debate about the nature of mumia. It is unclear now whether it was simply bitumen or the exudate of preserved corpses, mixed with myrhh, aloe and bitumen. Certainly, the best mumia was always said to come from Egyptian tombs in which corpses were preserved in this way, but relatively few people in Egypt were mummified with such expensive products. Several centuries of medical cannibalism might have been caused by a misunderstanding.

The beginning of the end of exotic antidotes might have been a simple experiment said to have been carried out by Paré. A cook who had been sentenced to death for theft agreed to be poisoned with corrosive sublimate (mercuric chloride) mixed with bezoar powder so that King Charles IX would have evidence that his expensive bezoars were worth what he had paid for them. The cook died in agony after a few hours. In Paré's *Discours de la mumie, de la licorne, des venins et de la peste*, he questioned whether anyone knew what mumia really was, given the debates over whether it was rock asphalt or preserved flesh. The ancient Egyptians, Paré wrote, had embalmed their dead "not for that end that they should become medicines for such as live, for they did not so much respect or imagine so horride a wickednesse". He also said that Egypt would not permit the export of its mummified nobles and therefore what was sold by apothecaries was at best the "basest people of Egypt" or at worst bodies collected by apothecaries in France, cured with salt and spices and then dried in ovens. Paré said he had administered mumia more than a hundred times but that "this wicked kinde of drugge doth nothing helpe the diseased".

There was always some questioning of the efficacy of these products and considerable scepticism over the motivations of doctors who prescribed ruinously expensive cures.

The emergence of direct observation and experimentation evolved through a process of conversations held in *spezieria,* a combination of salon, pharmacy and scientific lab that were mostly based in royal palaces. These were often linked to the production of theriac. Francesco Redi, a pharmacist and doctor to the Medici court in Florence, conducted a series of critical observations on vipers that offered a direct challenge to Galen's ideas. The bile of vipers was believed by Galen to be poisonous but Redi fed it to a variety of animals to show that this was not the case. Vipers were also said to be fond of wine. Their medical effect was said by Galen to have been discovered when one drowned in a jug of wine, producing an antidote that a farmer used to treat a snakebite. Galen believed that vipers quickly drowned when submerged in wine. In fact, Redi showed that vipers could swim in vats of wine for quite some time and even survive under the surface for more than an hour. Perhaps the greatest challenge to Galen's views came from Jacopo Sozzi, a snakecatcher from Naples who delivered vipers to Redi for the manufacture of theriac. Listening to the men of the *spezieria* debate the various positions of the ancients of the effects of snake bile, Sozzi simply took:

> one of the largest, liveliest and angriest vipers and made him spurt into half a glass of wine not only all the liquor that the sheaths held but also all the foam and saliva that this excited, provoked, overpowered beaten serpent could shoot forth; he drank that wine as if it had been so much pearly julep. The next day with three vipers twisted together, he played the same game, without a fear in the world and he was right.

The greater understanding of the actions of poisons on the body did not end the desire for a universal antidote, but it did make scientists more cautious about adopting ancient cures. Many of the antidotes recommended by Mathieu Orfila in his 1818 work *A Popular Treatise on the Remedies to Be Employed in Cases of Poisoning and Apparent Death* are still in use today. Many are remarkable for their simplicity and wide availability. "We shall reject such as are useless or dangerous and recommend on those the efficacy of which has been demonstrated by reiterated experiments. These are the whites of eggs, milk, common salt, vinegar, lemon juice, soap, gall nuts and some other substances which may be procured with the greatest facility."

Most of the treatments involved neutralising the poison based on observations of chemical reactions in vitro. Unfortunately this did not always consider some of the results of those reactions – including the creation of heat or carbon dioxide, both of which would cause patients additional problems. The widespread use of emetics was also a problem. Tartar emetic – a compound of potassium and antimony – was used to induce vomiting even though it was somewhat unreliable in producing the desired effect as an emetic and could kill if taken in excess. Even table salt and Epsom salts (magnesium sulfate) were dangerous, as a fatal dose could be absorbed before the patient vomited. Orfila was dismissive of the uses of charcoal on the basis of his experiments on dogs but others, such as the French doctor Michel Bertrand, believed in it so strongly that he experimented on himself, taking mercuric chloride and arsenic trioxide with a charcoal chaser without experiencing much more than indigestion.

The first recorded medical use of charcoal in a poisoning case was in 1818, but it was only at the beginning of the nineteenth century that it was discovered that activated charcoal – heated under carbon dioxide – was even more effective in absorbing toxins. Experiments in the 1940s showed that it could absorb a range of toxins but, surprisingly, it was not commonly used in poisoning cases until the 1960s. Nowadays, many cases of poisoning are treated with gastric lavage, followed by doses of activated charcoal to absorb any poison that lingers in the digestive tract.

Today, there is a vast and complex array of treatments for

symptoms brought on by poison. There are drugs that reduce the toxic impact of heavy metals, drugs that counteract the heart-slowing properties of opiates, drugs that can minimise the effect of drugs that act on neuron transmissions. There is also an extraordinary capacity to diagnose poisonings. But there are very few actual antidotes. The US Food and Drug Administration approves only five drugs as antidotes – atropine for nerve agents and some organo-phosphate pesticides, an anti-cyanide drug, epinephrine for allergic shock, Naloxone for opiates and oxygen for carbon monoxide.

But ancient knowledge may still have some uses. Going back to the days of Mithridates, it was believed that tiny doses of poison could actually be beneficial. This idea, known as hormesis, has often been associated with homeopathy and therefore was mostly disregarded by scientists. The idea is that substances that are regarded as toxic can act as drugs in low doses. A low consumption of alcohol lowers the risk of heart attack but a serious drinking problem raises it. Low doses of a range of poisons from dioxins to heavy metals such as cadmium can have beneficial effects.

Hormesis was first described in the nineteenth century when it was discovered that small doses of poisons enhanced the growth of yeast. The idea was out of fashion for much of the twentieth century, but re-emerged as a focus for scientific study when it was found that the paradoxical dose response was widespread. Plants grew more lush with small doses of herbicide; tiny amounts of antibiotics encouraged the growth of bacteria; and mice showed reduced rates of cancer after low doses of gamma radiation. The explanation is that these mild stresses trigger responses within cells that ultimately leave them stronger. A small amount of a chemical may produce a reaction to repair damage that will fix other problems within the cell. Exercise and very low calorie diets may create similarly beneficial responses; rats live longer when fed very low calorie diets. Different toxins trigger different mechanisms. Some produce more enzymes that repair damaged DNA; toxic metals provoke a response that removes the metals from circulation and also acts against free radicals that cause cell damage. However, small doses are not always beneficial. Studies have found that small doses of some toxins given to test animals produced worse problems than large doses.

Hormesis suggests that the relationship between toxins and health problems is not linear. This has a potentially profound effect on modern toxicology and the assessment of risk of the chemicals that surround us. Many chemicals that were believed to be toxic even at very low levels, notably dioxins and heavy metals, may be safe at those levels, allowing the reintroduction of an array of banned products. Other scientists maintain that precisely the opposite is true. We know so little about the possibly harmful impact of a great many chemicals at low levels that many products currently in widespread use may be harmful.

The hormesis effect may cause a range of problems. It can ramp up one type of immune response that provides protection to a certain virus or bacteria but undermine other responses and leave people more vulnerable to disease. You may fight off the common cold but be more susceptible to carcinogens. The study of hormesis tends to focus on narrow end points, such as protection again particular diseases, rather than seeing the full picture. Compounds can end up being classified as both harmful and beneficial at the same dose, depending on what is being measured. Hormesis is still widely debated among toxicologists, and studies of the suppression effect of chemicals are extremely difficult to research, in part because cancers rarely arise spontaneously in rats and therefore very large numbers are needed to be studied for any effect to be measured. There is caution about the idea that a small dose of poison may be the cure, for understandable reasons – it is counterintuitive, mechanisms have not been demonstrated and some of the science involves dubious extrapolations from limited data. It remains to be seen if a small amount of what ails you, does you good.

12. ARSENIC

And he followed her. The key turned in the lock,
and she went straight to the third shelf, so well did
her memory guide her, seized the blue jar, tore out
the cork, plunged in her hand and withdrawing it
full of a white powder, she began eating it.

Gustave Flaubert, *Madame Bovary*

Humans are highly evolved poison-detecting machines. We are far from infallible when it comes to avoiding toxins but we are surprisingly capable. Our senses of sight, smell and taste are remarkable in their ability to warn us away from danger as well as our semi-innate sense of fear around those things that might harm us. But humans are not simply animals. All experience of nature long ago became mediated through culture. Culture decides what we fear as poisonous and dangerous as much as our senses. As children, we learn what to avoid, that the juicy red berries may actually be deadly, that certain animals leave more than just tooth marks when they bite. As we grow older, our sense of taste grows more developed and sophisticated so that we are able to enjoy things that are not dangerously toxic but taste as if they might be. We learn to appreciate the bitterness of coffee and dark chocolate, or the bite of mustard, horseradish and chillies.

The word "taste" comes from the Latin *taxare*, to touch or evaluate. That evolved into *tasten*, a Middle English word meaning to test or sample. Taste is the last, most sophisticated and least understood line of defence against poisons. Our mouths are lined with taste buds, mostly on the tongue but also on the hard and soft palates, the oesophagus and epiglottis. The cells, which come in several forms, live for about ten days and have one of the most rapid turnovers of any human cells so they can remain fresh and alert. The cells can detect sweet, sour, salty and bitter tastes, each via a different mechanism. Salty tastes involve the passage of sodium ions into a cell, sour tastes are due to the concentration of hydrogen ions. Sweetness involves several receptors that polarise the cells when a sweet molecule binds to a protein. There is also said to be a fifth taste – *umami*, or the sort of savoury sense found in foods like broth, cheese or mushrooms. Taste makes up just one component of the senses we use; smell is equally, if not more, important in making judgements about the quality of food. Our bodies have evolved to eat an extensive range of foods. Being omnivores requires a digestive system that can cope not just with nutrients but a wide array of products that might be toxic. Our bodies are surprisingly adept at dealing with the poisonous substances we willingly ingest each day. But to keep out the acutely dangerous, we have developed a powerful memory of certain foodstuffs and we develop deep aversions to certain foods. Humans are programmed to avoid certain tastes that are associated with poisons. Just as children become mobile and thus able to forage for themselves, they also become extremely resistant to new tastes, as parents of two-year-olds can attest. This is an evolutionary mechanism to protect against accidental poisonings. Our fairly acute sense of taste and smell makes us quite difficult to poison. Anything could be a poison, but in most cases the tongue would detect the dose and the person would refuse to ingest the toxin. Salt is a poison but an average person would have to eat at least a gram per kilogram of body weight for it to be fatal.

Given the mechanisms we deploy against toxins, a perfect poison should be without any acrid or bitter taste. It should be colourless and odourless to avoid immediate detection, a chemical basilisk hiding under a cloak of invisibility. A small dose should be toxic. It would be better if it did not have an immediate reaction that might reveal the source of the poison and the identity of the killer. It should be readily available, almost undetectable, and easy to use by being water-soluble. It would be better if its symptoms mimicked those of a common disease that killed many people.

Between the fifteenth and nineteenth centuries, such a perfect poison existed. Arsenic could be administered easily and, if not given in a massive dose, it was not too obvious or immediate in its effects. It was also readily available and cheap, qualities that brought it within range of most people up until the nineteenth century, when controls on its sale were introduced in many European nations. As arsenic was a key ingredient in rat poison, no suspicion fell on those who bought it.

Arsenic is present in almost all of the food and water we consume. Some foods, notably shellfish, often contain large quantities of the metal in an organic form that is not very toxic and is mostly excreted in the urine. Inorganic forms – those not bound up with complex carbon-based molecules – tend to be more harmful. Metallic arsenic is mostly non-toxic but it is rarely found in this state, as it oxidises rapidly. Arsenic can combine with either three or five atoms to form trivalent or pentavalent compounds. It is the trivalent that is most toxic – arsenic trioxide is the common arsenic poison. A more exotic variation is arsine, a compound of arsenic and hydrogen that is a potent poison gas.

Most arsenic is turned into a less toxic chemical in the liver through a process called methylation; the resulting monomethylarsonic acid is passed from the body. However, in larger doses the liver cannot cope and the metal combines with haemoglobin and travels around the body. Within twenty-four hours of being ingested, arsenic can be found in all the major organs. Arsenic works on the body in several ways that disrupt the normal metabolism of cells. It stops the Krebs cycle in cells, leading to a depletion of ATP, a chemical that stores and transports energy – ATP is often described as a sort of molecular currency for the body. Arsenic also inhibits the transformation of thiamine, a vital amino acid. It acts by a process known as arsennolysis, in which it replaces phosphate in various metabolic pathways. The metal is incorporated into skin, nails and hair as it binds to the protein keratin.

In his *Treatise on Poisons* published in 1829, Robert Christison wrote that arsenic was "the poison most frequently chosen for the purpose of committing murder". The history of arsenic is the history of poisonings. Once a rare mineral, used mostly as a colorant and medicine, its use became widespread in the fifteenth century when a cheap and plentiful supply emerged as a by-product of smelting. Until the nineteenth century there was no reliable test for it. To get caught, one almost needed to be seen pouring arsenic into the victim's soup. Of the 404 criminal cases of poisoning recorded in England from 1750 to 1914, 237 used arsenic. The poison that Tofana sold to her clients was most probably a solution of white arsenic mixed with some probably harmless herbs, and possibly with the heart-slowing drug digitalis extracted from foxgloves.

Arsenic poisoning is not an easy death. Fictional victims of poisoning tend to slip away with a slight gasp, maybe clutching their throats to indicate what killed them. Death by arsenic would have been much more dramatic and unpleasant as the victim succumbed to terrible vomiting and diarrhoea and suffered intense pain. About half an hour after receiving a large dose of arsenic, the victim would develop a metallic taste in the mouth and garlic-smelling breath. Shortly after this they would suffer from severe nausea and vomiting, pains in their stomach and rapid diarrhoea. Damage to tiny blood vessels throughout the alimentary canal would result in blood loss. In cases of large doses the skin would appear blue, the extremities would be cold and clammy, and the victim would be racked with convulsions. In lower doses, they would have a severe headache, swelling around the eyes and muscle cramps. Chronic poisoning with the metal would show itself in kidney problems, dermatitis (described as looking like dew drops on a dusty road), the loss of hair, Mee's lines – white horizontal bands on nails – and peripheral neuropathy that causes pain and weakness in the hands and feet.

The pioneering Spanish toxicologist Mathieu Orfila described the symptoms of acute poisoning in his book *A General System of Toxicology*:

> An astringent taste, fetid mouth, frequent ptyalism (excessive salivation), continual spitting, constriction of the pharynx and oesophagus, the teeth set on edge, hiccups, nausea, vomiting of a matter sometimes brown and sometimes bloody, anxiety, frequent fainting, heat of the praecordia (heartburn), inflammation of the lips, tongue, palate, throat, and oesophagus, the stomach painful to a degree as not to be able to support the most emollient liquid; the alvine discharge blackish and of a horrible fetor, the pulse small, frequent, concentrated and irregular; sometimes slow and unequal; palpitation of the heart, syncope, unquenchable thirst, pungent heat all over the body, sensation as of a devouring fire, sometimes an icy coldness, breathing difficult, cold sweats, urine scanty, red and bloody, change of the features of the countenance; a livid circle around the eyelids, swelling and itching over the whole body which is covered with livid spots and sometimes with a miliary eruption (tiny small spots), prostration of strength, loss of feeling, particularly in the feet and hands, delirium. Convulsions often accompanied with an insupportable priapism, falling off of the hair, detachment of the epidermis and lastly death.

This was how Tofana's victims must have died. Blotchy and dehydrated, their hair coming out in clumps, vomiting up anything they were given, succumbing to delirium and convulsions as their anxious relatives kneeled by their beds, rosaries in hand, gagging on the stench. And all the while some of these men, perhaps being punished by their wives for their priapic pasts, might have had to endure a permanent and painful erection.

THE WORD "ARSENIC" comes from the Greek *arsenikon* meaning "potent". As well as being a poison, it has been a powerful cure for millennia. Aristotle refers to sandarach, or arsenic trisulphide, in the fourth century BC. This chemical was found in gold and silver mines, mostly from those near Mount Sandracurgium in Asia Minor. Two types of arsenic compound were known – arsenic trisulphide (known later more commonly as yellow orpiment) and arsenic sulphide (red realgar). Pliny the Elder describes orpiment as a brittle earth the colour of gold. Both are mildly toxic but are often mixed with the deadlier white arsenic. This substance, arsenic trioxide, was not generally known in classical times, but Arab chemists were known to have extracted it from about the eighth century. Dioscorides recommended realgar for asthma and coughing, as did Galen and Pliny.

White arsenic was not common in Europe until the fifteenth century, when the first industrial production began in Austria as a by-product of silver smelting. The powder collected in the chimneys and was scraped off to keep them clean. It was sold to traders for sale in Venice where it was used in glass production; a small amount makes glass opaque and white so that it resembles porcelain. Some was sold in the Levant as a rat poison but there was little interest in the chemical in much of Europe. By 1570 industrial arsenic production made it a cheap and plentiful substance that was not regarded as useful. Few books on metallurgy mention it, or consider it only as a by-product. By the seventeenth century output was so great that it was used as ballast on ships.

By the sixteenth century, arsenic was often prescribed as a drug. A shift to Paracelsian medicine created both great interest and anxiety about the medical value of metals, including arsenic and mercury. They were known as harsh cures with severe, even deadly, side effects, but at the same time emergent diseases such as syphilis were seen to require new approaches. Arsenic was also a component in "plague cakes" – amulets worn around the neck or in the armpits to ward off the dreaded illness. The chemical had been used as a cure and tonic in Central Europe, most famously by the arsenic eaters of Styria, a mountainous area of what is now Austria. In the hills around the town of Graz, people were said to consume arsenic each day in doses sufficient to kill a dog. The habit began with stable boys who fed it to their horses to keep their coats in good condition. Doctor Fritz Pregl, a German physician, noted in the nineteenth century that people preferred to eat orpiment but would also consume white arsenic, cutting off a thin slice from a block and eating it with bacon or bread. "The respiration and the heart are facilitated and nutrition is improved, the appearance is brightened and the hair of horses and men made smooth and glossy," he wrote. Pregl was unable to find out how much people ate as it was a controlled substance to be consumed quietly. He did note that the area was prone to more sinister uses of the chemical. "In a year, I have had to undertake and pass judgment upon in legal proceedings more arsenic poisonings than in all of Germany in ten years." Although Pregl believed that the use of arsenic as a poison was widespread, trials tended to falter on the view that people may have taken the drug of their own accord in what became known as the "Styrian defense".

In Britain in 1768, Doctor Thomas Fowler of Staffordshire invented his eponymous and enduring solution – a 1 per cent solution of potassium arsenate. It became a popular cure for fevers although Fowler was accused by many of being a fraud. Fowler's solution was regarded as being useful against some cancers, and in India doctors adopted it to treat a range of illnesses from angina to whooping cough and rabies, and even to ensure women gave birth to boys. In the nineteenth century, the metal oxide was a common drug, used in pills, inhaled in vapours, injected into muscles and used in enemas. It was used in skin treatments, mostly against eczema and psoriasis. Dentists applied it to kill the nerves in damaged teeth and dull the pain of dentistry. It was also prescribed for ulcers, migraine, neuralgia and fever, for asthma (curiously mixed with tobacco) and for cholera, where it was believed that the stronger poison would get rid of the weaker one.

Arsenic's real value emerged as a treatment for syphilis in the early twentieth century when the German chemist Paul Ehrlich, in collaboration with the Japanese scientist Sahachiro Hata, developed

Salvarsan, also known as arsphenamine or "606". A later arsephe-
namine derivative, known as "914" (it was Ehrlich's 914th experiment),
became the first remotely effective drug against syphilis. Ehrlich
had been searching for what he called a magic bullet, a chemical
that would harm the microbes that caused the disease but not the
surrounding cells. There was an initial euphoria about the treatment,
which appeared to be successful and caused patients to suffer fewer of
the terrible side effects of earlier treatments. But arsenic would soon
fall out of favour again with the introduction of antibiotics that ended
the dread of the disease and its associations with poison, witchcraft
and dangerous sexuality.

Arsenic is again a subject of medical interest. The chemical has
been involved in possibly the largest environmental mass poisoning
of people ever – the arsenic in water from wells dug in Bangladesh. In
the eighteenth century, a form of arsenic known as paste of Rousselot
was used in the treatment of some cancers, and from the 1860s it was
used as one of the first forms of chemotherapy in the treatment of
leukaemia. Researchers in the United States have been examining the
role of arsenic in the human metabolism. It appears that arsenic is
an essential nutrient in small quantities but it remains unclear why
exactly. It appears possible that it plays a role in a process called DNA
methylation, which involves the addition of a molecule at specific
locations in DNA. Low DNA methylation is associated with an
increased risk of cancers. Very low levels of arsenic in the diet have
been shown to reduce the amount of a chemical in the liver that is vital
for this process. Too much arsenic kills; too little may as well.

FROM THE 1840s onwards there was a growing number of reports
that suggested that arsenic was being used not just to kill rats but
to kill people. Britain was late to the idea of regulating the sale of
arsenic. In 1682, a French law introduced some controls – although
doctors, colour makers, metalworkers and other "officials" could
buy arsenic from dye makers or apothecaries without any problems.
Peter Navier, a French chemist writing in 1777, was critical of the

unregulated spread of arsenic, suggesting that many people died because of its misuse. A meeting of the English Provincial Medical and Surgical Association in 1849 declared that "the practice of secret poisoning had prevailed of late years to a most fearful extent". A doctor from Torquay in southwestern England told those at the meeting that arsenic was "the agent usually employed to this diabolical purpose".

Arsenic came to the fore because of the emergence of a test for the metal in 1836. The Marsh test was the first way of detecting the chemical in poisoning cases and was the star of several key trials in the nineteenth century. The test often gave contradictory or uncertain results but it was a step ahead of earlier tests, particularly those that relied on burning arsenic-tainted material with charcoal and trying to detect a garlic smell. The chemist James Marsh was at the Royal Arsenal in Woolwich developing a recoil break for naval guns. He was sent pots of coffee and the intestines of a prosperous farmer whom a local magistrate suspected had been poisoned by a grandson. Marsh followed the current German research and showed that a yellow precipitate could be extracted from the remains of the poisoned coffee. A jury, however, rejected this evidence and freed the man, who later confessed to the murder. The acquittal stung Marsh, who was determined to develop a test that produced white arsenic that could then be shown to sceptical juries as proof of poisoning.

The key breakthrough was in 1836 when Marsh, expanding on work done by the Swedish scientist Carl Scheele in the late eighteenth century, added acid and zinc to samples. Hydrogen was given off, mixing with arsenic to produce arsine gas, which has a characteristically garlicky smell. A small residue of metallic arsenic is also left behind. The test was first put on trial in the case of Marie Lafarge in 1840. Repeated tests failed to find arsenic traces in the organs of the alleged victim. Orfila was brought in to give a definitive answer and declared that there was arsenic in the victim and that it could not have come from any other cause except poisoning. Orfila's evidence established the Marsh test as valuable evidence, although it actually has problems. The test does not always distinguish between antimony and arsenic, and it is unreliable. Jöns Jacob Berzelius refined it in 1837 and Hugo Reinsch and Max Gutzeit developed new tests later in the century that provided greater precision in the detection of the metal.

The sudden anxiety that erupted about arsenic in the mid-nineteenth century stemmed from new tests for the metal that had been used in high-profile murder trials such as that of Lafarge. By the middle of the nineteenth century it was commonly accepted that arsenic was a menace, that it was widely used in poisonings, notably by women and that it presented such risks that it must be more successfully controlled. The fact that it could be easily bought in grocers and chemists for use as rat poison made people anxious. Its use in high-profile murders, or at least allegations of murder, meant that there was always an intense interest in the poison. Given that some rural grocery stores sold up to a tonne of the substance each year, what is remarkable is how few, not how many, died from it. Statistics on poisoning are always doubtful as many cases of poisoning may go unrecognised. Some reports had as many as six hundred people a year dying from arsenic poisoning, but this is not borne out by the records that do exist. In 1841 there were 3,939 recorded violent deaths in Great Britain, of which just 128 involved poison. Of these only fifteen involved arsenic. Despite the fact that the poison was cheap, almost universally available, easy to administer and disguise, it was used in only a small number of recorded killings.

A problem in the nineteenth century in Britain was the prevalence of cholera, which hit the country for the first time in 1831 and then another four times until 1866. Some toxicologists feared the disease was being used to cover for poisonings and provided detailed ways in which the symptoms differed. Symptoms occurred in a different order, with those poisoned by arsenic experiencing pain followed by vomiting and diarrhoea, while cholera victims tended to vomit first. But only chemical tests could really distinguish between deliberate death and disease. Cholera epidemics were often accompanied by accusations of mass poisoning that provoked riots and lynch mobs. In Paris in 1832, a cholera outbreak led to the mob killings of accused poisoners. Stories circulated of bread being poisoned and of wells being tainted. Most of those singled out for attack were foreigners or itinerant workers, most commonly wet nurses and masons who came in from the countryside.

In Britain, the Arsenic Act of 1851 tried to put a halt to unregulated sales but only by confining purchases to adults and asking vendors to keep registers. Arsenic bought to kill rats had to be mixed with soot or indigo to ensure that it could not be used undetected. Regulation made little difference. Arsenic was common in insecticides, sheep dips, treatments for grains to control worms and fungi and most commonly in dyes. Scheele's Green (emerald green or copper arsenite), discovered by Scheele in 1775, was used in many items found around the home including toys, flowers, packaging, furnishings, wallpaper, medicines, confectionary, jellies and other foods. If it was green in Victorian England, the likelihood was that it contained arsenic.

Green was the highly desirable colour of the nineteenth century. Victorians sat in rooms coloured with arsenic while wearing clothes dyed with arsenic. People lived in what must have been a miasma of this poison in quantities that must certainly have contributed to the chronic ill health of the time. Domestic bliss may have been one of the aspirations of Victorian society but in fact they lived in a world of self-inflicted toxicity. The widespread use of arsenic in products that could leak into the environment may undermine many of the accusations that arsenic was widely used as a deliberate poison. Many of the victims could simply have received chronic doses of the chemical from their environment. "There appears good reason for believing that a very large amount of sickness and mortality among all classes is attributable to this cause and that it may probably account for many of the mysterious diseases of the present day which so continually baffle all medical skill," according to one forensic scientist.

One of the biggest arsenic killers may have been wallpaper. In 1836, the government reduced duties on imported wallpapers and they surged in popularity. Gas lighting replaced candles and the fashion turned to darker coloured rooms. The development of machine-printed paper in the 1840s made wallpaper more available than ever before and most homes across the country were decorated with wallpaper, much of which was coloured with arsenic-based dyes. When the paper became damp or mouldy, as was often the case before central heating, arsenic compounds were released. By 1860 arsenic wallpapers had become a matter of public concern, in part because of articles published in *The Lancet*. One called for all arsenical wallpapers to be printed with a skull-and-crossbones pattern. Ironically *The Lancet*'s founder, Thomas Wakley, was himself a victim of poisonous wallpaper after his office was redecorated.

In one study, it was found that a thirty-eight centimetre square of paper contained enough arsenic to kill two adults. A room covered with a hundred square meters of paper would contain 2.5 kilograms of arsenic. The problem was both dust coming off the paper and the volatility of arsenic compounds caused by dampness leaching through the paper. The first fictional murder by wallpaper appeared in the *Chambers's Journal* in 1862. In "Our Best Bedroom", the heir to a great fortune is pressured into sleeping in the "Green Room", the best bedroom but one unfortunately hung with arsenical wallpaper. The problem lingered as long as rooms retained their arsenic-laden decorations. Clare Boothe Luce, the playwright and politician, suffered arsenic poisoning from the paint on the ceiling above her bed when she served as the American ambassador in Rome in the 1950s.

The most famous possible victim of deadly wallpaper was Napoleon. The French emperor died in St Helena aged fifty-one in 1821 and since then, his death has been an endless source of speculation. The poisoning theory was revived in 1961 when locks of Napoleon's hair, cut from his head after his death as keepsakes, were analysed and were found to contain high levels of arsenic. This was immediately taken as a sign that he was deliberately poisoned but it was not necessarily so. The metal could well have been given to him medicinally. There is also the possibility that the hair might have been treated with arsenic to prevent mites or that over the years the locks had been stored in a container with a high level of arsenic. Samples of his hair taken long before he was held by the British also show high levels, leading to speculation that he used a hair tonic containing arsenic or had taken medicines, possibly for syphilis. The official autopsy at the time declared that he had died of stomach cancer, an illness that had killed his father and several other close relatives, and recent studies back up this claim. Scientists have studied the water, earth and coal of St Helena to determine if the arsenic could have come from those sources. Coal seems to be a likely source of contamination, along with wig or hair powder or medicines such as Fowler's solution.

Analysis of the wallpaper at Longwood House where Napoleon died showed that it did not contain much of the metal. It may have come from the curtains around his bed, which were dyed a colour known as Paris Green that contained copper-aceto arsenite. Hairs analysed in 1982 and by the FBI lab in 1994 did not show enough arsenic to have poisoned him, while his symptoms suggested cancer rather than any acute or chronic poisoning. Those who believe he was poisoned say that the hairs analysed by the FBI may well have been taken from Napoleon before his stay at Longwood. They also maintain that the well-preserved state of his corpse when it was exhumed from a grave in St Helena in 1840 attests to the presence of a high level of arsenic in the body, as it can reduce decomposition. Supporters of the poison theory believe that arsenic was used to make Napoleon sick, to break down his body until it would be vulnerable to other illnesses and poisons. He was then fed calomel (mercurous chloride, used as a purgative) and orgeat, a drink made from the kernels of bitter almonds. These combined in the stomach to form mercuric cyanide, which may have finished him off in his weakened state. This may not have been deliberate. Medical treatment at the time was as dangerous as any disease.

Part of the appeal of arsenic was said to be its history. Its alleged use by the great and the bad provided it with an aristocratic provenance, although it is hard to see that this existed outside the imagination of writers. *The Poison Trail*, a breathless work of true-crime stories published in 1939, declared that arsenic was a popular choice for murderers because it had "the endorsement of history". The use of it by the Borgias and the Marquise de Brinvilliers led to its popularity, according to the author William Boos. "And since their days – to a large extent, also, because of their activities – a veritable tradition of arsenic poisoning has grown up in the world to seize the popular imagination as only a tradition can." Not only were the traditions of the Borgias still alive but they represented a threat in the United States, according to Boos.

> Now among Anglo-Saxons, poisoning is by no means the usual way of getting rid of an enemy or inconvenient relative – other methods are distinctly preferred. But this is not so true of Central and Southern Europeans. To many of them killing means to poison and to poison means to use arsenic, simply by a force of familiarity and tradition. Thus a majority perhaps of the poison murders in this country are committed by foreign immigrants – Poles, Hungarians, Czechs, Italians, French Canadians – and arsenic is almost always the universal agent they employ.

The main dangers from arsenic were accidental rather than deliberate poisoning or the general illness induced by sustained exposure at home or in the workplace. Arsenic often found its way into foods and other products, killing or sickening many people. The Violet Powder incident in Essex in 1896 involved baby powder that was contaminated with large quantities of arsenic. Some twenty-eight children fell sick and thirteen died. Towards the end of the nineteenth century ideas about environmental health and the dangers of workplaces were being developed and food safety became a key area of reform. Sir Thomas Oliver's *Dangerous Trades* prompted broader investigation into the issue of arsenic poisoning. There was in the mid-nineteenth century a considerable scare over toxins – an arsenic crisis – but this had passed by 1890 without significant government action. Consumers and publicity forced voluntary reforms between 1855 and 1875 that led to a massive reduction in the number of products that were made with poisonous arsenic salts.

AFTER AN OUTBREAK in Marseilles in 1722, the plague no longer presented a serious problem in Europe. After centuries in which it had killed millions of people across the Continent, it no longer inspired such intense fear. Plague broke out elsewhere, in northern Africa, Russia, the Levant and many other countries but after the eighteenth century, it was very rare in Europe. Several explanations for the end of the plague have been suggested but few of them are satisfactory.

The last occurrence of plague in any place was often as devastating as any other, so it appears there was no slow increase in immunity that might have explained why it died out. The idea that this period saw improvements in sanitary conditions is unlikely, given the rapid expansion of urban populations and the continuation of illnesses like cholera and typhoid. There could have been a change in the ecology of rats. The brown rat, or *Rattus norvegicus*, did become dominant around this time and was less efficient than the black rat at transmitting the plague. The problem with this as an explanation is that the dominance of the brown rat rather than the black rat began a long time before any decline in the plague and ended a long time afterwards. It has been suggested that better methods of quarantine may have slowed the spread of the disease but given that there was very little enforcement of regulations, it is hard to imagine why this particular area of law and policing – one that is fraught with problems to this day – should have suddenly experienced a sudden improvement. In the eighteenth century, the early medical investigator John Howard travelled some of the plague routes and found no more policing of quarantines than would have been expected. Controls were lax and officials easily bought off.

All the common explanations are either not proven or would have taken effect too slowly to explain the rather abrupt end of plague in Europe, which came to a near halt with the Marseilles epidemic of 1720–22. Around this time, arsenic trioxide became widely and cheaply available as a rat poison. Arsenic was the first poison that was truly effective against rats. As it was colourless, tasteless and odourless, rats, surprisingly picky eaters with intensely cautious senses of taste and smell, would actually consume it. Traps were not very effective and cats rarely attack full-grown rats as they are quick and dangerous fighters. Cats probably accounted for only a fifth of all rat mortality, not enough to cut the population given their high fertility rate. Before arsenic, wolfbane was the most common poison used against vermin but it is less toxic than arsenic and tends to dry out and become ineffectual. Reducing the rat population was vital in dealing with the plague as it cut the number of fleas geometrically. There needed to be a critical number of rats for plague to spread and keep on spreading. "It might just be that the beginning of industrial-scale toxic pollution brought about a vast improvement in the health conditions of Europe," according to the historian Kari Konkola. The use of arsenic became commonplace just as plague disappeared, suggesting that the large-scale killing of rats with an effective poison may have disrupted the ecology of plague. Arsenic may have once been the perfect poison, but it possibly saved far more lives than it has ever taken.

13. THIS ALL BLASTING TREE

Fierce in dread silence on the blasted heath
Fell upas sits, the Hydra-tree of death
Lo; from one root, the envenom'd soil below,
A thousand vegetative serpents grow.
Erasmus Darwin, *The Botanic Garden*

By the eighteenth century, the basilisk had lost its fearsome reputation. No longer accepted as a real animal, it appeared only as a device in heraldry or occasionally as architectural decoration. A basilisk sits perkily above a doorway in Fort Belvedere in New York's Central Park. Just as the basilisk faded, it was replaced by a new fear of something equally deadly and mysterious. Reports arrived in Europe of a deadly tree to be found on the island of Java that gave off toxic emanations that killed everything around it. Upas trees were so deadly that only a handful of people had ever seen them, standing alone surrounded only by saplings of the same species as no other plants or animals could survive in its shadow. J. J. Stockdale, an English writer, said in his *Sketches of Java* that criminals who were to be executed could win a reprieve by volunteering to collect the poisonous sap of the tree. If they approached upwind, it was possible to get to the tree without being affected by its deadly miasma. Only one in ten prisoners would survive the ordeal.

Malayans, as Stockdale called all people from what is now Indonesia, were adept at using the poison. Javanese princes used it to kill off rivals among the vast numbers of half-siblings in their polygamous families. Soldiers used it to poison their rivals. The Dutch, the colonial power, supposedly had lost many men to the toxin. In Europe the story of the tree goes back to the early travel writers such as Friar Odoric and Sir John Mandeville, a later author who collected exotic stories from various sources and told them as if he had witnessed the miraculous events and objects on his own travels. *The Travels of Sir John Mandeville* is a compendium of sixteenth-century male fears of the unknown, most graphically illustrated in a passage in which he describes an island where men do not sleep with their wives on the first night of marriage but:

> instead employ a man known as a gadlibiriens or 'fool of despair' who take their virginity. The origins of the custom came from the belief that men had died sleeping with their wives for they had snakes within them which stung the men's penises within and so they follow the custom there to make other men test out the route before they themselves set out on that adventure.

The Travels of Friar Odoric, the stories of a Franciscan monk who travelled across Asia about twenty years after Marco Polo in the early fourteenth century, includes a mention of a place called Panten or Thalamasyn, which is believed to be Borneo. "Here can be found trees that produce flour, and some that produce honey, others that produce wine and others a poison the most deadly that exists in the world." He describes men going into battle with poisoned arrows that kill almost instantly. The only antidote was to "temper the dung of a man in water" and "so drinke a good quantitie thereof". Mandeville, who stole with abandon from Odoric, also described a poison tree. "An yle that is called Pater and some call it Salmasse are trees that beare venym againe the which is no medicine but one that is to take of the leaves of the same tree and stampe them and tempre them with water and drinke it or else he shalle dye sodainly for Treacle may not help."

The naturalist Georg Eberhard Rumpf, better known as Rumphius and as the "Blind Seer of Ambon", wrote of the toxin:

Up to now I have never heard of a more horrible and more villainous poison coming from plants than that which is produced by the milk-tree. I call it a milk tree because it yields a reddish brown sap and this sap is the pride of the Indians throughout the Water Indies and with it they dare provoke Dutch Arms. The soldiers were erstwhile more afraid of this than they were of canon or musket.

Rumphius never saw the tree himself which, as he put it, "nature had wisely placed away from human dwellings" on the sides of high mountains. It was also hidden away by locals who refused to divulge any information about it. Dutch officials were ordered to find out more. The prefect of Macassar sent him some branches of a female tree. It was, according to Rumphius, possible to:

approach the tree if one came covered in cloth so that nothing from the tree dropped onto the skin. To extract the poison, you take a thick bamboo and pierce the trunk from which flows a sap that eventually turns a dark russet colour. Nothing grows around the tree and the earth beneath it is covered with feathers from birds that alighted in the tree and then died. The only thing that can live under the tree is a horned snake that cackles like a hen or crows like a cock and has eyes that shine in the dark. Others describe it as a basilisk – a snake with a cock's comb and legs at the front.

In the late eighteenth century, the tree gained widespread notoriety. The German surgeon J. N. Foersch visited Java in 1774 and described a tree he found about seventy-five miles from Batavia, now Jakarta. "It is further surrounded on all sides by a circle of high hills and mountains and the country round it, to the distance of ten or twelve miles from the tree is entirely barren," he wrote in a London journal. An elderly priest who lived nearby provided directions and consolation to the condemned men who were sent to collect its poison. He had sent more than seven hundred men down the path to the blighted area where the tree stood and fewer than one in ten had returned. Foersch claimed to have witnessed an execution in which thirteen concubines of the "emperor" who were accused of infidelity were wounded with a lance infused with the poison and all died within fifteen minutes.

Foersch's article captured the imagination of many a scientist and writer. Erasmus Darwin, physician and grandfather of the theorist of evolution, included it in his *Loves of the Plants*, describing it as the "Hydra-Tree of Death". Lord Byron used it to describe the nature of original sin in his *Childe Harold's Pilgrimage*: "This eradicable taint of sin, This boundless upas, this all blasting tree." The upas became a metaphor for deviancy and destruction. Pushkin devoted an entire poem to it. "Deep in the desert's misery/ far in the fury of the sand/ there stands the awesome Upas Tree/ lone watchman of a lifeless land." The poem describes the fate of a man sent to collect the poison, which is then used by a king to poison all his enemies. In 1820 the painter Francis Danby submitted a vast canvas called *The Upas* to an exhibition in London. The painting showed a bleak, rocky chasm littered with corpses felled by the tree. By the early nineteenth century, the upas had replaced the basilisk as an image to conjure up pervasive destruction.

But Foersch's story was overblown. It is not even clear that he was anything other than an imposter who had never served as a doctor in the East Indies. He had probably never seen the upas, nor is the reality of the tree as dramatic as he portrayed it. The upas is *Antiaris toxicaria*, known in Malay as *bohun upas*. It produces a poisonous white sap that was used as an arrow poison for hunting monkeys and birds; it contains a glycoside that acts in a similar manner to digitalis, the poison found in foxgloves. But the tree represents little danger unless someone takes in a considerable quantity of sap. Birds nest in the trees and plants thrive around them.

The upas shared many of the same characteristics as the basilisk. It came from a mysterious land, grew only in the most isolated conditions and all those who approached it died, making it conveniently difficult to verify its existence. Its appearance at a time when there was intense European interest in Java, allowed it to be used to confirm the views of the place at the time.

FROM THE TIME of the sixteenth-century explorers onwards, poisons used by the people they encountered were an immense source of fascination. The first mention in English of curare is in Sir Walter Raleigh's pamphlet *The Discovery of Guiana* published in 1596. He mentioned a people called the Wakiri:

> as black as negros but have smooth haire and these are very valiant or rather desperate people and have the most strong poison on their arrowes and most dangerous of all nations of which poison I will speak somewhat being a digression not unnecessary.
>
> There was nothing whereof I was more curious than to finde out the true remedies of these poisoned arrows for besides the mortality of the wound they make, the partie shot indureth the most insufferable torment in the world and abideth the most ugly and lamentable death sometimes dying starke mad, sometimes their bowels breaking out of their bellies and are presently discolored, as black as pitch, and no man can endure to cure or to attend to them. And more strange to know that in all this time there was never Spaniard either by gift or torment that could attaine to the true knowledge of the cure although they have martyred and put to inuented torture I know not how many of them. But everyone of these Indians know it not, no not one among thousands but their southsaiers and priests who do conceale it and onley teach it but from the father to the sonne.

The British Empire was advancing, establishing new outposts across the world. The colonial fascination with poisons extended beyond the meticulous collection of poisoned arrows – a vast number of which are stored, and still handled very carefully, in the Victoria and Albert Museum in London. The interest with new poisons led scientists to experiment with them in ways that would establish the basic techniques of toxicology, dragging the study of poisons away from its classical roots and into the realm of modern science. Meanwhile the cultural anxieties of colonialism and enslavement and the encounter with an array of new experiences led to a new array of fears. Poisoned arrows and upas trees all reflected concerns about the brutal realities of colonial rule; they were a strenuous effort to deflect the real deadliness of European colonialism. Natives were depicted as masters of poisons, adept at using these hidden means of death against unsuspecting Europeans. The reality was that poisons were rarely used. The novelist Michael Ondaatje wondered in his personal history *Running in the Family* why the people of Sri Lanka had not used their supposedly formidable knowledge of poisons to kill more of the foreigners who invaded their land. They may have resisted colonialism in other ways but poisonings remained rare.

Poisoned arrows excited the European imagination but there is little evidence that they have been widely used. Anthropological research on warfare and its technology is extensive and yet rarely mentions the use of poison. In the *Encyclopaedia of World Cultures*, a compendium of a thousand cultural groups, there are only a handful of mentions of the use of poisons, mostly relating to hunting. In dense forest, hunters need to paralyse animals rapidly so they fall to the ground and can be found, rather than disappearing, wounded, across the forest. Certainly, cultures have used poisoned arrows in war – Juan Ponce de Leon, the Spanish conquistador credited with discovering Florida, is said to have died from a poisoned arrow fired by a member of the Calusa people in 1521 – but there are few other recorded deaths from this means.

The early eighteenth century was a critical moment in the history of science as experimentation took hold, and the emerging disciplines of chemistry, physics and biology interacted in new ways. Experiments with poisons started to take consideration of the dose given to animals and also began to examine them postmortem. These basic steps in experimentation were critical to developing an understanding of how poisons worked on the body – the basis of modern toxicology. Experimentation often meant contradicting the work of established authorities – these were contentious, sometimes even dangerous times to be a scientist. Working on poisons was not always safe. One was exposed not only to dangerous chemicals but also to accusations of sorcery and mischief. Moses Charas, an early experimenter with poisons, was the Royal Demonstrator of Chemistry at the Jardin des Plantes in Paris until the Edict of Nantes drove him to exile in England. He was later called to Spain to the court of Charles II, who was ill and required treatment. Charas happened to offer the opinion that Spanish vipers were poisonous, which they are. He was thrown in jail as a heretic because an archbishop had exorcised all snakes in Spain, rendering them harmless, according to Church doctrine.

Snakes and poison arrows were critical in the development of modern toxicology, a process begun by Felice Fontana, an eighteenth-century empiricist and pioneer of experimentation. Debates raged over the nature of viper poison. Charas believed that it was not the liquid that came from their fangs when enraged that was deadly but spirits they conjured up. Fontana showed that death did not result from a universal inflammation as some has supposed – it happened too quickly, while it took much longer for the blood to coagulate. Testing how much it took to kill animals, Fontana concluded that it would take the venom of five or six vipers to kill a man. "And here I am not afraid to advance freely that the bite of the viper is not absolutely mortal to man, and that those have been mistaken who have regarded the disease caused by the viper as one of the most dangerous and from which it was impossible to recover." His argument was based on the size of the animal – he noted that it took several bites to kill a fifty-pound dog and that a man was much bigger. He was also unable to find much in the way of anecdotal evidence of deaths. He found only two and in one case the man died of a gangrenous arm several weeks after being bitten.

In London, Fontana was able to experiment on some arrow poison

– some of it contained in a pot labelled "Indian Poison", brought from the banks of the Amazon by Don Pedro Maldonado. This poison, then known as ticanus and now called curare, attracted much interest. It was viewed as so poisonous that even its odour was thought deadly. The mythology around the poison was that it was prepared by women condemned to death as they died just at the point in the process when it became powerfully toxic. Fontana wafted some around a pigeon, which survived the experience. He found, however, that it was a strong poison – it was most effective when introduced intramuscularly but was not fatal when swallowed. Fontana's work established some of the basic ideas of modern toxicology – dose is critical; poisons act in different ways, some directly attacking the nervous system and others disrupting other physiological processes; and observation of experiments was more important than the regurgitation of ancient knowledge.

For many Europeans, colonialism meant living amidst unfamiliar cultures and enduring an array of tropical diseases (although it was indigenous people who suffered truly awful losses from unfamiliar diseases such as smallpox and measles brought in by colonists). Nevertheless, despite their toxic presence, the fear of poison hung over colonial communities. The dread was acute in India, a land associated in the European imagination for centuries with exotic poisons, deadly snakes and terrible plagues. From the fifteenth century, stories circulated in Europe about the Hindu practice of sati in which widows burned themselves to death on their husband's funeral pyres. This in itself made a dramatic addition to the traveller's tales from the exotic East, but it was frequently paired with the idea that it was the law that women would die alongside their husbands to stem an epidemic of poisonings. The Reverend John Ovington, writing in his *A Voyage to Surat in the Year 1689,* blamed the eagerness of women to find new lovers. Indian widows were burned alive "because of the libidinous disposition of the Women, who through their inordinate Lust would often poison their present Husbands to make way for a new lover".

The story became a commonplace understanding of why women might be treated in this way and appears in an array of stories written by early European travellers to India. The same story of the lustful, poisoning wife is curiously absent from Indian culture at this period. Other travellers from China and the Middle East who visited India also failed to pick up on it. Al Biruni, an Islamic commentator who wrote of his travels to India in the eleventh century, mentions sati but explains it as the honourable death of a widow. None of the Indian epics contain any warnings against poisonous wives nor do any law books mention poison as the reason for the practice of sati. The practice is not justified anywhere in Hindu scriptures, nor is any specific punishment set out for women who poison their husbands. The *Laws of Manu* are clear about the punishments for other transgressions: "If a woman... should prove false to her husband, the king should have her devoured by dogs in some much frequented place." It might be expected that if men lived in fear of poisoning women at this time, they would have laid down some sort of punishment, and yet the Laws are silent on the issue. At the time these stories were popular in Europe, India was ruled by the Mughals. Records show only two cases of poisonings being tried in Mughal courts from 1526 to 1707 and both were done by men. The Mughals had outlawed sati and there are records of trials for the burning of "unwilling widows" but there is no evidence of an epidemic of poisonings followed by illegal incidents of widow burnings.

The story appeared in European writings on India just at a time when poisonings were an obsession. The witch hunting manual *Malleus Maleficarum* had warned that if men lost faith in their wives, the women would "prepare poisons for you and consult seers and soothsayers; and will become a witch". It was also a time when works on murderous women had gained immense popularity. John Reynold's compendium of "true crime", published in 1621 under the title *The triumphs of God's revenge against the crying and execrable sinne of murther,* details six poisonings by wives in just a few years. These crimes were quite rare in comparison with violence by men but they had a powerful hold, being endlessly recycled in pamphlets, sermons, ballads and in plays. Notorious cases had a grip on the public imagination up to a century after the crimes had occurred. The often-repeated stories of sati appeared to echo European anxieties about poisonings. In Europe, poisoning wives were perhaps at least as common as they were in India, and in England in the seventeenth century their punishment was similar to their imagined fate in India. Women convicted of killing their husbands were given the same sentence as those convicted of treason: they were burned at the stake.

IN OCTOBER 1856, a dispute over the Chinese crew of a vessel moored outside the new colony of Hong Kong blew up into a minor conflict, which became known as the Arrow War after the name of the vessel in dispute. The Chinese authorities had seized the Chinese crew of the ship, charging them with collusion with barbarians. The British accused the Chinese of violating the Nanking Treaty, the one-sided agreement forced on China that allowed the British trading access to their ports. Relations were tense anyway as the Chinese had been blocking access to Canton and the British were determined to open up trade with the country, an enormously lucrative market for British opium.

The governor of Hong Kong, Sir John Bowring, ordered Viceroy Yeh Mingchen to apologise and release the crew or face punishment. The British started seizing Chinese vessels. Yeh dismissed Bowring's demands, prompting the British to shell forts along the Pearl River estuary. Even as more buildings were shelled by the British, Yeh remained intransigent. In retaliation for the British aggression, mobs burned British warehouses and homes in Canton and Whampoa. Yeh announced a price of a hundred dollars per head for British killed by Chinese and urged the colonists to leave Hong Kong immediately.

Hong Kong, declared a British colony in 1841, was a rapidly growing entrepot with more than seventy thousand people, mostly Chinese, arrayed along a narrow stretch of the island and in smaller settlements in Kowloon. Like many new colonies, it suffered from the strains of rapid growth with regular outbreaks of disease such as smallpox, cholera and diphtheria. It occupied a precarious position as a tiny outpost of traders set up against a politically and militarily weak but massive China. It was vulnerable to blockades and the allegiance of many of its residents was in doubt. There were also frequent rifts between the merchant community, essentially adventurers and drug dealers with few scruples, and the moralising colonial administrators who sought to bring their British Christian order to the local heathens.

Tensions with China and the prospect of another war caused dismay among the British population, who were there to profit from China trade. They had had little interest in punishing the Chinese for some slight, particularly if it cut into their income. Hong Kong was already in a feverish state when posters began appearing on streets calling on Chinese to avenge slights by the British. The colonists appealed to the Governor to expel all Chinese whose honesty could not be vouched for, but he refused to adopt such a measure. It was hardly practical to expel most of the people living in Hong Kong. Yeh remained as stubborn as ever and the British fleet retreated from Canton, unable to make him move. The British were feeling isolated and vulnerable.

On 15 January 1857, a similar scene occurred at breakfast tables across Hong Kong. After eating bread bought by the servants at the city's main bakery, dozens of people began to sweat, feel nauseous and finally vomit. The first stricken were members of the Indian community who tended to rise earlier than Europeans and, unlike Chinese, ate bread for breakfast. More than four hundred Europeans had been poisoned by arsenic, it was later determined, that had been put in the bread sold at the E-Sing Bakery on Pottinger Street, owned by a Chinese man known as Ah-Lum. Among those who had eaten the bread was Lady Bowring, the wife of the governor, who was to die a few years later from the chronic effects of the poison.

Later, analysis of loaves showed that the bread contained nearly 1 per cent arsenic by weight, a massive dose and one that probably led to the vomiting up of much of the poison. Had it been a lower dose, it might have actually killed more. Although nobody is believed to have died immediately from the poison, several deaths over the next few years were attributed to the long-term effects of arsenic. Ah-Lum had left that morning for Macau but he and his family had also been poisoned. He returned to Hong Kong and was arrested. More than fifty of his workers were arrested and were held in a tiny police cell for more than a month before being released. A clamour arouse in the local newspapers for them all to be hanged and their bodies displayed outside the bakery as a lesson to other Chinese who might defy British rule. Ah-Lum was put on trial in front of the chief justice.

He was prosecuted by Thomas Chisholm Anstey, a former English MP known for being a garrulous bore and later notorious for being the author of the longest letter ever submitted to the *Times*, the 116 pages of which were published as a pamphlet. Anstey pompously appealed to the British jury by saying, "It was better to hang the wrong man than confess that British sagacity and activity have failed to discover the real criminals." The judge and jury disagreed and Ah-Lum was found not guilty on 6 February 1857. He remained in detention and was finally expelled from the colony. The poisonings led to much criticism of Viceroy Yeh, mostly from residents of countries other than Britain who had no quarrel with China. Yeh merely responded that he did not see the difference in barbarity in putting arsenic in bread or lobbing shells into the streets of Canton as the British had been doing.

Four people were said to have died from the effects of the poison, although they all died between six and fifteen months after the ingestion of one large dose. The likelihood is that they died of other causes, although it is not inconceivable that a large dose of arsenic might have left them more susceptible to the many other illnesses that struck people in the tropics. The incident was grabbed on as a sign of resistance to colonialism. Writing in the *New York Daily Tribune* in June 1857, Karl Marx sounded almost approving:

> There is evidently a different spirit among the Chinese now to what they showed in the war of 1840 to '42. Then, the people were quiet; they left the Emperor's soldiers to fight the invaders, and submitted after a defeat with Eastern fatalism to the power of the enemy. But now, at least in the southern provinces, to which the

contest has so far been confined, the mass of the people take an active, nay, a fanatical part in the struggle against the foreigners. They poison the bread of the European community at Hong Kong by wholesale, and with the coolest premeditation. (A few loaves have been sent to Leipzig for examination. He found large quantities of arsenic pervading all parts of them, showing that it had already been worked into the dough. The dose, however, was so strong that it must have acted as an emetic, and thereby counteracted the effects of the poison).

While those out establishing an empire on which the sun never set faced an array of poisons – real and imaginary – at home in Britain, deliberate poisonings were having a dramatic impact on legal proceeding and the development of toxicology. Part of this obsession came from the industrialisation of Britain and the often catastrophic environmental impact. It was also the moment that toxicology emerged as a science and its development took place under an intense public examination because of the role of scientists in high-profile trials. The 1840s were a period of political and economic turmoil in Britain with the Chartist movement, mass migrations from the land to urban industrial jobs and revolution afoot in Europe. Domesticity was a refuge from all this and anything that disturbed the emerging Victorian cult of the home was regarded as a potent threat. Science was changing, as were policing techniques. There were new tests for chemicals such as arsenic. Many communities now had police forces that were more inclined to investigate crimes, mostly those in the home that in earlier days might have been ignored. Rising levels of literacy, which improved markedly through the century, along with advances in printing and distribution of newspapers, magazines and books, created the means by which poison fears were spread.

A series of prominent trials that were widely reported in the press – in a much more sensational manner than would be possible even in today's British media – meant that poisoners, particularly young women who may have resorted to the method, were some of the first criminal celebrities. Newspapers created a new public sphere in which crimes were reported in intense detail. Investigations and police work was criticised, forensic scientists questioned, witnesses doubted and the accused often found guilty before the courts had ruled. Prosecutors in the United States today refer to the "*CSI* effect" – the fact that the television procedural created a popular sense that forensic evidence will provide the answer to all crimes. Prosecutors who fail to provide forensic details often find themselves facing sceptical jurors. A similar effect occurred in the nineteenth century as intensive newspaper reporting on trials produced a public that was both fascinated by and often intensely critical of the emerging role of science in the courtroom.

The trial of William Palmer of Rugeley was one of the most sensational criminal cases of the nineteenth century. Palmer was accused of using strychnine to poison his wife, his brother and a friend. In each case, Palmer stood to gain from insurance money. He had clear motive and means to have carried out the murders. He was found guilty at the Old Bailey and was hanged on 14 June 1856, in front of a crowd of thirty thousand. "I am innocent of poisoning Cook by strychnine" were his last words. He might well have been, as the evidence against him was circumstantial. Strychnine was not found in the body of his alleged victim but the similarities in the way he died to the known effects of the poison were enough to convince a jury. Palmer was certainly a rogue and was certainly guilty of gambling, adultery, forgery, fraud and medical malpractice. But the doctor was convicted in the court of public opinion of being a mass murderer even though he was only ever tried for one killing. His waxwork stood in Madame Tussaud's in London for 127 years.

Palmer's trial came at a moment of profound changes in the media and public interest in poisonings. Newspapers doubled their subscription numbers during the trial. Copies sold for more than four times the normal price. Local passions rose so high in Staffordshire that Parliament passed a special law allowing the trial to be moved to the Old Bailey in London. Later, the town of Rugeley even requested that it be allowed to change its name. The move to London meant many of London's grandees could attend. The Lord Chief Justice presided over the court. William Gladstone, later Liberal prime minister, visited, writing in his diary, "I liked ill both the looks and the demeanour of the prisoner." Charles Dickens wrote about the case, describing Palmer's "complete self-possession and constant coolness". Crowds gathered outside even though they were unable to follow what was going on inside. Palmer was one of the first celebrity murderers and one of the figures who precipitated the Victorian obsession with poisonings.

Palmer's refusal to admit his guilt before his execution was a disappointment to many. Alfred Swaine Taylor, the leading British expert on toxicology and jurisprudence, testified that Palmer had used strychnine. However, he was unable to find any trace of it in the victim and had based his testimony on descriptions of the death. Taylor had written in his work *On Poisons* that toxicology "demonstrates at once the means of death; while symptoms and postmortem appearances are, as we have seen, fallible criteria". In this critical trial he was working only with symptoms and postmortem appearances, leaving him vulnerable to attack on the stand. Palmer's trial focused substantially on the scientific and medical evidence, with more than half the witnesses addressing these issues. Scientists were pitted against each other "like rats or prize fighters", according to an article in *Lloyds* magazine. There was considerable disquiet over the verdict; science had been expected to prove the man's guilt but had failed to meet the standard set by the public.

Newspapers were convinced that there was an epidemic of poisonings in the nineteenth century. *Household Words*, a magazine edited by Charles Dickens, reported that between 1839 and 1849 there had been two-hundred-and-forty-nine deaths by poisoning but only eighty-five convictions. It did not say how it had reached these figures but the conclusion was clear – people were poisoning each other at an alarming rate and getting away with it. Shops had to have licences to sell tea, coffee, tobacco or snuff but anyone could sell arsenic. Poisoning was worrying because it happened at home. "The crime of murder by poison admits more readily of a fiendish sophistication in the mind of the perpetrator than any other form by which murder is committed." Journalists were intensely disturbed by the idea of well-off or middle-class women who poisoned their spouses. They were less easily dismissed as hatchet-faced and unfeminine – as working-class female murderers were often described. Crimes of well-presented women murdering men had an almost aristocratic bearing to them and almost no trial of this kind passed without reference to Lucrezia Borgia.

The risks faced by Victorians were greatly overstated. Quite detailed records were kept in Britain of births and deaths and these give us one picture. Poison is difficult to quantify in comparison with other forms of violence such as stabbings or burns and is probably underrepresented in the data but the figures show remarkably low numbers of deaths. The number of deaths range from 1 to 3 per cent of murders each year across the century. William Farr, who wrote extensively on the collection of statistics in nineteenth-century Britain, wrote to the Registrar General in 1839 to say that coroners were probably missing large number of deaths by poisoning. Coroners had little experience of either diseases or of poisons and did not often carry out extensive postmortems that included analysis of stomach contents. "The result of this negligence is that little is known positively of the causes of sudden death; and the facility of procuring all the more intense poisons, as well as the prospect that the effects of poisoning may be confounded with natural causes, offers a strong temptation to the commission of this dreadful crime."

Certainly, there must have been many opportunities. Opium, mostly mixed with alcohol in the form of laudanum, was imported in vast quantities and was widely used in calmatives for children. "Baby farmers" – people who took over the care of abandoned children and were paid by parishes – found it easier to cope if children were all sedated. If they died, nobody was much concerned. Although statistically, poisoning was extremely rare, it was certainly under-reported, mostly in the cases of children who were highly vulnerable to accidental or deliberate poisoning. The existence of "burial clubs" – cooperative societies that gave a pay-out on the death of a relative to pay for funeral expenses – might also have encouraged the speeding of infants along their way. By registering a child in multiple clubs, parents could get quite a windfall from their death. *The Lancet* reported in 1861 that while the children of working men died at a rate of 36 per cent before reaching the age of five, those registered in burial clubs died at a rate of 62 to 64 per cent.

Although poisonings were seen as "A Crime of the Age", as the *Illustrated Times* put it, Victorian writers also saw the crime as a sign of civilisation. Victorians were alongside the Romans and the poisoners of Renaissance Italy in their use of a subtle and skilled way of killing. As the historian Mark Essig wrote in his study of nineteenth-century American poisoning trials: "Poisoning was fascinating because it was a crime with a heritage, and an aristocratic heritage at that."

The great Victorian crime was the domestic poisoning. A survey of poisoning crimes in Britain, based on court records from 1750 to 1914, showed that of 540 criminal trials linked to poisons, 277 involved women. These cases also included ones of poisoning animals and attempted murder. Of 342 people charged with murder involving poison, 210 were women, showing that courts disproportionately charged women with the most serious offence while men were often charged with manslaughter, mostly in cases involving physicians or pharmacists. Overall just over 50 per cent of these cases involved women and a large share involved infanticide. In nineteenth-century England, women accounted for about 40 per cent of all murder charges but once infanticide was classified separately, the share fell to below a quarter.

Murders by women arouse such passion in part because of their rarity and also because they almost always involve members of their families. Women are most likely to murder their children and then their husbands. Occasionally it involves lovers but very rarely strangers – from 1957 to 1962 in England there were ninety-nine women tried for murder – in eighty-eight cases victims were related to the killer. Of the sixty-eight women executed for murder in England since 1843, thirty-seven used poison in their crimes. A fifth of all women executed for poisoning came from two small towns in East Anglia – an area previously known for its large number of witchcraft accusations and trials. In the United States the number of women poisoning men appears to be no higher. Just twelve women were executed in the United States from 1682 to 1962 for poisonings. The number in Britain is higher – from 1843 up until the abolition of the death penalty in 1966, thirty-seven women were executed for poisonings. When women do kill, they are more likely to use poison than guns or knives. This may of course reflect a bias in terms of documenting the trials of women.

Court records are often one of the best avenues for research on the history of crime but they have limitations. Even the most detailed records contain only the evidence as presented by a court in a very specific manner. Much of what goes on in the courtroom is never recorded. Some crimes never reach courts; until the nineteenth century it was very rare to be tried and even more rare for a record to survive. Crimes committed in the domestic realm were often covered up. Infanticide would often be hidden, the deaths of other family members disguised as sickness. Courts can provide us with a useful history of the judicial system but only a partial history of crime, much of which will also remain beyond the possibilities of history.

More than a third of murders carried out by women in the United States involved knives or other household implements, and many murders took place in the kitchen or elsewhere in the home. Cesare Lombroso and Guglielmo Ferrero's study of women and crime, *The Female Offender*, published in 1895, understood the nature of crime in biological and Darwinian terms. Those who committed crimes were less evolved than the law-abiding. Women committed less crime than men because of a natural conservatism that came from the relative immobility of the eggs compared with sperm. Women criminals were doubly condemned – they were not just criminals, but by being criminals they were not quite women and were biologically and sexually abnormal. "As a double exception, the criminal woman is consequently a monster," they wrote. Lombroso came up with a way to detect women with potentially criminal mindsets. Their physical appearance was similar to men's, they had heavy brows and strong jaws like men. They also had darker skin and more hair than other women. Lombroso had mostly studied women from southern Italy and his work did little but reinforce the heavy prejudices of the time.

The criminologist W. I. Thomas wrote later that crime came down to a difference in energy. Men were ketabolic and expended energy; women's bodies were anabolic and designed to store energy and thus they lacked some of the drive to commit crime. The pioneering criminologist Otto Pollak thought that women's crimes were different because of their essential deceitfulness. He even suggested that they chose work such as nursing, teaching and domestic service because these gave them access to realms in which they could carry out crime undetected. Pollak believed that women's crimes were underrepresented in statistics and histories because of their essentially secret nature. Pollak wrote that this was because women were able to disguise their sexual feelings in a way not available to men and were therefore lacking in sincerity. It was a short step, he believed, from faking orgasms to slipping something toxic into the soup. Women rarely do kill but when they do, it tends to be people they know – men kill strangers, women kill intimates. In Britain, 60 per cent of poisonings involved family members and the figure rises to 70 per cent if servants are included. Only 10 per cent of poisonings involved someone who could generally be regarded as a stranger – these crimes were often similar to the use of date-rape drugs today in that they mostly involved people spiking drinks with laudanum or chloral hydrate before robbing them.

A French scientist, Rene Fabre, in the introduction to his work on toxicology, said that 70 per cent of poisonings were carried out by women and that most took place in the countryside. However, there is little evidence for such assertions. We do know that overall men kill more frequently with poison and yet the idea of poison does not attach itself to ideas about masculinity or male bodies in the same way it does with women. The use of poison undermines and destroys notions of femininity, nurturing and family. The notion of women as resorting to poison due to their physical weakness does not stand much examination – it has been used just as widely by men and it does not necessarily take great strength to shoot someone. The main motivation for the use of poison is the possibility of the crime going undetected rather than an aversion of violence. Death by poison is rarely the placid, calm demise that is often imagined but can be a harrowing and prolonged process.

As providers of food and medicine, women have more opportunities to get an unsuspecting and trusting victim to consume the poison – always the killer's greatest challenge. The historical record of trials and their incorporation into the wider cultural life through journalism, fiction and movies have all created a focus on women and in mostly aristocratic and middle-class women. This enduring image of the female poisoner is not upheld by the historical record. The historian Katherine Watson, author of *Poisoned Lives*, showed that most women who used poison in England from the eighteenth to twentieth centuries were poor, often caught in some desperate situation and often striving to maintain respectability in the face of unwanted children, divided families or unfaithful spouses. When women acted together, it tended to be mothers and daughters killing an unwanted child, although there was one reported "thrill killing" in which two teenage girls who were obsessed with poison killed a female lodger apparently for no reason other than the desire to poison someone.

In seventy-five of the cases recorded by Watson, the killer murdered her husband. Infanticide and the murder of spouses account for the vast majority of poisonings by women. In eighteenth- and early-nineteenth-century England, divorce was almost unheard of and was available only to those who could afford a private act of parliament. Divorce through the courts was introduced in 1857 but was only granted on the grounds of adultery that had to be proved in court. Up until 1923, women who wanted a divorce also had to show themselves to be victims of some additional offence such as desertion, rape, sodomy or cruelty. On top of all this, divorces were only available in London and the court rarely awarded women custody of children. Very few women were in a financial position to endure the scandal and cost of divorce. Suffocating Victorian standards of respectability drove people to crime and those same standards intensified the spectacle that was made of them if they were caught.

In eighty-two cases, mothers or stepmothers killed their children. A third of all poisonings cases collected by Watson involved children under sixteen. Younger children have always been vulnerable to poisoning, but were very much so in Victorian times when babies under the age of one made up an astounding 61 per cent of victims. In the 1970s, children under one made up about 6 per cent of murder victims.

Women who use poison are categorised as unusually evil, but so also are women who concoct poisons – a condemnation not suffered by the forgers of blades or the manufacturers of guns, both of which are entirely respectable, even celebrated trades. Even the men who manufactured mass poisons have tended to be termed weapons scientists rather than poisoners. But women like Tofana, who herself was never known to have administered poison to anyone, were conflated with those who carry out the crime. It is an idea that dates back to the Lex Cornelia, the Roman law that laid down punishments not just for poisoning but for the manufacture and provision of poison.

The toxicologist John Trestrail examined 679 documented cases of poisoning and found that 46 per cent were male, 39 per cent were female and in 16 per cent of cases the sex of the poisoner was unknown. These figures have to be regarded with some caution; we do not know if cases of women poisoning are more or less frequently documented than men and Trestrail's poisoning cases are drawn mostly from the United Kingdom and the United States. Certainly, there have been suggestions that women tend to get tougher sentences for murder involving poison because of the malice aforethought involved, which meant that they could not claim to have acted in the heat of passion. Trestrail's analysis also suggests that a substantial number of women who poison might be serial killers with multiple victims.

Clearly many poisonings must have gone undetected, mostly those involving small children. Life expectancy in nineteenth-century Britain was lower than in Afghanistan today and child mortality was very high. Sudden deaths were not uncommon and many illnesses had similar symptoms to the signs of poisoning. Up until the end of

World War I in England, most prosecutions were brought either by private citizens or by senior police officers rather than a government prosecution service. Prosecutions only tended to come about if someone reported suspicious circumstances or if the coroner decided to press a case. Until the twentieth century there was almost no routine application of forensic investigations or regular investigation into suspicious deaths.

Shifts in legal views of poisonings have also occurred. In 1800, poisoning animals was a crime punishable by death. By 1900, it was barely considered a crime. In 1800 it was not a crime to administer poison with intent to do harm. Later, administering poison with intent to cause death or abortion was an extremely serious offence, punishable by death. Trials were extremely short – most were over in a scant few hours with a handful of witnesses appearing for the defence – and the laws were confusing. The Offences Against the Person Act of 1861 made it an offence to administer poison with the intention of inflicting harm. Up until then it had been almost impossible to prosecute unless a person had consumed a poison and died. The change in the law was driven by a case in which a coachman had dosed the maids in the household with Spanish fly, a powerful irritant made from blister beetles, that killed them. His aim had been to use this aphrodisiac to arouse passions that he hoped might be directed his way. He had not intended for them to die. He was brought before a magistrate who decided there was no law that covered the offence and in response to the case, the law was changed to allow for a crime of administering poison.

The idea of women poisoning men clearly creates a sense of wider anxiety beyond the real risks of it happening to any individual. The historical record from the past two centuries shows poisoning to be rare and a mostly domestic crime. Poison has become indelibly associated with women, but the American journalist H. L. Mencken put his finger on the real issue when he wrote: "One sometimes wonders that so few women, with the thing so facile and so safe, poison their husbands."

Most murderers are men and most of these murderers kill their spouse, most often by beating or stabbing them. The handful of women who murdered their spouses, particularly those who poisoned them, aroused extraordinary levels of public interest. Sentencing Anne Merritt to death for poisoning her husband, the judge commented on "the strange and horrible frequency of the crime with which you are charged". But the realities of nineteenth-century poisoners was far from the mythology as the detailed study by Katherine Watson has shown. In 1851 Dickens was writing at the end of spate of poisonings that would have been alarming, but the trend was somewhat short-lived. What lingered was the fear of the violation of domesticity, the concern that these murders had become so much easier in an age of widening scientific knowledge. There were also profound fears that the intensive media coverage of the time was fuelling the urge to poison.

In "The Decline of the English Murder", George Orwell noted that of the nine most famous murders that took place between 1850 and 1925, a period he described as "an Elizabethan age of murder", six were poisoning cases. These were the murders that captured the public attention most closely. Eight of the ten criminals involved were members of the middle class and social position was one of the main motivators in these killings. They were also profoundly domestic – of twelve victims, seven were husbands or wives of the killers. Orwell's perfect murder was a middle-class solicitor, living a life of quiet desperation in the suburbs, who is seduced into killing by some great passion. The man would be a paragon of the local community, head of the local Conservative Party, a member of a Non-Conformist Church and perhaps a strong advocate of temperance. He uses poison, always the means of choice for the better murder, but makes some tiny mistake that leads to his discovery. "In the last analysis, he should commit murder because this seems to him less disgraceful, and less damaging to his career, than being detected in adultery."

Watson's study of Victorian poisoners confirms what Orwell suspected. Poisonings were most often about maintaining respectability in the face of imminent scandal. They were frequently by poor women who would not be able to divorce a spouse or endure the scandal of a child born out of wedlock. Roger Lane, a historian of American murder, estimated that poisonings accounted for less than 1 per cent of murder cases that involved the justice system. And yet poisonings provoked more discussion in the emerging literature of medical jurisprudence than any subject except insanity. More than a hundred works on poison trials were published in the nineteenth century, most of them "true crime" tales with a desire to promote the idea that poisons were a considerable risk. Hand in hand with the growing interest in poisonings as a filler for newspapers and magazines was the emerging fascination in the role of the expert in these cases. The science of toxicology grew up in an environment of intense public scrutiny. Trial reports were the pre-runner of police procedurals and the coverage devoted to what witnesses told courts was intense.

Toxicologists such as Mathieu Orfila, generally regarded as the founder of the science, and the Scottish scientist Robert Christison gained their fame as expert witnesses in trials that received over-whelming press coverage. They played a critical role in the way doctors and experts were viewed, establishing the idea of expertise in very public trials. Modern toxicology emerged from the scientific developments of the eighteenth century, notably the new understanding of chemistry and the nature of matter. Orfila's prominence came from his testimony in the Lafarge trial in France in 1840 in which he used the Marsh test to prove that Madame Marie Lafarge had poisoned her husband with arsenic. The trial and the use of the controversial test in securing a conviction created a storm of fascination in France. Marsh tests were carried out as entertainment at fashionable dinner parties in Paris.

14. TO TRAFFIC IN SOULS

I will not give a fatal draft to anyone if I am asked,
nor will I suggest any such thing.
        The Hippocratic Oath

T he inclusion of this line in the oath sworn by doctors suggests that a few have been inclined to administer fatal draughts or at least suggest one. This has happened more at the behest of others to alleviate pain or speed up an inevitable end, but medical poisonings also occur for reasons of greed or a lust for killing. The most prolific poisoner of recent times was a British general practitioner working in a small town in the north of the country. Doctor Harold Shipman is believed to have killed at least 250 of his patients and possibly more between 1971 and 1998. He was convicted of the murder of fifteen before he killed himself in prison in 2004. His motives for murder were unclear, according to the inquiry into the deaths, which pointedly did not ascribe one. He just liked killing. Death gave him a sense of importance in an otherwise mundane life.

With most poisoning cases there are some doubts. Essentially all cases before the twentieth century that were brought to trial are surrounded by legal and forensic questions. Despite new techniques, problems still remain in determining whether someone has been deliberately poisoned unless there is a confession by the killer. Even the Shipman case is clouded with uncertainty, despite the trials and the subsequent extensive judicial inquiry. We do not know how many people he killed or why he killed them.

Shipman's case may have been extraordinary, but there were signs of problems that might have been picked up. Doctors who kill often have records of disciplinary issues. In many cases, they have repeatedly lost their jobs or been under the suspicion of colleagues. Many have abused drugs or experimented on patients and colleagues with poisons. In some cases they are even caught or suspected of practising poisonings before they move on to more deadly crimes. And yet these disciplinary issues rarely end with doctors being forbidden to practise. There is a strong inclination in the profession to ignore the danger signs. Harold Shipman was convicted in 1976 of illegally obtaining the powerful painkiller pethidine. At his trial he asked for an astounding seventy-four similar cases to be taken into consideration. The case was reported to the General Medical Council, the medical disciplinary body in Britain, but no action was taken against him. No special controls were implemented and he was allowed free access to drugs, once obtaining enough morphine in the name of a terminally ill patient to kill nearly four hundred people.

Explaining the action of doctors like Shipman is a challenge. The inquiry into the deaths could not determine a motive and Shipman never offered one. He refused any psychological examination. Money was not a driving motive. In only one case – the murder of Kathleen Grundy, which led to his discovery – did he attempt to forge a will. Few of his patients left him anything. There are themes in the personalities of these killers; a sense of superiority and an overtly aggressive attitude to those regarded as intellectual inferiors (although letters written by Shipman show him to be a poor writer). He had an exaggerated sense of his own worth and a tendency to dramatise his skills as a doctor and the cures he effected. Even though he was generally regarded as having a good and attentive bedside manner, he would turn oddly brusque after the deaths of some of his victims, treating relatives curtly. He was an adept and fluent liar. In one case, he arrived at a patient's house and killed her, unaware that a friend was in the next room. When he discovered her there, he was remarkably unflustered although he had clearly believed he was alone in the house when he carried out the murder.

The judge who investigated the killings speculated that he might have transferred his 1970s addiction to the painkiller pethidine to a similar reliance on murder. The only evidence that killings gave him some sort of thrill or psychological release was a conversation he had with a patient who was a care assistant for elderly people. He asked her if she ever discovered them dead and when she said that she had and that it was a very distressing experience, he asked her whether she had ever felt a little "buzz" from it. It is also clear that his murders were not all opportunistic. He sometimes sought out patients to kill, even doing it when he was off-duty.

"I think the only valid possible explanation for it is that he simply enjoyed viewing the process of dying and enjoyed the feeling of control over life and death, literally over life and death," said South Manchester Coroner John Pollard. A forensic psychiatrist, Richard Badcock, who talked to Shipman for ninety minutes while he was in police custody, said: "He could either be in a state of complete control, in which case he was relaxed and normal, or he was in a state of collapse… He is not doing it for excitement. He is trying to get rid of an anxiety, but an anxiety which he might not even let himself think about at the time."

The sense of power that goes with controlling life and death is a key part of the killing of patients by medical staff. By the end of his murders, Shipman was quite blasé about the deaths he had just caused and became increasingly sloppy about covering his tracks. He was unconcerned about the impact of his behaviour on the living. One man, misdiagnosed by a local hospital with cancer, was kept for months on large doses of morphine, thinking he was going to die any moment, even though Shipman knew he did not have cancer. It might have been a way of getting his hands on more morphine without arousing suspicion. It turns out the man's father might also have been killed with the morphine that Shipman prescribed for him.

"God complex" killings affect healthcare workers the world over, cropping up with a somewhat alarming regularity. In 2005, a German nurse, Niels Högel, was reported by a colleague for giving a patient an unauthorised injection. The patient died the next day along with another; hospital authorities finally investigated and found Högel responsible. He was charged with attempted murder and jailed for seven and a half years. Authorities started looking at his work record and he admitted he had injected some ninety patients, resuscitating sixty of them while thirty died. He would induce heart failure with drugs that lowered blood pressure or cause arrhythmia in order to show off his skills in resuscitation.

At a subsequent trial in 2015, Högel was convicted of killing another two patients and jailed for life. Further investigations involving the exhumation of ninety-nine bodies found he had killed thirty-five more at a clinic in Oldenburg before leaving with glowing references even though other staff had suspicions about his behaviour. An additional forty-eight were believed to have died in the northern German town of Delmenhorst. In 2019, he was convicted of the murder of eighty-five people all because bringing people back to life made him feel important.

UNDERLYING THE WIDESPREAD anxieties of being poisoned by unlikely sources is the lingering and more possibly realistic concern of being poisoned by those who more than anyone else do have access, means and sometimes motive – doctors and other medical workers. Plato took a hard line on the issue, saying that any doctor who administers a drug that harmed someone must be executed. For him, the doctor must be fully responsible for any treatment – if he gives a poison to someone then, as he knows about drugs, he must have administered it deliberately and therefore is guilty of murder. There was no room for accidents or mistakes in Plato's world. Pliny and others would question the abilities and truthfulness of doctors, saying that we are all vulnerable when sick to being "lulled by the sweet promise" of being cured. "Their experiments lead to deaths; and yet for doctors, and only doctors there is no penalty for killing a man. In fact they pass on the blame, reproaching the deceased for his lack of moderation and self-indulgence."

Beyond the risks of incompetence and quackery was the anxiety of malevolence. Libanius, a Greek writer famous for his rhetorical skills, wrote a speech purporting to be that of a prosecutor addressing the jury in the trial of a doctor accused of poisoning. It is not believed to have come from a real case but it is thought to reflect common views of the time. Poisonings were clearly not an uncommon occurrence at the time – physicians were frequently accused of poisonings and the issue was also a common theme of these sorts of rhetorical exercise. The tension of the healer turned poisoner had a dramatic resonance that it retains to this day. "Behold gentlemen, a man abusing the noblest office, who calling himself a doctor, has actually murdered more people with drugs than those who practice poisoning for a living." Libanius is unstinting in his praise of the medical profession – after being struck by lightning when he was young, he was plagued by illnesses and depression for most of his life and had considerable experience of doctors. But he is blunt in his condemnation of this doctor "the most maleficent, the most defiled, the most hated by men and Gods – who mixed the most dreadful with the most noble, the destructive with the healing, the deleterious with the beneficial, the lethal with the protective".

Libanius points out that in other cases of poisoning, doctors were often able to identify signs that a toxin had been administered but when the doctor himself was the poisoner, he was quick to point out the weak health of the victim. He describes the doctor as passing through "more harshly than the plague", saying that many had lost their teenage children to his care. "Slain by whom? Not by enemies nor by the cold steel of robbers, nor by slaves rebelling at night nor by the sorcery of an old woman secretly stirring poison into the wine." Libanius suggests the motive was money, a desire by the doctor to "surpass, lazily and quickly, the revenues of tyrants… You discovered therefore the least honourable route and brought about a kind of traffic in souls". Why do doctors poison people? It seems, rather than for purely financial gain, there is a deep thrill, a deep sense of power gained, in the traffic of souls.

AT THE START of World War I, Agatha Christie trained as a ward nurse and dispenser. During these years the British writer developed an enduring interest in poisons and a suspicion of doctors. During World War II, she returned to voluntary service as a dispenser at University College London. As an author, she liked to write what she knew. Guns were a mystery to her and raised problems of storytelling as one had to get the gun into the hands of the killer and then dispose of it or have it discovered. Poisons were much closer to her heart.

There are eighty-three poisonings in Christie's novels – in more than half of her sixty-six novels, at least one of the corpses has been poisoned or been a victim of an overdose. She also uses a wider range of toxins than any other author. Sir Arthur Conan Doyle killed off only around ten people with poison and involved no real science in the detection of any chemicals. By contrast, Christie was the crime writer of a much more toxic and cynical age, in which the biggest dangers were no longer the spurned lover or the jealous spouse but a doctor or nurse. Her characters would not dream of killing with arsenic but instead used cyanide, morphine, digitalis, strychnine, nicotine or barbiturates. She even employed some remarkably exotic poisons – ricin, the poison derived from castor beans; gelsemium, the mixture of toxins from yellow jasmine; and taxine, the chemical from yew trees that is now used today in chemotherapy. Only twice does she violate the unspoken rule of the mystery writer and invent a poison, in both cases prescription drugs that are actually similar to existing medicines. By the 1960s, Christie's novels are a veritable *Valley of the Dolls* of murder, such is the tumble of toxic pills and wayward doctors scrawling prescriptions with abandon.

More than a dozen of Christie's killers were doctors, nurses or those involved in medicine in some way. Her characters often join in her pursed-lipped disapproval of the state of medical care, complaining about the way young doctors hand out medicines so freely or are indifferent to the feelings of their patients. Her own experiences as a nurse in Torquay had shaped her view of doctors as egoists who demanded the obeisance of underlings. For her, they commanded too great a status in society. "Nothing had prepared me for the need to fall down and worship physicians," she wrote in her autobiography. "Even those doctors who were, by secret nursing opinion, despised as below standard, in the ward now came into their own and were accorded a veneration more appropriate to higher beings." They were thus the perfect murderers for her novels, as their position sheltered them from suspicion but they had access to drugs, victims and a great height from which to fall. As Orwell wrote in "The Decline of the English Murder", the enjoyment of crime stories was greatly enhanced when an otherwise upright citizen commits a terrible act. It is the distance of the fall that intrigues us.

Christie's own experiences of poor health and her father's long and slow death from an illness that was never satisfactorily diagnosed amplified her suspicions. Her work as a pharmacist undermined any remaining confidence. She was surprised to find the degree of vagueness in prescriptions, that doctors simply followed fashions for certain drugs or prescribed unnecessary drugs. She was also a frequent critic of the jargon doctors employed "to baffle laymen", as she put it in *A Caribbean Mystery*. At the time, British pharmacology textbooks still urged doctors to prescribe in Latin as it was better to keep the patient in ignorance of the drugs they were receiving.

In eight of her novels, physicians are murderers, mostly using poison to dispatch their victims; although in *Four and Twenty Blackbirds*, a doctor uses the less sophisticated method of pushing his uncle down the stairs to gain an inheritance. In *The Adventure of the Sinister Stranger* published in 1929, the physician claims that someone is trying to steal his research into what he describes, very conveniently for anyone planning a murder, as "certain obscure alkaloids (that) are deadly and virulent poisons and are... almost certainly untraceable".

When they are not killing people, doctors were often the subject of ridicule in Christie's works. Their access to drugs is often mentioned as giving them the perfect means of murder. In *Murder in Three Acts* she writes of "something rather amusing about a doctor being poisoned... wrong way around. A doctor's a chap who poisons other people." Pharmacists are more sympathetically treated by Christie, perhaps because of her own experience. She passed the Apothecary Hall Examination in chemistry, *materia medica* and compounding, which qualified her to dispense medication for a doctor or pharmacist. In her autobiography and in other writings she often mentions the possibility of mistakes by pharmacists – whereas she regarded doctors as ill-intentioned and the subject of anxious humour, she found the risks of mistakes in prescriptions to be a very serious subject, perhaps because it had been a constant worry for her while working as a dispenser. The pharmacist in *Poirot Loses a Client* observes that they have to be especially careful making up drugs because if doctors make mistakes in their prescriptions, the druggist is often blamed.

While Christie creates several doctor-murderers, she only wrote of one pharmacist who used his skills to kill. Zachariah Osborne in *The Pale Horse* of 1961 is a cherubic but sinister pharmacist who carries around a dark lump of curare in his pocket, as it makes him feel powerful. In the novel, he is obsessed with the idea of being the star witness in some major trial, as a pharmacist friend of his father's had been in a famed poisoning trial. In *The Adventure of the Egyptian Tomb*, Hercule Poirot discovers that the victim has been poisoned with strychnine rather than died of tetanus caught from a cut received while opening a tomb. The poisoner in this case – a doctor – kills himself with cyanide after being uncovered by the wily detective. In *The Murder of Roger Ackroyd*, the doctor kills both his victim and himself with barbital.

Christie was certainly inventive in her use of poisons, employing several long before they became notorious. The doctor in *Cards on the Table* kills a victim by putting anthrax spores on his shaving brush. Another victim is injected with an unspecified poison while being told it is an inoculation against typhoid. Bioweapons are used in some of her works – in *The Labours of Hercules*, a cult leader injects his followers with microbes to kill them, while in *Easy to Kill*, a woman induces a fatal case of septicaemia in her doctor by scratching him with some scissors and then dressing the wound with a bandage infected with some pus from a cat's ear. In *Sad Cypress* from 1940, a nurse drinks of a pot of tea with a patient that contains oral morphine. She immediately injects herself with an emetic to expel the contents of her stomach while she leaves the patient to die, thus creating an alibi for the poisoning. Zachariah Osborne in *The Pale Horse* uses one of the most unusual poisons – thallium. Christie's detailed description of the symptoms of thallium poisoning, notably the acute hair loss that goes with it, were said to have saved the life of a child who was brought into hospital with convulsions but could not be diagnosed. A nurse who had recently read *The Pale Horse* identified the symptoms and it turned out that the child had consumed a thallium-based cockroach poison.

When one considers these works – the prevalence of doctor poisoners, the scepticism of many about their health care and the expenses involved, it is clear that the doctor as villain has a resonance with the reading public. There is always an ambivalence about doctors that is mostly commonly expressed in literature and satire and perhaps nowadays more often in lawsuits. On one hand, when we are sick we are grateful and willing to submit to their knowledge and power; however, once cured, a certain resentment sets in that is clearly a reflection of an enduring ambivalence in the relationship. Doctors have been engaged in a constant struggle to maintain a positive image. Poison trials may have established them as repositories of expertise and professionalism, but fiction tends to present them as possible murderers, driven by greed, vanity and often the desire to play god.

The literary doctor normally falls into one of two types, the bumbling idiot spouting incomprehensible jargon or the dark and malevolent murderer. Shakespeare was the rare writer from his time who portrayed most doctors as professional and efficient. This may have had something to do with his son-in-law being a doctor. He was less kind to lawyers ("The first thing we do, let's kill all the lawyers", *Henry VI Part 2*) or schoolmasters, but doctors and apothecaries were rarely sinister or ridiculous figures in his plays. Personal experiences had shaped the views of some other authors. The French playwright Moliere never forgave the doctors who dosed his son with antimony and probably killed him. The result is several scathing portrayals of doctors as corrupt and murderous.

Renaissance literature had as many stock characters as cop shows or sitcoms today. The pompous lawyer, the self-important idiot, the conniving Jew, the ageing harpy, the shrewish wife, the cheerfully oblivious fat man. In Thomas Middleton's *A Fair Quarrel*, first performed in 1613, the surgeon is the comic idiot, coming out with streams of jargon that would have audiences in the aisles. "Now I must tell you his principal dolour lies i' th' region of the liver; and there's both inflammation and tumefaction feared; marry, I make him a quadrangular plumation, wher I used sanguis draconis, by my faith, with powders incarnative, which I tempered with oil of hypericon and other liquors mundificative."

In Thomas Nabbe's *The Bride*, from 1638, the surgeon, Mr Plaster, speaks in an incomprehensible mix of English and Latin in order to impress the patient's cousin and to undercut a common herbalist who has been treating him up to then. It is a problem mentioned by authors from this time. Thomas Dekker urges readers of his *The Gulls Hornbook* to "send them [doctors] packing, therefore to walk like Italian mountebanks, beat not your brains to understand their parcel-Greek, parcel-Latin gibberish".

Occasionally the comic and the sinister were merged. Doctor Julio in Webster's *The White Devil* is modelled on Roderigo Lopez, the attempted assassin of Elizabeth I. After his sensational trial, there was a rash of characters that bear some resemblance to him, being either Spanish, Portuguese or Jewish doctors. "He will shoot pills into a man's guts, shall make them have more ventages than a cornet or a lamprey; he will poison a kiss, and was once minded for his masterpiece, because Ireland breeds no poison, to have prepared a deadly vapour in a Spaniard's fart that should have poisoned all Dublin." As the author Dominic Green notes in his work on the Lopez plot, it insinuates itself into much of the drama of the times – Middleton's *A Game of Chess*, Dekker's *The Whore of Babylon*, and Thomas Nashe's *Lenten Stuff* all have references, and Shylock in *The Merchant of Venice* is widely believed to have drawn on the events surrounding Lopez.

The sixteenth and seventeenth centuries provided the greatest array of literary doctors and poisoners. Debates between Galenists and the emerging power of Paracelsians, new diseases and new cures from the New World, the emerging understanding of the body and anatomy all add up to a world in which there was an extraordinary literary interest in doctors. It was the harshness of medical treatments that worried many patients and encouraged the idea of doctor as poisoner. The use of mercury in the treatment of syphilis – leading to the adage "A night with Venus, a lifetime with Mercury" – left people sicker than ever and did little to cure the multiple horrors of what Shakespeare called "the infinite malady". One of the most common complaints against doctors was the dangers their cures represented. In *Timon of Athens*, Shakespeare inserts one of his rare criticisms of doctors when Timon warns some thieves: "Trust not the physician / His antidotes are poison, and he slays, more than you rob." Dekker also warns against falling into the hands of doctors "and so consequently into the Lord's". Ben Jonson's *Volpone* – one of the great works of fictional toxicology – is replete with cautions against doctors, including "Most of your doctors are the greatest danger / And worse disease to keep out."

Attitudes towards doctors were far less reverential than they are today, in most part because they were distinctly less successful. The omniscient, kindly doctor, worthy of respect, is something of a Victorian conceit and was certainly not seen in earlier centuries. They are poisoners, grasping cheats who con their patients into buying outrageously expensive but useless medicines, or ill-informed and pompous idiots. Each trend in medicine, be it sword-salves (potions that were put on weapons in the belief that this would cure any wound it had caused) to the urolscopy, the analysis of urine by doctors scornfully known as "piss prophets", were all treated with contempt by various authors. Scientific debates were far more common in literature than they are today and were far less removed into the realm of specialisation and expertise. The medical knowledge in Shakespeare alone is astounding. Not only was he familiar with a wide array of drugs and medicines, but he was up to speed on the latest developments in anatomy such as the discovery of the Eustachian tubes that link the ear to the throat.

The establishment of the Royal College of Physicians in 1518 was an attempt to allay fears of doctors and exclude the unlicensed, but it did little to improve the situation; concerns about doctors intensified over the century. The herbalist Nicholas Culpeper was so contemptuous of the college that when he published his translation of the dispensatory, he addressed its fellows as "so Proud, so Surly and so Covetous". The college's restrictions were, for many, part of the problem, while others complained of the wide range of people calling themselves doctors and trying to heal the sick. Nearly a century after its founding, a critic wrote: "All sorts of vile people and unskilled persons without restraint make gainfull trafique by botching in physicke and hereby number of unwitting innocents daily enthrall and betray themselves, their lives and safety to sustain the riot, lusts and lawlesse living of their enemies and common homicides." Unwitting innocents certainly had a hard time distinguishing between good and bad doctors, a problem since the beginning of medical practice.

THE HEALTHCARE MURDER is quite a rare crime. There have been, according to a count by the historian John Trestrail, just 118 cases since 1832 that have reached the courts. But 76 per cent of all known healthcare murderers have been detected since 1980, and 59 per cent since 1990. It is unclear if there has been a marked increase in killings but there clearly has been an increase in the detection of doctors and nurses that kill. It might be that the professional secrecy and lack of scrutiny that protected some medical killers is waning. Trestrail's study is based on court documents from Britain and the United States and it almost certainly underestimates the number of medical killers, who mostly go undetected. Healthcare killers tend to fall into three categories – those who kill for money, normally to gain from an inheritance; those who put their knowledge of medicine to use for those with murderous intent, particularly in cults; and those who kill numerous patients over a long period. It is the medical serial killers that have presented the greatest challenge to the medical profession and possibly represent the most common use of poison by individuals for murder.

Of the 540 case of poisoning crimes examined by Katherine Watson in England from 1750 to 1914, only thirty-four were "medical attendants", a category that includes doctors, druggists, herbalists and healers. Six were nurses and seven abortionists, which in the days before this was legalised in the 1960s normally meant anyone who carried out abortions without any medical training. But prosecutions of doctors were rare and they generally remained above suspicion.

Murder by doctors and nurses is far from being a modern phenomenon, but we may have become more aware of it because of shifts in the way people are treated and how they die. In the twentieth century, it became more normal to die in hospital rather than at home. The final act of poisoning someone who may be fatally ill shifted from the domestic sphere to a more public realm. Detection also became more likely due to registration of deaths, postmortems and more demanding families who wanted to know exactly how relatives died. While women make up only 14 per cent of non-healthcare poisoners, they represent more than 40 per cent of those involved in hospital poisonings. New studies of medical care killers show traits they seem to exhibit – twenty-two have been identified in all. Secretive personalities; a preference to work alone, often at night; an undue belief in their own skills; a dislike of intervention by other professionals; a history of mental illness; drug problems; and disciplinary issues were among the red flags. Unusually, studies have shown that there is an almost equal divide between healthcare murderers, although many are nurses – a profession dominated by women.

There has been little external scrutiny of doctors and few controls to limit their access to drugs or to monitor how those in their treatment die. Harold Shipman could murder his patients because he worked alone and treated mostly elderly people whose deaths evoked little suspicion. In most cases when doctors are suspected of killing patients deliberately, they are moved to another unsuspecting hospital but no action is taken. Employers have few means of checking the past of a doctor if a previous employer prefers to cover up a crime rather than face unwanted publicity and the potential for lawsuits. In her conclusions to the Shipman inquiry, Dame Janet Smith wrote:

> The system appears to have operated on the assumption that all doctors were essentially decent and strove to do their best for patients; a few would commit some form of misconduct and some might fail to provide an adequate standard of care. Those would be reported to an appropriate authority and would be dealt with. Shipman's was not the only case that demonstrated that these assumptions could not be made. Nor was his the first.

PART THREE

15. A COLD-BLOODED CALCULATION

But someone was still yelling out and stumbling,
And flound'ring like a man in fire or lime
Dim, through the misty panes and thick green light,
As under a green sea, I saw him drowning.
Wilfred Owen, "Dulce et Decorum Est"

The first deadly chemical weapon attack of modern times took place on 22 April 1915 near Ypres in Belgium. As the sun was going down, a low cloud of pale blue haze rose from the German trenches and drifted across the pockmarked landscape, turning into a thick green cloud as it filled craters pounded into the earth by the incessant shelling. The gas billowed silently into the French trenches across an area four miles wide, striking the 87th Territorial Division and the 45th Algerian. From the Canadian and British positions nearby, soldiers watched in horror as French soldiers staggered out of their trenches, clawing at their throats and eyes, gasping for breath, their skin turning the colour of livid bruises. The victims were drowning from within as their lungs filled with fluid. Those with horses rode furiously away from the cloud, others shed their rifles and uniforms as they ran. The Germans moved cautiously across the gassed landscape, avoiding low-lying pockets where the chlorine had accumulated. Within a few hours they had captured two villages, taken more than two thousand prisoners and seized fifty artillery pieces.

It was said that more than five thousand men died that day and ten thousand more were horribly wounded, their skin coloured "a shiny grey black" as they rocked back and forth gasping for breath. There was no actual body count in the chaos and the number of deaths were probably exaggerated, but it was certainly a horrifying attack. A day later, another attack was said to have claimed the lives of five thousand Canadians, although this is also probably an overstatement. A line had been crossed; these were the first mass poisonings of the modern age and they occurred despite efforts to outlaw these weapons before they could be used.

In 1868, a group of twenty nations agreed in St Petersburg that the use of certain types of explosive bullets would be against the laws of humanity. The weapons would cause unnecessary suffering and were therefore beyond the pale. It was a key statement in the development of the modern laws of war – laws that have too often been broken but nevertheless remain an important constraint. In 1899, the first Hague Convention laid down the first restrictions on chemical weapons, banning the use of projectiles that could only be used for dispersing toxic gases. For the first time, mankind had banned a weapon before it had been used. In 1907 the Second Hague Convention included the clause: "It is especially forbidden to employ poison or poisoned weapons." Apart from Serbia and Turkey, all the combatants in World War I had signed and ratified the Convention. Many of them used the weapons.

William of Malmesbury, author of the *Gesta regum anglorum* ("Deeds of the English Kings"), wrote in the twelfth century that when someone "uses poisoned arrows, venom and not valor, inflicts death on the man he strikes. Whatever he effects, then, I attribute to fortune, not courage, because he wars by flight and by poison." The philosopher John of Salisbury also denounced the use of poison in the twelfth century. In 1589, Alberico Gentili, the Italian who was a pioneer of international law, called for a ban on such weapons. The Dutch jurist Hugo Grotius, in his seventeenth century *De Jure Belli ac Pacis* ("On the Law of War and Peace"), said of the poison taboo:

> Agreement upon this matter arose from a consideration of the common advantage, in order that the dangers

of war, which had begun to be frequent, might not be too widely extended. And it is easy to believe that this agreement originated with kings whose lives are better defended by arms than those of other men but are less safe from poison, unless they are protected by some respect for law and by fear of disgrace.

Grotius has no doubts about the ancient nature of the poison taboo and that it had been widespread. "From old times the law of nations – if not all nations, certainly those of the better sort – has been that it is not permissible to kill an enemy by poison."

There has been a taboo against the use of chemical weapons that goes back millennia, but as with all such taboos, it has been broken all too often. There is little agreement on the origins of such a taboo and why it has endured. Poisoning has always been a special case, singled out in the rules of war going back to the Hindu *Laws of Manu*. There is far less anxiety about swords, or even small arms, which, of course, kill significantly more people each year than chemical weapons. Bombs arouse far less sense of moral outrage despite their terrible, often indiscriminate toll.

Humans are instinctively afraid of poisons. There is some evidence of a deep fear of snakes among primates and almost all people in all cultures have some anxiety about them. People display the same "mixtures of apprehension and morbid fascination" with snakes that primates show in the wild, according to the biologist Edward Wilson. Studies show that there is a degree of innate fear: monkeys bred in a laboratory were shown videos of other monkeys that gave the appearance of exhibiting equal fear towards both toy snakes and toy rabbits. Later, the lab monkeys showed fear about the snakes but not the rabbits. They did not learn to fear rabbits from watching the videos.

Translating primate studies into models of human behaviour is always a stretch; the idea that the taboo is connected to an innate sense of repugnance is only suggested, not proven, by this evidence. All human activities have an overlay of culture, and anxieties about poisons may be as much about nurture as nature. Poisons were bound up with healing and mysticism; access to the gods or their messengers often came through hallucinogenic drugs that were also known to be poisons. For example, in the Amazon, arrow poisons used in hunting are also used by shamans. The blurring of the worlds of potions, poisons and spirituality might have made people wary of the profligate use of toxic weapons against enemies. As misfortune was often seen as the result of upsetting spirits or gods, there were serious risks in dabbling in this mysterious realm.

After several hundred years in which the use of poisons in warfare had been very limited – by technology mostly, as there were simply not many good poisons around, but also by social opprobrium – those limits were breaking down in the middle of the nineteenth century. Lord Dundonald, a British admiral, requested permission to use sulphur dioxide against the defences of Sebastopol during the Crimean War between Britain and Russia. He got the idea from witnessing the deadly effects of gas released from sulphur kilns in Sicily. The plan was to light a fire of coal, wood and sulphur when the wind was blowing in the right direction. The prime minister, Lord Palmerston, initially went along with the plan, which died while wending its way through interminable government committees – each of which rejected it as too horrible for use in warfare. Commanders in the American Civil War asked to use chlorine gas shells but the proposal was ignored. A political philosopher named Francis Lieber drafted a code of conduct for the Union secretary of war. In it he was quite unequivocal about poisons: "The use of poison in any manner, be it to poison wells, or food, or arms is wholly excluded from modern warfare. He that uses it puts himself out of the pale of the law and usages of war."

Most new weapons prompt initial outrage that generally fades into acceptance. The Second Lateran Council of 1139 banned crossbows, a technology that clearly worried elites because it could, and did, lead to large and effective armies outside of feudal control. The ban was limited to wars between Christians. Outrage over crossbows soon waned. This has not been the case with chemical weapons, which have never come to be fully accepted. Their use initially showed a pattern of terrible escalation. Germany dismissed British and French outrage over the violation of the Hague Conventions and responded with the sort of hairsplitting that would become a feature of all such agreements; the convention banned artillery shells containing gas but it had not specified a ban on gas being released from cylinders. The Germans first attempted to use chemical weapons at the Battle of Bolimov in Russia in February 1915, but the winter temperatures caused the gas to become ineffectual. The attack at Ypres two months later was much more deadly. Once provoked, the British and French took revenge, gassing enemy lines in September 1915. After using chlorine for about two years, the Germans deployed mustard gas in July 1917, massively worsening casualties. Mustard gas was invisible and could contaminate and penetrate clothing, making it much more difficult to avoid. A gas mask was no longer enough.

Around eight million people were killed and twenty-one million were injured in World War I. Gas deaths made up less than 1 per cent of the total. Chemical weapons killed around 100,000 people in the twentieth century, a lower death toll than during Tajikistan's largely unnoticed civil war in the 1990s, but about the same number as casualties from a single night of bombing with high explosives and incendiaries in Tokyo in March 1945. It was just a tiny share of the tens of millions who died in conflicts, pogroms, acts of genocide and famines created by government policies in the twentieth century, and yet those deaths have aroused more fear than almost any other. Gas was not decisive in the Great War. It worsened the stalemate and made fighting and logistical support more difficult for both sides. It rarely created a decisive victory, in part because of the cautious tactics of the German army, which never took advantage of their first attack to press through and capture Ypres.

Two arguments have been made as to why there has been such outrage around chemical weapons. The first is that they are not useful on the battlefield and therefore the military establishment did not work to legitimise them in the way it had with other weapons. The other is that the weapons are so intrinsically cruel, coming upon the victim without warning and leaving them choking to death in a horrible way, that some sense of human decency won out. Neither reason is very satisfactory. Most weapons fail in their early forms. Muskets and cannons took many centuries to evolve into effective weapons. Likewise, the cruelty argument does not stand up well. Bullets arrive without warning and cause immense pain and trauma. High explosives can cause terrible, indiscriminate injuries and yet there are no real limits on their use in war. Chemical and biological weapons arouse our outrage, but it is the cheap and commonly available AK-47 rifle that takes the greatest toll on lives around the world – increasingly of non-combatants.

It might have appeared that The Hague Conventions had done nothing to limit the use of chemical arms, but they had created the idea that chemical weapons were abhorrent, that if they were to be used, their use had to be restrained and that Germany would ultimately be punished for being first to violate the ban. The conventions have often been dismissed as ineffectual, which they clearly were during World War I, but the establishment of societal norms and taboos requires the reinforcement of laws after there have been violations. The use of chemical weapons in that war strengthened the chemical weapons taboo in Europe and the United States. The Treaty of Versailles prohibited the use, manufacture or import of chemical weapons. In 1922, at the Washington Armament Conference, the US government insisted on a ban on first use of chemical weapons despite the warnings of experts that the best that could be expected was restrictions on their use against civilians. An article was drafted forbidding the use of chemical weapons but it never came into force because the French refused to sign the document over a dispute about submarine warfare.

In 1925, the Geneva Protocol laid down restrictions on using, although not on making or storing, chemical weapons. The Protocol is still in force, enhanced by further restrictions over the years, but more than thirty nations attached reservations saying they would use chemical weapons if attacked with them first. One of the justifications made at Geneva was the tradition, albeit one that dated back less than thirty years, of agreements banning chemical weapons. Treaties have been an important part of the process of putting weapons beyond the pale. At each step, political leaders worked to cultivate a public moral revulsion at the idea of the use of gas, much as later in the twentieth century various public figures cultivated the same disgust about landmines, weapons that met with few moral qualms until the 1990s. At least this was the case in most countries. The US Senate did not ratify the Geneva Protocol until 1974, nearly fifty years after it had been signed and only after the government had already decided to give up its chemical weapons stocks. Chemical weapons were developed just at a time when people were starting to question the idea that warfare was the natural and inevitable state of nations. They were also developed at the start of a century that would see terrible obliterating wars in which, amazingly, chemical weapons were rarely used. The history of chemical weapons in the twentieth century is mostly a history of restraint rather than unfettered use.

AT THE START of World War II, there was a widespread expectation that gas would be used. Some forty-four million gas masks were issued across Britain and schoolchildren were trained in their use. Not only were memories strong from the trenches of World War I, but Britain had used chemical weapons near Archangel in Russia during an ill-fated attempt to intervene in the Russian revolution and then later against tribal insurgents in Iraq. Winston Churchill had been quite an enthusiast. As Secretary of State in the War Office in May 1919, he requested gas for use in the Middle East. "I am strongly in favour of using poisoned gas against uncivilised tribes. The moral effect should be so good that the loss of life would be reduced to a minimum. It is not necessary to use only the most deadly gases; gases can be used which cause great inconvenience and cause a lively terror."

Fears had also been stoked by Italy's use of chemical weapons in Abyssinia, now Ethiopia, in 1935. Chlorine gas had been used against defenceless civilians as part of Mussolini's efforts to build an empire, and the response from the international community had been muted and ineffectual, a pattern that would be repeated for many decades. Chemical weapons use was by no means rare: Spain used them in

Morocco between 1921 and 1927; Turkey in 1937; Japan in China from 1937 to 1945, Portugal in its colonies in the 1970s and Egypt in Yemen in the 1960s.

Germany was banned from making chemical weapons but it still managed to take important steps in developing even more deadly gases. Many chemicals were used in industrial production or had legitimate uses as pesticides, so control regimes faced the problem of "dual use" products that could be used as weapons and in legitimate processes. By the mid-1930s, the German government under the Nazis was repudiating what they viewed as an unfair treaty and there were no hindrances on the development of new weapons. In 1936, Doctor Gerhard Schrader, a chemist with I. G. Farben who was searching for new pesticides, discovered a nerve agent that was given the codename "tabun". This chemical, colourless and odourless, was fatal if inhaled and could poison through the skin. It was a cholinesterase inhibitor, a nerve agent that prevented the body from breaking down acetylcholine, the chemical that causes messages to be sent to muscles to contract.

The age of nerve agents had begun. Two years later, Schrader discovered sarin, a similar chemical that was ten times as toxic. In 1944 German scientists developed soman, an even more deadly chemical. The Germans began production in 1936 and did not stop until the end of the war. By 1942, they could produce hundreds of tonnes of nerve agents a month. But the Nazis, so bestial and unrestrained in their violence, never used them in war. The Allies did not even know about these chemicals until Germany was defeated.

Gas attacks in World War I might have deepened the taboo against their use. Political leaders who were young men, sometimes soldiers, in World War I, held back from their use in World War II. Hitler had been a victim of gas just before the end of the war in October 1918. In an episode he recounts in *Mein Kampf,* he was caught in an attack just south of Ypres. The chemical seared his eyes; after the battle, he was taken to a hospital in the eastern part of Germany to recover. The impace this episode had on him is unclear, but during the war, he repeatedly refused to use the chemical weapons that the Nazis had eagerly developed. After the Battle of Stalingrad in which the German Sixth Army was surrounded and decimated, many in Hitler's inner circle, including the propagandist Joseph Goebbels and his secretary Martin Bormann, wanted to use the new chemical weapons against the Soviet Red Army. In February 1943, Hitler ordered preparations for a chemical attack to begin, setting 20 April – his birthday – as the target date. When the date came, he would not authorise an attack.

Hitler gathered a group of officers and scientists at the Wolf's Lair, his Prussian headquarters, on 15 May 1943 to discuss using nerve agents. By this stage, Germany had stockpiled more than 45,000 tonnes of chemical agents and could produce 350 tonnes of tabun a month at the plant at Dyhernfurth. Germany had been constrained by a lack of resources, particularly chemicals derived from petroleum. The sense of those attending the meeting was that whatever Germany had, the Allies probably had more of it. Asked if the Allies had nerve agents, the experts equivocated and Hitler left the meeting without deciding to use the stockpiles. It might have been a fear of massive retaliation that held Hitler back. In fact, the Allies had nothing equivalent in potency to the new German nerve gases.

Albert Speer, Hitler's architect and planning chief, maintained in his testimony at the Nuremberg trials that Goebbels had urged Hitler to use gas in retaliation for the devastating firebombing of Dresden in February 1945. On 19 and 20 February, Hitler held two days of meetings with his top generals on how to respond. Admiral Karl Dönitz, the commander of the Germany navy, warned against the use of chemical weapons, saying they would not delay the end of the war.

The order for their use was never given. The German weapons programme was plagued by anxieties about production and the possibility of Allied retaliation. It was never able to meet its ambitious targets as the war effort faltered and resources had to be diverted elsewhere. Hitler was told that the Allies probably had twice the German stockpiles. By the end of the war the United States alone had around 146,000 tonnes of chemical arms against a high estimate of 78,000 tonnes for Germany, and so the German estimates were not far off. Goebbels had warned before the end of the war that the Nazis would deploy "a wonder weapon" in their last stand, a phrase that caused alarm among the Allies. It turned out to be the V-1 rocket, a weapon of fear but not an especially deadly one; indeed, more enslaved people died making them than were ever killed by them. Although obsessed with maintaining control over his chemical weapons up to the last moments of the conflict, Hitler never used them.

Before the declaration of war on 2 September 1939, France and Britain said they would abide by the Geneva Protocol. Germany followed with a similar pledge. But even so, with memories of Ypres and the Somme still fresh, most of Europe lived in terror of German gas attacks. British fears of what war would mean had been expressed a year earlier by Prime Minister Neville Chamberlain. "How horrible, fantastic, incredible it is," he said on the BBC, "that we should be digging trenches and trying on gas masks because of a quarrel in a faraway country between people of whom we know nothing."

Major General J. F. C. Fuller pictured London drenched in mustard gas, a common fear as the country moved towards war. "London for several days will be one vast raving Bedlam, the hospitals will be stormed, traffic will cease, the homeless will shriek for help, the city will be pandemonium... the government will be swept away in an avalanche of terror." When the attack did not occur, London breathed a collective sigh of relief, although right up to the end of the war, the fear endured that Hitler might carry out some sort of "mad dog" final attack on the city.

Over and again during World War II, the use of chemical weapons was discussed by the Allies. Churchill offered to send mustard gas to help Stalin on the eastern front. In the Pacific, the use of gas would have aided the US forces in their island-hopping campaign against Japanese forces. Britain considered using gas in retaliation for German V-1 rocket attacks on London in 1944. Military leaders advised against it, although Churchill was clearly in favour:

> I want a cold-blooded calculation made as to how it would pay us to use poison gas... principally mustard...
> I should be prepared to do anything that would hit the

enemy in a murderous place. We could drench the cities of the Ruhr and many other cities in Germany in such a way that most of the population would be requiring constant medical attention... I want the matter studied in cold blood by sensible people and not by that particular set of psalm-singing, uniformed defeatists which one runs across now here, now there.

Churchill did not know of the new German nerve agents and yet the restraint still held. The main anxiety in World War II was the fear of retaliation. Both sides knew that to launch a chemical attack was to suffer one in return. On 8 June 1943, President Franklin D. Roosevelt made clear the US policy. "Use of such weapons has been outlawed by the general opinion of civilised mankind. This country has not used them and I hope that we will never be compelled to use them." But he warned US enemies that any country that used gas would face "the fullest possible retaliation upon munitions centres, seaports and other military cities through the whole extent of the territory of such Axis country". Britain, feeling more vulnerable to gas attacks than the US, was concerned that a German attack would precipitate an immediate response from Washington and a further round of attacks. By 1943, the two commands had agreed to consult before launching any retaliation.

The other arguments against chemical warfare were not nearly so powerful. Moral constraints were hardly applicable in what became a "total war" in which civilians were seen as legitimate targets. Legal limitations mattered little when both sides did not hold to the Geneva Conventions. The military concerns that weapons of mass destruction somehow violated conventions of professionalism and honour among soldiers fell away with the preparation and use of nuclear arms. Political restraints on countries fighting as allies or the moral repugnance leaders felt did not matter. And yet in Europe, chemicals were not used deliberately, although at least eighty-three people died and many hundreds were injured in Bari in Italy when a German bomb hit an American ship carrying a hundred tonnes of mustard gas bombs.

President Roosevelt had decided that the US would not be the first to use the weapons, a policy that upset many in the military. By the end of 1943, Japan no longer maintained air superiority and therefore gas could be used without significant risk of retaliation. Japan had never ratified the Geneva Protocol and had used mustard gas during its invasion of China. By 1939, the Chinese complained to the League of Nations that Japan had carried out 886 gas attacks. The League, true to form, offered no help. As the balance tipped against Japan, Major General William Porter, the commander of the Chemical Weapons Service, called for the use of gas to shorten the war against Japan. His idea was quickly rejected as it would give the Germans reason to use gas against troops involved in amphibious landings in Europe.

Although they were winning the war in the Pacific, the US command was anxious at the thought of fighting island by island, having to wage ferocious battles against suicidal Imperial soldiers dug in across the region. The chief of staff, General George C. Marshall, spoke of the possibility of using gas in a limited way on outlying islands as a way of dislodging die-hard troops without losing too many American servicemen. Closer to the end of the war, when the threat of retaliation by Germany had retreated, plans were drawn up to gas large areas of Japan in preparation for an invasion in November 1945. Around fifty targets would be hit with 120,000 tonnes of gas – the study prepared by the Chemical Warfare Service and the US Army Air Forces suggested that 250 square miles of urban areas might be blanketed in gas and that this could result in the deaths of five million people. The attacks would start fifteen days before the landings of US forces and would drench Tokyo and twenty-four other target cities including Osaka, Kobe, Nagoya and Kyoto. In Tokyo, the plan was to attack an area of 17.5 square miles north of the Imperial Palace. This area was estimated to have a population of nearly one million people and the attacks would start at eight in the morning when the greatest number of people would be in the area. Bombers would drop large quantities of phosgene into the area. Mustard gas was to be used in urban areas because it penetrated wood, making it almost impossible to decontaminate Japanese buildings.

On 21 June 1945, the order was given for the massive expansion of the manufacture of poison gas in the United States for use in Japan. Planners of the attack did not know about the Manhattan Project to produce atomic weapons. By the time this study was produced, the US had a prodigious armoury of toxic weapons: 4.4. million artillery shells, 1 million mortar rounds, 1.25 million aerial gas bombs and 112,000 canisters for spraying gas from planes. They would not be used.

Had it gone ahead, it would have almost certainly been the deadliest military attack. The London Blitz led to just over forty thousand deaths, 135,000 died in the firebombing of Dresden, and the bombing of Tokyo in March 1945 killed about 83,000. The eventual death toll at Hiroshima was 140,000, and at Nagasaki 74,000 died. The planned gas attacks on Japan might have killed millions.

World War II was the first time in history that a weapon used widely in a previous conflict was not used by the same parties in the next war. The statements early in the war that the two sides would abide by the Geneva Protocol set the tone. Neither side had a believable margin of superiority that might have prompted the use of the weapons. For most of the war, neither side would gain by initiating the use of gas; early on it would have hampered the rapid Axis victories in Europe and Asia, later they faced the risk of massive retaliation when they had lost air superiority. Both sides also faced immense logistical obstacles. It is hard to make and transport chemical weapons in peacetime; it was very difficult for both sides to expand production in wartime when neither had a target for the use of the weapons. The one moment when the US leadership would have likely bent to those who wanted to use the weapons – the invasion of Japan – never came.

ON 13 JANUARY 1993, representatives of 130 countries gathered in Paris to sign the Chemical Weapons Convention. This agreement was the first ever to outlaw an entire class of weapons. It banned the development, manufacture, storage and use of these weapons. Four years later, when sixty-five countries had ratified the convention, it came into force. These weapons are now illegal, although that is not to say they will not be used. Several countries – Egypt and North Korea – did not sign the Convention and still possess chemical weapons. Due to that threat, and from countries such as Russia and Syria that have maintained programmes illegally, militaries still have to maintain the cumbersome equipment and procedures that protect against chemical weapons. This is, in fact, the main tactical use of the weapons: to slow down armies, to ensure they waste time and effort protecting their troops against weapons that may not be deployed.

The legal regimen against chemical weapons has been considerably strengthened with the Convention. But far more countries than ever, perhaps more than one hundred, have the necessary industrial capacity to make chemical weapons. The military restraints have all but vanished. Few soldiers would question the use of chemicals in warfare nowadays, given the widespread use in the latter half of the twentieth century of defoliants, tear gas, so-called non-lethal gases and napalm. Political restraints against their use have also evaporated. In the case of terrorists, there is no real concern about retaliation.

16. THE SONG OF SARIN THE MAGICIAN

I will with engines never exercised,
Conquer, sack and utterly consume
Your cities and your golden palaces,
And with the flames that beat against the clouds
Incense the heavens and make the stars to melt
As if they were tears of Mohomet
For hot consumption of his country's pride.
> Christopher Marlowe,
> *Tamburlaine the Great, Part II*

Death came suddenly. Some were able to stumble to their doors, gasping for breath and clutching at their throats. They dropped in the earthen streets, their bodies unmarked by wounds or blood. Later, a grey slime would ooze from their mouths and their milky eyes would sink back into their sockets. The skin on their hands, feet and faces was burned a livid red. Across the impoverished town of low, flat-roofed houses, mothers lay clutching their children, old men tried to shelter their families. They were like the bodies captured in ash in Pompeii, their hands clasped over their mouths in a vain effort to keep the scorching gas from their lungs. Families were found in a final embrace, parents desperately trying to shield their children.

On the morning of 16 March 1988, the town of Halabja in northern Iraq was bracing for an attack by Saddam Hussein's Revolutionary Guards. Kurdish guerrillas, backed at the time by Iran, had driven the Guard from their posts around the town and a counter-offensive was expected. Many of the town's Kurdish residents were in their cellars, having taken refuge from artillery fire and napalm bombs. The attack had begun in the late morning and lasted until two in the afternoon before dying down. Then the sounds started to change. No longer were shells exploding as they landed but, instead, the incoming weapons were quieter. Soon people began to notice unusual smells; first a stench of garbage, then a sweeter smell that some described as apples, cucumbers or perfume, and soon the deep rankness of rotten eggs. The gases, heavier than air, were sinking into the cellars where they had taken shelter.

When people emerged, they saw that birds had fallen from the trees. Cows and goats were lying on their sides, breathing rapidly, mucus running from their eyes and mouths. Then the gas hit those who had come out of their shelters. Many vomited as they clawed at their throats. People desperately scrabbled to find water, to wash the scalded faces of children and to quench an intense thirst. Covering the ground was a white powder that caused skin to blister and redden.

Urfiya Hama Ahmad was a walnut farmer and mother of five children who lived in a small village called Zarin, near the border between Iran and Iraq. In 1978, the Iraqi military dynamited her house and forced her family to move to Halabja. Her orchards and animals were left to die. They were paid a pittance in compensation and given a half-built, roofless house outside Halabja. Urfiya's husband Muhammad scraped together a living working on construction sites.

On 16 March, Urfiya and her family were sheltering in a basement because Iranian forces had been shelling Halabja. Around four o'clock she ventured outside but immediately smelled chemicals and rushed back into the shelter. About twenty people were packed into the cellar, desperately trying to keep out the chemicals with wet blankets and towels. After the building was hit by a shell, they feared the shelter might collapse and so they fled. Urfiya's two cows and two calves were dead in the street. There were bodies all around, in the street, in cars with the engines still running. They set off towards Iran, taking nothing with them, heading into the mountains. Urfiya's youngest child, a thirty-day-old boy called Chwestan Muhammad Abd-al-Qader, died that night. His body was left by the road.

After walking all night, they reached a village under Iranian control where people from Halabja had gathered. Iranian helicopters flew them across the border where they were treated with injections of atropine, an antidote to nerve agents, and eye drops. After several months in refugee camps in Iran, Urifya and Muhammad heard that the Iraqi government had rebuilt Halabja and were encouraging people to return. About two thousand people decided to return and were taken by bus to a border crossing. From there Iraqi forces took them to a prison. The young men were separated from the others and were not seen again. The rest were driven south to a prison in a remote area. All the Halabja refugees were kept together in the prison, which housed some six thousand to seven thousand people. People died at rate of ten to fifteen each day and were buried in a shallow trench outside the camp, where dogs would dig up the remains at night. Urfiya and her family were in the camp from August to October 1988. Finally they returned to a town called Zara'in in Kurdistan, but in 1991 after the uprising against Hussein, they once again fled to Iran.

Halabja was one of around fifty towns and villages in Kurdish areas that were victims of poison gas attacks carried out by Hussein. We know more about the town than any others because shortly afterwards, Iranian forces retook the area and invited journalists to record what they had found. Not only were there press reports, but the Iranian Red Crescent Society documented the massacre with hundreds of photographs. A mixture of mustard gas and the nerve agents sarin, tabun and VX killed an unknown number of people in the city. The death toll was somewhere between two thousand and five thousand people. Ten thousand more were injured and many suffered long-term illnesses from the gases. It was the worst ever poison gas attack on civilians.

Saddam Hussein first used poison gas on Iranian troops in 1983 when the war he had launched three years earlier settled into a gruesome stalemate. Tehran sent "human waves" of child soldiers into Iraqi lines, some of them carrying plastic keys that would admit them to paradise. The United States was wary of an Iranian victory and wanted to lure Iraq away from its alliance with the Soviet Union. President Ronald Reagan dispatched Donald Rumsfeld, who had served as defence secretary in the Ford administration, to meet Hussein. He was the first US representative to visit since the countries broke diplomatic relations in 1967 over American support for Israel in the Six Day War. The United States already knew that Hussein was using chemical weapons, but Rumsfeld did not raise it in two meetings with the dictator. As a token of their support for Hussein, the Reagan administration offered critical battlefield intelligence including satellite imagery that allowed Iraq to target their chemical weapons more effectively.

In 1984, United Nations Secretary-General Javier Perez de Cuellar sent a team to investigate allegations of Iraqi use of chemical weapons. It was the first of eight UN missions that would confirm that Hussein had used poison gas, but the Security Council did nothing until 1986 when Iraq was finally named in a resolution criticising its use. The weapons had a devastating effect on Iranian morale; on several occasions the Iraqis only had to use smoke bombs to cause their opponents to flee. In 1988, Tariq Aziz, the foreign minister, finally admitted that Iraq had used chemical weapons but said that Iran had used them first. There is no evidence that Iran ever used chemical weapons in the conflict despite massive provocations. Joost Hiltermann, the author of an exhaustive study of chemical weapons use by Iraq, was unable to find any evidence supporting allegations that Iran has also used them. The order against using them may have come from the top. Ayatollah Khomeini, in a rare moment of restraint, issued a fatwa condemning the use of chemical weapons even when fighting a just war.

It was almost inevitable that the Iraqi government would go on to use poison gas against civilians. Halabja was close to the Iranian border; it was a base for Kurdish rebels and had been occupied by the Iranians on 13 March. The town was important for the Iraqis, as it was near to a reservoir that supplied water to Baghdad. A cable from Iraqi military commanders directed a "firm escalation in military might and cruelty".

From March to July, the city was under Iranian control. Tehran, realising that this was a public relations coup for them, brought in foreign journalists to report on the massacre. The pictures of Halabja provided proof that Hussein had used poison gas against civilians. Later, when Iraqi forces retook the area, they destroyed most of the buildings, razing the city with bulldozers and dynamite. Across northern Iraq, tens of thousands of Kurds would die in the Anfal campaigns, an act of genocide by Hussein that killed more than 100,000 people. Iraqi forces bombed dozens of Kurdish villages, many using chemical weapons.

Reaction by Western governments to the massacre at Halabja was muted. Although shocking photographs were published around the world, they sought to downplay their significance. There was concerted effort in Washington DC to fog the windows by suggesting that Iran might have been to blame or might have staged the incident to discredit Iraq. A week after the attack, a State Department spokesman suggested that Iran may have also used chemical shells in Halabja, a piece of misdirection that was picked up by a credulous media. This hid a major degree of culpability by the United States for chemical attacks on Iran. The United States never provided Hussein with chemical weapons but they did sell him precursor chemicals, as did many other countries, as these chemicals have many uses. But the satellite imagery Washington provided was critically important. Without it, these weapons would never have been as effective. Between 1983 and 1988, Iraq is believed to have used 100,000 munitions filled with chemical weapons against its neighbour and the Kurdish people.

The United Nations sent eight missions to Iraq and Iran from 1984 to 1988 and each concluded that Iraq had used chemical weapons. The US government had first-hand knowledge of the use of these weapons. Senior military officers had toured Iraqi battlefields and had seen for themselves evidence of their use including areas cordoned off with warning signs and discarded atropine injectors used by Iraqi forces at risk from their own nerve agents. Despite knowing that the Iraqis had used these weapons and despite the barrage of evidence from Halabja, the US and British governments refused to allow a UN report to say that there was "conclusive evidence". A Senate bill aimed at punishing Iraq was blocked, with farm lobbyists eager for business in Iraq pushing hardest against sanctions. No measures were taken by the Security Council, which passed four resolutions that mentioned chemical weapons but never named Iraq as the culprit. The Reagan administration even opposed a resolution at the UN Human Rights Commission condemning Iraq's use of chemical weapons. The United States, alongside European nations and the Soviet Union, ensured that Hussein faced no punishment for using poison gas on his people.

In the run-up to the invasion of Iraq in 2003, the less than honourable role of the United States in the development and use of Hussein's weapons was brought out into the light. George W. Bush justified his war in part by saying that Hussein has used chemical weapons against his own people, but ignored how the United States had helped. The State Department put up a page on its website detailing the Halabja massacre and pointing out that not only had thousands been killed on the day of the attack, but that thousands more died later or suffered terrible enduring after-effects. This was the same government department that had earlier urged its diplomats to pin blame for Halabja on Iran and to muddy the waters of any condemnation of Iraq by always sharing the blame with its opponent. Even up until 2003, former CIA agents and others were saying that Iran was responsible for the attack, claiming that it had used hydrogen cyanide gas at Halabja. However, for this gas to be fatal to the same degree as the attack on the Kurdish town, it would have had to have been used in enormous concentrations. A better explanation for the presence of the chemical is that the Iraqis may have used tabun gas contaminated with cyanide, a by-product of the manufacturing process. There is no evidence that Iran ever used chemical weapons in the Iran-Iraq War. There is no dispute that there were many Iraqi chemical attacks.

Hussein went on trial in August 2006 for his use of chemical weapons against Kurdish civilians during the Anfal campaign. There was ample evidence linking him and his senior officers with the use of the weapons during their war against Iran and against the Kurds. On 30 December 2006, he was hanged for other crimes before the trial could be completed.

FOR SHOKO ASAHARA, death was a step towards enlightenment, part of the journey to a higher state of being. Destruction of the world was part of that journey, a move to the next stage in the cycle of death and rebirth. But he was not quite so keen on dying when the reality of death presented itself. During the wedding of Crown Prince Naruhito in May 1993, Asahara drove across Tokyo to the Imperial Palace in a car equipped with an aerosol sprayer filled with botulinum toxin, one of the deadliest substances known to man. As the botulinum toxin was released, Asahara panicked and ordered the driver to stop the car, as he was terrified that the toxin was leaking back into the vehicle. He jumped out and fled into the crowd.

This release of a cloud of botulinum toxin was just one experiment carried out by members of Asahara's Aum Shinrikyo cult. Their actions culminated in the deaths of twelve people in the Tokyo subway gas attacks, an event that provoked a wave of anxiety around the world about the future of terrorism. The subway attack on 20 March 1995 was the first in which terrorists used a chemical weapon indiscriminately against civilians. The attack injured at least a thousand people, some of whom were left with long-term neurological damage. The taboo on the use of chemical weapons in a terrorist attack had been broken, and a religious group had shown itself capable of manufacturing and using weapons of mass destruction without the help of a government.

Aum Shinrikyo had carried out at least fourteen biological and chemical attacks before the Tokyo subway killings brought them to global attention. They sprayed crowds with toxins, tried to kill the Imperial family and attacked US military bases in Japan. Aum was founded in 1984 by Asahara, a yoga instructor who preached a syncretic mix of doctrines drawn from Buddhism, Shintoism, Christianity and other diverse sources such as Nostradamus, Isaac Asimov, Japanese science-fiction movies and New Age thinking. Asahara was obsessed with violent leaders such as Hitler and Hussein. As with many cults, it developed an apocalyptic vision of the world, believing current society was beyond redemption and needed to be replaced with a new order.

Asahara believed the world was living through a period known as *mappo,* during which people had forgotten Buddha's laws and were headed for disaster. He was also intrigued by the imagery of the apocalypse in the Revelations of Saint John. He believed the cult would need biological and chemical weapons to provoke Armageddon and to survive threats against them from the Japanese government and a global conspiracy, which inevitably involved Jews and Freemasons. He also drew from popular culture, from manga and animation series such *Space Battleship Yamato* and *Genma Wars*, as well as from disaster movies such as *The Day After*, about a nuclear confrontation between the United States and the Soviet Union. He was taken with an image of Tokyo being swamped with water from a 1970s disaster novel called *Japan Sinks*.

Asahara's appeal came from his much publicised associations with some of the key figures of Tibetan Buddhism, including the Dalai Lama himself. He was also intelligent, lively and charismatic; many of those who met him described him as outgoing and open. In just a few years, Aum expanded, eventually signing up around thirty thousand members in Russia and ten thousand in Japan, although there was probably only a core membership of 1,400 mostly Japanese adherents who had renounced all other aspects of their life. Many were well educated and prosperous, and the cult benefited both from their money and expertise.

Aum was one of many new religions catering to young Japanese who found their society soulless and stultifying. In 1993, despite being one of the most secular countries on earth, Japan had 231,000 registered religious groups together claiming a membership that was larger than the entire population. Many were tiny congregations that often had overlapping membership. Most tried to find ways to make traditional beliefs more relevant to modern life and this was part of the appeal of Asahara. Many of those who joined Aum were attracted by the idea of being more than just a small cog turning in the machinery of Japanese life. Within the cult, they were given important titles, even running so-called ministries in a parallel government, and enjoyed opportunities to exercise creativity that were lacking in their lives.

The cult's profile was greater than most of these often tiny religions in part because it became a successful business, running yoga studios and printing presses. Members were pressed to hand over their possessions to Asahara and were forced to cut connections to their families. This inevitably spawned concern from families and critical media coverage. In 1990, Asahara and twenty-four other members of the group stood in national parliamentary elections under the banner of a political party they named "Truth". They were ardently covered by the media, as their campaign involved wearing white robes and Asahara masks while chanting outside train stations – a more arresting image than the drab ranks of grey-suited Japanese politicians. None were elected, but the publicity was overwhelming, leading to both new members and increasing tensions with families. As journalists grew increasingly curious and intrusive, the group ran into problems buying property around the country. Local administrators and landowners were no longer keen to sell to them.

That same year, prompted by a growing sense of persecution, notably from lawyers acting for families of cult members and from journalists, Asahara ordered his followers to start developing chemical and biological weapons in a secret laboratory in the village of Kamikuishiki at the base of Mount Fuji. The effort was led by Seiichi Endo, a biologist who was one of several highly qualified scientists who had joined the cult and was named by Asahara as the "minister of health". Aum was organised as a para-state with twenty ministries like those of the Japanese government, so that when the apocalypse came – Asahara made frequent predictions that it was due sometime in 1995 – the group would take over world government.

Endo first concentrated his efforts on developing botulinum toxin. He took a sample of the clostinium botulinum bacteria from soil in Hokkaido and began trying to weaponise the toxin by developing a sludge that could be sprayed through an aerosol. The first tests were done around the Diet building in the centre of Tokyo in April 1990. A vehicle was fitted with a spray and drove around the parliament, releasing the solution of botulinum poison into the air. Similar experiments were carried out at a US naval base in Yokohama and near Narita airport. Nobody was killed, probably because the botulinum sludge was not sufficiently potent and most of it fell harmlessly to the ground. In May 1992, cult members had set up a modified cooling tower on the roof of their building in Tokyo as a means of spreading a culture of the anthrax bacterium. Neighbours complained of a foul smell from the building and said strange jelly fish-like blobs of mucus appeared on surrounding streets. Local health officials demanded access to the building, but cult members had already spirited away the equipment used for making and spreading the anthrax culture.

Nobody died in this attack and no cases of anthrax, an extremely rare disease, were reported, despite Aum's indiscriminate spreading of the bacteria. Anthrax occurs widely in nature but infects few people. The strain of the bacteria used by Aum was weak – it had been used to inoculate animals. It also lacked any protective capsule and was rapidly rendered harmless by sunlight. The few cases of the disease that occur naturally tend to be cutaneous anthrax, in which the bacteria penetrate the skin through a cut. Gastrointestinal anthrax can occur when people eat infected meat. Inhalational anthrax, in which spores enter the lungs and then multiply, is by far the rarest and most deadly form of the disease. It is the form that engineers have harnessed as a weapon. In 1992, Aum, attracted by some overblown press coverage about an outbreak of Ebola haemorrhagic fever, decided that the deadly African virus might make a more effective weapon. Asahara and a group of around forty followers travelled to Congo, officially as part of a relief effort to provide aid to hospitals, but suspicions have lingered that they had attempted to gather samples of the virus, which has killed up to nine out of ten of those infected in some outbreaks. Ebola is a dangerous virus but not necessarily one that would be effective as a bioweapon, as it is spread only through bodily fluids and not through the air.

By 1993, Aum leaders decided that bioweapons were too difficult to develop and the group turned its attention to chemical weapons – notably sarin, the gas developed by the Nazis. The first signs that the group was interested in these weapons came when Asahara publicly urged the Japanese Self-Defense Forces to arm themselves with sarin. He followed this up with several statements alleging that Aum members had been victims of sarin attacks. In fact Aum members were probably falling ill from leaks from the sarin production facilities near Mount Fuji, but in the enclosed paranoid world of the cult, many believed the threat came from outside. One leak in the compound may have injured more than fifty people, including several children. Former members of the cult believe that these releases of the gas may have been deliberate attempts to cultivate the idea within the cult that it was under attack. From 1993 onwards, Asahara frequently complained that his enemies were using sarin against his followers, blaming the CIA, a rival religious group called Soka Gakkai, and people he believed were "Jewish Japanese" including Emperor Akihito and then Crown Princess Masako.

Sarin became an obsession, a way in which Asahara's paranoia was acted out and reinforced, a means by which the cult could protect itself and a way in which it could bring about the apocalypse that he was predicting. The poison became a central part of the cult's life. Not surprisingly, given Japanese predilections for saccharine sweet children's imagery and karaoke, the police even found two songs written in praise of the poison. One was called "Song of Sarin the Magician". It was sung to the theme from a popular children's TV show about sarin and the lyrics went:

> *It came from Nazi Germany*
> *A dangerous little chemical weapon*
> *Sarin, Sarin.*
> *If you inhale the mysterious vapour,*
> *You will fall with bloody vomit from your mouth,*
> *Sarin, Sarin, Sarin*
> *The chemical weapon.*

Producing the poison gas was far from easy. Aum was able to obtain the recipe from open sources and from scientists in Russia, but manufacturing it proved complex and dangerous, even for the reasonably skilled scientists in the cult. Aum spent more than ten million dollars building a three-storey production centre complete with corrosion-proof reactor vessels, heat exchangers, pumps and ventilation. The group is estimated to have spent thirty million dollars on sarin production, using the skills of more than thirty scientists. Asahara even purchased a Russian Mi-17 helicopter that he planned to use to spread the gas over Tokyo. But Aum faced immense difficulties dealing with the corrosive and dangerous precursors and by-products. Chemical leaks were common from the start; farmers in the area complained of foul smells coming from the compound and said that their cows had stopped producing milk.

Masami Tsuchiya, the cult's most talented chemist and one of its most fervent believers in the coming apocalypse, was the first to manufacture the gas but he often managed to make himself ill doing so, even losing consciousness on several occasions. During an attempt to test the weapon by assassinating the chairman of a rival religious organisation, the head of Aum security, Tomomitsu Niimi, was sprayed with sarin and was only saved by a quick injection of an antidote. He would later be sentenced to death for his role in the attacks.

ON 27 JUNE 1994, a team of Aum members drove a refrigerated truck into a residential neighbourhood in Matsumoto City, about 330 kilometres northwest of Tokyo. Parking by a stand of trees, they activated a computer-controlled system that released a cloud of sarin gas into the air. Their target was a dormitory in which judges stayed during trials. Aum had become embroiled in a real estate dispute with a local landowner. The group had purchased land in the area through a front company, but when the owner found that Aum was behind the deal, he pulled out of the sale. Aum decided to sue. A panel of three judges, all staying in the dormitory while hearing cases in the town, were set to rule against the cult.

Yoshiyuki Kono, a former chemical salesman, woke in the night to find himself choking and his wife unconscious. He called for an ambulance, as did others in the building, before being overcome by gas. Seven people died, five hundred more required hospital treatment, including the three judges, forcing a lengthy delay in the case. Kono immediately fell under suspicion from the police, who believed that the fumes had come from chemicals that he mixed to make insecticide. In fact, none of the chemicals he had in his house could possibly have produced sarin. The police released his name, creating a media storm that diverted attention from Aum. Asahara had noisily blamed the attack on the US military, but this did not draw the attention of investigators. Japanese authorities insisted on calling the attack an "accident" and speculated that the gas had been created by the chance mixing of otherwise innocent chemicals even as chemists said this was an impossibility. Only nine months after the Matsumoto City attack did the police acknowledge that they thought it was deliberate and had probably been an experimental run carried out by a terrorist group. The day after they made that announcement, sarin was released in Tokyo's subway.

Immediately after the Matsumoto attack, Asahara had ordered the weapons production facility at the headquarters to be disassembled and turned into a hall for worship, as he was fearful that Aum would be identified as the source of the gas. Despite many years of reticence about investigating a religion and police incompetence, the authorities were closing in on Aum, fuelling Asahara's paranoia. He ordered Tsuchiya to restart sarin production, but without the full weapons lab it was a difficult process. They managed to produce a small amount of the poison but it was impure and less potent than earlier samples.

Asahara's readings of Nostradamus and his eager consumption of disaster fiction led him to predict that from 1995 until the end of the millennium, the world would be plagued by a series of apocalyptic events. Japan would be destroyed by nuclear weapons and Tokyo would be engulfed by the sea. According to Asahara, this was all due to the government's persecution of him. Aum members would survive the coming destruction and the world they created would be a perfect spiritual community. "After the Third World War, I imagine that this world will be filled with love. Every person will overcome his or her suffering and work for the good of others." Paranoia in the cult worsened with the Great Hanshin Earthquake on 17 January 1995. Aum followers believed it had been predicted by Asahara and caused by new quake-inducing machines from the United States.

The attack on the Tokyo subway came just a few months later, on 20 March. The main target was the Kasumigaseki subway station at the heart of Tokyo's government district. The station, a major hub, has exits by the national police headquarters and the foreign ministry. Five members of the cult boarded three different trains and converged on the station, each carrying a solution of sarin contained in two plastic bags and wrapped in newspaper. Using sharpened umbrellas, the men stabbed the bags, releasing pools of the nerve agent into the carriages. They fled out of the station to safe houses around the city.

The sarin, even in a relatively diluted solution, took a significant toll. More than five thousand people ended up at hospital but, according to the Japanese police, nearly three quarters of them had no symptoms except anxiety. Some 493 people were admitted, sixty were seriously affected and around a thousand suffered mild symptoms, mostly headaches, runny noses and vision problems caused by the contraction of the iris. The wider impact on Japanese society was enormous. It came at a low moment for the country. The government was struggling with a recession and had just bungled the aftermath of the Kobe earthquake. The response to the Tokyo subway attack was little better. Emergency services misidentified the toxic chemical, believing it was acetonitrile, a chemical Aum scientists had mixed with the poison gas to make it evaporate quicker. Patients were not intubated and drips were not inserted until they had arrived at hospital. Sarin that had soaked into the clothing of victims affected ambulance workers and medical staff, thirty-three of whom required hospital treatment.

After the attacks, when Japan indulged in soul searching about Aum, the media tried to paint the group as an aberration, focusing on the many admittedly very bizarre aspects of the group and the powers of "mind control" that Asahara was said to possess. But Japanese scholars saw in it echoes of earlier traumatic events: militarism in the 1930s, the violent left-wing student movements of the 1960s and the dystopian popular culture of the 1980s. Commentaries often blamed the group on whatever most troubled the writer, be it Western influences, rampant consumerism, political corruption or the ossification of religious values. Aum became the symbol of whatever was feared; it represented whatever people saw as wrong with the society. Japan's right-wing establishment saw Aum as the geek generation's version of the left-wing Red Army terrorist group. It was equated with communism, in part because of its links to Russia, and also with politicians on the left whom the right believed had forced pacifism on the country. Aum took on an extraordinary array of associations, far beyond its real impact in society, showing how the use of poison could amplify political meaning way beyond other forms of violence.

Questions linger over how Aum Shinrikyo crossed over from fearing the apocalypse, a feature of many religions including most forms of Christianity, to trying to provoke the apocalypse. The psychologist Robert Jay Lifton, an expert on cults, found characteristics that Aum shared with other violent groups. A critical component was the paranoid guru at the heart of the cult. Asahara was at the centre of everything; there was no deity beyond him, no other source of knowledge or power within the group. Violence lay at the heart of Aum. As many as eighty people, including several children, may have been murdered by its members. The first murder committed by cult members – a killing to cover up an accidental death that was under investigation – launched the group down a path of increasing violence. The cult developed an idea of "altruistic killing" in which murder was seen as a merciful release for someone low on the spiritual hierarchy. When a spiritually superior person killed an inferior person, the act moved both of them along the path to enlightenment. It was a short journey from this to the idea of purifying the world on a larger scale.

The lure of weapons of mass destruction came out of pervasive anxieties about them in Japan following the use of nuclear weapons at Hiroshima and Nagasaki. Disciples were also numbed by a process of having to tackle tests put forward by the guru and by his insistence that they rise to cope with these challenges. Killing with poison gas was part of this process.

The scenes of people stumbling out of subway stations blinded and gagging opened the eyes of the world to the horrors of chemical attacks. It also showed the limitation of these sorts of weapons. A group that was disciplined, well-organised, extremely wealthy and able to procure equipment and materials from around the world killed far fewer people than the Madrid train bombings in 2004 or the London underground attacks a year later. Although Aum was able to make sarin, it was not able to produce it at a strength or in sufficient quantities to cause mass casualties. They were not able to kill anyone with botulinum toxin or spread the Ebola virus. A committed group that existed almost entirely outside of government scrutiny or control was unable to come up with a truly effective chemical or biological weapon. The subway attack certainly had a profound impact on Japan, but it did so with relatively few casualties, fewer in fact than were killed by Palestinian suicide bombs on buses in Israel.

ON THE 6 and 26 July 2018, more than twenty years after the attack, Asahara and twelve other members of Aum Shinrikyo were executed by hanging. The impact of the attack had been profound, setting off decades of anxiety about whether other terrorist groups would use chemical or biological weapons. It sparked waves of fear, particularly in Japan, where new religions came under increasing scrutiny.

The thousands who ended up in emergency rooms on that spring morning were mostly suffering from anxiety, showing how widespread the effects of such an attack can be. Fear spread far wider than Tokyo. We are once again in an age in which nameless, faceless poisoners haunt us. Chemical and biological weapons have considerable appeal to terrorists. In some cases, there is almost a mystical allure. They are weapons of the apocalypse that could create more terror than anything that has come before.

17. MEN IN BLACK

He gives me no rest night or day,
My man in black. He's everywhere behind
Me like a shadow. Even now he seems
To sit here with us as a third.

Alexander Pushkin, *Mozart and Salieri*

On 23 November 2006, Alexander Litvinenko, a former officer in the Russian security force known as the FSB, died of massive organ failure caused by radioactivity from polonium-210, a rare isotope that is one of the deadliest substances known to man. The isotope emits alpha radiation made up of intensely energetic and heavy particles that only pass through a few close cells if ingested in the body. A dose the size of a barely visible dust mote is deadly. Weight for weight, polonium-210 is about 25 million times as toxic as cyanide. It must be swallowed, inhaled or absorbed through a cut in the skin to kill. Once in the body, it spreads rapidly to cells and destroys them. The first to die are those fast-growing cells that make up hair, blood and the lining of the intestine. After a day of vomiting, the Russian defector was admitted to hospital. He soon lost his hair and his digestive problems worsened. Radiation depletes the bone marrow and causes a sharp drop in white blood cells. The liver absorbs much of the toxin, leading to jaundice. As the cells that line the intestinal tract die off, the patient suffers from toxic shock and organ failure. Death is slow and terribly painful.

Litvinenko lived longer than his poisoners probably expected.

He was young and fit, and was cared for in one of London's best hospitals. But he was probably doomed from the start. His poison was only identified on the day he died, but there is little doctors could have done if they had known earlier. It had been suspected that he had been poisoned with thallium, but tests at the Atomic Weapons Establishment found a spike in polonium. That was regarded as too unlikely to be anything but an error in the equipment. Another urine sample was taken that showed conclusively that it was this isotope that had killed him.

Litvinenko was a relatively unimportant man, something of a fantasist who exaggerated his knowledge of secret matters and traded in fevered speculations about Vladimir Putin and his sex life. Litvinenko left Russia in 2000 and by 2006 he would no longer have had useful secret information. He had however betrayed the FSB, first by publicly accusing it of involvement in a plot to kill the Russian oligarch Boris Berezovsky and by blaming it for the bombing of apartment buildings that Putin had used as a rationale for restarting the war in Chechnya. He made some unsupported accusations against Putin, suggesting, without any proof, that he was a paedophile. Litvinenko may also have been a minor blackmailer and his role as a "security consultant" on the fringes of the Russian community in London brought him into contact with many unsavoury characters. But it was his death by means of an extraordinarily exotic poison that ensured him a place in history. He was the first person known to have been murdered through the administration of a radioactive poison. A Russian strangled, shot or defenestrated in London might have been a story for a day; the use of a poison like polonium ensured worldwide headlines for weeks. We may never learn the true intent of the assassins given the opaque nature of Russian politics, but by using this chemical they tapped into almost universal fears of an invisible, inescapable threat.

Polonium occurs naturally and extensively but the 210 isotope is present only in the most minute quantities; the main artificial sources are nuclear reactors which can produce around 100 grams, or less than a teacup, of the substance each year. Marie and Pierre Curie discovered the chemical in 1898, naming it after their homeland, which was occupied by other powers at the time, including Russia. Alloyed with beryllium, polonium has some uses as a component in nuclear weapons, as a fuel for some satellites and to eliminate static electricity on machines that spin threads or produce paper. Where the isotope might come into contact with people, it is contained in beads that minimise its threat to life, as alpha radiation cannot move far. Whoever carried out Litvinenko's murder had access to well-trained chemists and probably to the resources of the state, as the chemical is very expensive. Polonium-210 can also be carried across borders without attention; alpha radiation is not picked up by the detectors installed at airports and border crossings.

The murderers – evidence points to two or three people involved in the killing – managed to spread the poison around buildings in London and Hamburg, as well as on three British Airways jets. The staff at a London hotel were contaminated, although fortunately none ingested any of the poison. The Russians might have underestimated the traceability of polonium-210, as it is so rare that standard toxicology tests would be unlikely to pick it up. They may have intended to spread fear as well as radioactivity, or they might have been kept in the dark about what they were handling. They were somewhat cavalier in their approach to the poison, most likely slipping it into a cup of tea that Litvinenko drank at a hotel. The poisoners turned into walking dirty bombs, spreading traces of it around Europe – a warning perhaps to other former members of the FSB who might want to defy Vladimir Putin.

THE TWENTY-FIRST CENTURY began with as much anxiety about poison as any other century. Terrorism, war, biological and chemical threats, food poisonings and genetic modification of animals and plants have created an inchoate sense of vulnerability. This may not be as pervasive as the fears of previous centuries but it billows up in similar ways. In November 2004, reports circulated that Yasser Arafat, the president of the Palestinian Authority and chairman of the

Palestinian Liberation Organization, had been poisoned. Toxicology screening at the French military hospital in Percy, near Paris, where he died apparently showed no poisons in his blood. Public opinion in the Arab world was not convinced. His death, possibly explained by his age, health and the stresses of his imprisonment in his compound in Ramallah, was not accepted by those who had seen him as a powerful and immortal father figure who had survived numerous close brushes with death. For many true believers, it was not credible that he might have died of natural causes.

One of the first to mention poisoning was Ashraf Al-Kourdi, a doctor to the Jordanian royal family who had treated Arafat. The stories were stoked by Arafat's wife Suha, who accused some of his closest colleagues of trying to kill him but later refused to authorise the full autopsy that might have ended speculation. French privacy laws, which do not allow for the public release of medical records, compounded the problem. Those around Arafat saw similarities in his symptoms with those suffered by Wadi'a Haddad, the military commander of the Popular Front for the Liberation of Palestine whom they believed was poisoned by Israeli agents in the 1970s.

Conspiracy theories always proliferate in a vacuum of information. Arafat's medical records were later leaked to several journalists. The verdict of French doctors was that he was neither poisoned nor died of AIDS, as some had suggested, but had succumbed to idiopathic thrombocytopenic purpura, in which the immune system reduces platelets in the bloodstream, leading to uncontrolled bleeding. A brain haemorrhage was followed by gradual organ failure and death. But doctors at the Percy Hospital could not say what killed him. "Consultation with a large number of experts of various medical specialties and the results of many tests do not allow us to offer a medical cause that would explain his symptoms," Doctor Bruno Pats, chief of staff at the hospital, wrote in a confidential report. Immediately there was speculation of lurid causes. An Israeli doctor suggested Arafat had died from AIDS, an odd diagnosis given that people suffering from AIDS tend to die of an opportunistic infection. Others suggested poison, possibly ricin fed to him at dinner on 12 October 2004.

After the murder of Alexander Litvinenko in London two years later, Arafat's widow requested that samples taken in hospital be tested for polonium-210. Doctors at the Lausanne University Hospital in Switzerland found levels of polonium that indicated he might have been poisoned, even though he did not exhibit all the symptoms evident with Litvinenko. Arafat's body was exhumed and high levels, up to 20 times what were expected, were found in bone samples. The Lausanne doctors determined that Arafat might have been poisoned by repeated tiny doses of polonium that eventually prompted his medical emergency and death. To date nobody knows who might have poisoned him.

STORIES OF ARAFAT'S poisoning had a particular currency in the Middle East because of an attempt by the Israeli secret service Mossad to kill a top official of Hamas in 1997 under orders from Prime Minister Benjamin Netanyahu. Khaled Meshal was walking from his car to his office in the Jordanian capital Amman when two men approached him. One pressed a metal object up to his neck, which turned out to be an air pressure injector loaded with poison. Another man opened a heavily shaken can of coke as a distraction. Meshal, who had been alerted by shouting from his driver, said later that he felt as if he had been electrocuted. The two men ran off, pursued by security guards, and were cornered. The assassination attempt, which was supposed to be a covert effort, blew up into an international incident, with President Bill Clinton forcing Israel to hand over an antidote to the poison, believed to be Lofentanil, a variant of the powerful synthetic opioid fentanyl. Israel was also forced to release Sheikh Ahmed Yassin, the spiritual leader of Hamas, and seventy other prisoners. The Mossad agents had used Canadian passports, a fact that soured relations with Ottawa, while King Hussein of Jordan, one of Israel's few friends in the Arab world, remained furious for years. Danny Yatom, the head of Mossad, was forced to resign over the botched operation. The Israeli government released a heavily bowdlerised report on the affair that admitted guilt, but offered few insights into how it went so spectacularly wrong.

In the same month as Arafat's death, the Italian film director Franco Zeffirelli claimed that the opera singer Maria Callas had been poisoned twenty-seven years earlier. She had died alone in her Paris apartment aged fifty-three, by then a somewhat diminished figure, sadly often remembered more for being the abandoned mistress of Aristotle Onassis than for her extraordinary talent. No autopsy was carried out, but her death certificate said she had died of a heart attack. Zeffirelli, perhaps keen to publicise a new movie about the diva, said that in the months before her death she had been surrounded by "a group of strange Greeks" who later ended up benefiting from her quite considerable estate. "It's possible they filled her with substances that killed her. After all her body was cremated immediately after the funeral service." It has all the perfect components of poison lore – a rich and famous woman who dies young, a group of "strange Greeks" who were not identified, jewels, last-minute changes to wills and a hasty cremation. But the sad and ordinary reality, perhaps too sad and too ordinary for her admirers, was that she had been in poor health for some time and probably died from a heart attack, a broken woman.

The likelihood is that neither Arafat nor Callas were poisoned. But Viktor Yushchenko certainly was. The Ukrainian opposition leader and later president ingested a massive dose of TCDD, a form of dioxin. These are isomers of 210 mostly man-made organic compounds that are soluble in fat and can accumulate in living tissues. They are known to be extremely toxic in animals but as only a limited number of people have been poisoned with dioxin, knowledge of how it kills in humans is somewhat sketchy. Its toxic effects in large doses are not well understood. Its carcinogenic effects are disputed and the small number of people given massive doses suffered disfiguring chloracne and some stomach problems but few other ill effects.

The mystery has yet to be unravelled but toxicology tests carried out in the Netherlands showed that Yushchenko had six thousand times the expected amount of dioxins in his blood. This was the second highest level ever recorded after the unsolved poisoning of a secretary who worked in the chemical lab of an Austrian research institute. Dioxins left Yushchenko with chloracne, his normally youthful features now distorted, reddened and bloated. Even months later his complexion was an unnatural grey-green and he'd had to wear make-up during the campaign to look less sick. Chloracne can last for years and can be accompanied by porphyria cutanea tarda, a disorder of haemoglobin metabolism that leads to fragile skin and unusual pigmentation.

A dioxin is a strange choice of poison. It does not kill rapidly, although it might lead to death from soft tissue cancers or lymphoma many years later, so it is unlikely that someone intended a rapid assassination to remove him from the election. Its symptoms only emerge slowly, often taking weeks to produce the terrible facial disfiguring suffered by Yushchenko. It is easily detected – it may never be fully flushed from the body – and can be fingerprinted in a way that makes it possible to trace the source of the chemical. Far from being the perfect poison, it is one of the worst ways to kill someone if trying to avoid detection.

Yushchenko fell ill shortly after attending a dinner with the head of the Ukrainian secret service SBU, Ihor Smeshko, and his deputy, Volodymyr Satsyuk. He complained of a headache and an upset stomach a few hours after the meal and soon his symptoms worsened. A few days later he was taken to hospital in Vienna suffering from intense back pain and a digestive tract that was ulcerated from top to bottom. But the secret service head was known as something of a reformer and politically aligned to Yushchenko, so it was hard to understand why he might attempt to poison him. The food had been served from common plates and so it would have involved a risky process such as poisoning part of a dish. It might have been done by someone to implicate and embarrass Smeshko. Initially the story that made the rounds in the Ukrainian press was that the poison had been put into soup served at the late-night meal in the dacha, but those who attended the meal said no soup was served.

It is unlikely that Yushchenko was poisoned at the meal with the SBU. Dioxins normally take several weeks to produce the symptoms that the Ukrainian politician suffered shortly after the meal. Although dioxins are said to be tasteless, nobody knows where this assertion comes from, as it would be a brave and foolhardy scientist who tasted such a toxic chemical. Nobody knows if it would have been possible to have given him such a massive dose in a single meal.

Dioxin poisoning on this scale is so rare that nobody even knows how it might play out. The acne normally breaks out some time after the initial poisoning so Yushchenko might have been given the chemical months before. What his case does illustrate is how difficult it is to poison someone. It is unclear whether whoever poured the flask of dioxin into his food planned to kill him or just to sicken him enough to make it difficult for him to campaign. Political murders have not been unknown in Ukraine but they have been obvious and brutal with little attempt to hide the violence. But whoever carried this out managed to create an intense mystery, a sense of a hidden hand working in a way that leaves everyone in danger. Anyone who could get his hands on a phial of dioxin would probably know that it was unlikely to kill immediately. But it is hard to fathom why anyone wanted to induce cancer in a politician when it might not fell him for another twenty years.

The poisoning did do one thing that poisonings always do. It created a miasma of fear and doubt around almost everyone involved. Everyone came under suspicion. The SBU suggested it might have been someone in Yushchenko's inner circle who poisoned him. Those around the candidate thought it must have been the SBU, an organised crime figure or the Russian secret service, the FSB. Everyone denied involvement. Yushchenko refused to cooperate with the investigation, saying he did not trust those involved. He refused to allow a commission of inquiry access to blood tests and other information gathered by his doctors in Austria. His wife, a US citizen of Ukrainian origin, said she had noticed the smell of poison on his breath the night he came back from the dacha, but whatever she smelled, it was not dioxins, as they are odourless.

Another political murder by poison was also unfolding in the last months of 2004. Munir Said Thalib, a prominent Indonesian human rights activist, collapsed and died on a Garuda Indonesia flight from Singapore to Amsterdam. After falling ill three hours out of Singapore, a doctor on board examined him but thought he was suffering from an upset stomach. Just two hours from Amsterdam, he died. The mysterious and sudden death on board a plane of a thirty-eight-year-old man, albeit one who had suffered some health problems, was bound to arouse the concern of the Dutch authorities, who immediately opened an investigation. An autopsy performed at the Netherlands Forensic Institute found that Munir's body contained three times the fatal dose of arsenic. He had been poisoned sometime after leaving Jakarta.

Munir had been a persistent critic of Indonesia's human rights record, notably the violent suppression by the military of separatist movements in Aceh and Papua. He had often stood up to intimidation, saying he had lost count of the number of death threats he had received over the years. But his murder was extraordinary, and given the controversy it aroused, an enormous error of judgement by those in the military intelligence establishment who were believed to be behind it. Munir's death on the way to the Netherlands was bound to prompt an investigation. Arsenic would inevitably be detected in a toxicological screening. The number of people who could have administered the poison were clearly limited to those on board the plane. A murder on a jet is an almost perfect modern-day Agatha Christie setting. Killing him on board a jet would almost certainly lead to an investigation abroad by competent authorities. The prime suspect was Pollycarpus Budihari Priyanto, a Garuda Indonesia pilot who had invited Munir to sit in business class on the Jakarta-Singapore leg of the flight and had given him a glass of orange juice. In December 2005, Priyanto was found guilty of poisoning Munir and was sentenced to fourteen years in prison, although his conviction was overturned by the Supreme Court a year later. Munir's murder is as baffling as the poisoning of the Ukrainian president. Had he been killed in a staged traffic accident in Jakarta, it would have been easy to scupper the investigation. The Indonesian state has been unafraid to use violence against its critics but it has normally done this in far less elaborate ways.

The difficulties of using poison to kill raise the question of why it should appeal to secret services. The attraction partly lies in the secret nature of poisoning, which makes it appear more sophisticated and less thuggish than simply shooting someone. It also spreads fear more effectively, making enemies feel more vulnerable and less able to predict the threat against them. This was likely the motivation for its use by the KGB and its Russian successor. In 1921, the Soviets had established a division known as Laboratory 12 to develop poisons. A favourite target were Russians in exile in cities such as Paris and Zurich; the aim was more to let them know that nothing could protect them if the NKVD secret police wanted them gone.

A former KGB agent who now lives in London said that the agency regarded a poison as just a tool, "like a pistol". The FSB has the same view. A Saudi man fighting for the Chechens, known as Khattab, was said to have been killed in 2002 by a poisoned letter given to him by a Dagestani messenger. It has never been fully revealed how successful this unit was in terms of killing people without anyone being aware that the deaths were murders, but some significant failures are known and some of its activities existed in the realm of fantasy. Certainly, few of the claims made by former KGB agents come with much proof. A 1953 document from the NKVD details a plan to use a Soviet agent to kill Josip Broz Tito by spraying him with a cloud of pneumonic plague bacteria while the Yugoslav leader was in London. As a back-up, there was a suggestion of presenting him with some jewellery that would contain a rapid-acting poison that would kill him when he opened the box. Tito died in hospital of heart failure in 1980 at the age of eighty-eight.

The writer Aleksandr Solzhenitsyn was believed to have survived a ricin poisoning attempt by the KGB in 1971 when an agent stabbed him with a needle in a cathedral in Novocherkassk in southern Russia. He was left in terrible pain for days. An elaborate effort to poison the Afghan president Hafizullah Amin also went dramatically wrong. Mitalin Talybov, an Azeri KGB agent who could pass as an Afghan, infiltrated the presidential palace as a chef. On 17 December 1979, Amin's son-in-law was taken ill with acute food poisoning and flown to Moscow for treatment. Hafizullah was fearful of being poisoned and frequently switched foods – in this case it seems his relative ate the toxic food prepared for him. On 25 December, the Soviet invasion began and seven hundred KGB agents stormed the Darulaman Palace in the suburbs of Kabul where Amin was living. The Afghan presidential guards put up greater resistance than expected and more than a hundred of the KGB paramilitary died. Amin and his family were gunned down in the palace. Bullets proved much more successful weapons of assassination than poison.

The Soviets' was not the only secret service to attempt to use toxins in assassinations. France's intelligence agency, the SDECE, plotted to poison the Egyptian president Gamal Nasser in 1956. He died fourteen years later of a heart attack. Fidel Castro stands as an enduring testament to the difficulties of introducing a sufficient dose of a toxin to a wary enemy. The US government is believed to have made at least eight attempts to kill the Cuban leader in the 1960s, several of them involving poison. The Technical Services Division of the CIA developed different approaches to killing or at least harming Castro. One idea was to provide a Cuban military officer who was close to Castro but in the pay of the CIA with a pen fitted with a needle that was so fine you could inject someone without their being aware of it. The pen would be filled with Black Leaf 40, a pesticide that contains a lethal dose of nicotine. The aim was to scratch Castro with the needle hidden in the pen. The officer never got close enough. It was then decided that the easiest way of getting a poison to pass Castro's lips was on one of his cigars. Plans were drawn up to infect them with botulinum toxin, one of the most deadly poisons known. Another plan was to lace his tobacco with BZ, a super-hallucinogen, before he made a speech.

Killing Castro was clearly the aim but if that could not be done, then embarrassing him would be satisfying. There was an attempt to put the metal thallium in his shoes while he was staying in a hotel in New York for a meeting of the United Nations General Assembly. Thallium can act as a depilatory – the aim was that a bald, beardless, eyebrow-less Castro would lose the respect of his nation, his machismo receding with his hairline. Castro was fond of scuba diving, so it was decided that a lawyer who was negotiating with Castro for the return of those captured during the Bay of Pigs invasion would present him with a diving suit. The mouth piece had been infected with tuberculosis bacteria while the suit itself had been dusted with the spores of a fungus that causes Madura foot, a disease that leaves the skin covered with deep and suppurating ulcers. The poisoned wet suit was a novel take on the Nessus shirt, the toxic garment that killed Hercules. A range of other poisons were said to have been deployed against Castro. A former lover was sent with some pills that she hid in her face cream. They melted into the cream and she changed her mind about killing him anyway. The Cuban government claims there have been 638 attempts to kill Castro. All failed but those involving poison presented the least risk. Fidel Castro died of natural causes in 2016 aged ninety.

So why assassinate with poison when there are easier methods that are more likely to work? The most obvious reason is to go undetected and for a death to seem natural. Of course, we have no idea how many people might have been killed in such a manner but some of the most high-profile poisonings have been carried out with toxins that were quite readily detectable. The use of polonium-210 was either an act of extreme arrogance in that the perpetrators may have presumed it would go undetected, or it was a deliberate effort to sow fear among Russian exiles. Opponents of the Russian leader living abroad must now see every plate of sushi, every cup of tea as potentially lethal. Bodyguards and bullet-proof limousines will not protect against the tiniest amount of some exotic toxin that could be slipped into any meal. The point of using poison is not about the death of an individual. There are easier ways of killing. Poison is about making people live in dread of what might occur.

18. THE MEDIUM IS THE MESSAGE

If someone had wanted to poison him, they would have finished him off.
          Vladimir Putin

Very little happens in Salisbury, the somnolent cathedral city in southwest England, and so the underworked police and ambulance services responded rapidly to a report that a couple had fallen ill in The Maltings, a small shopping centre by the River Avon. A man in his sixties and a younger woman were sitting on a bench by the river, feeding the ducks despite the cold and dreary weather at the beginning of March. The man complained of feeling ill, but it was the woman who first collapsed, falling into the man's lap as he remained upright. Passers-by might have thought they were another pair of drug users, not unknown in British country towns, but an army nurse shopping there sensed something was wrong. Four minutes after calling 999, the police were on the scene and moments later an ambulance arrived.

Sergei Skripal and his daughter Yulia were saved by the proximity of the police, the speed of the ambulance crew and the fact that Britain's military chemical weapons research institute at Porton Down is just outside Salisbury. The ambulance crew at first thought the pair were drug users and administered the usual treatment for overdoses, but the police officers thought they were unusually well dressed for heroin users. A search of their wallets turned up IDs and a warning flag came up when Sergei Skripal's name was run through a police database. He was a former Russian military spy who had been recruited by British intelligence. Within an hour they were in intensive care, attached to a full array of life support systems. At that point they were so sick, there was very little expectation they would survive.

It was a mystery what caused them to fall so seriously ill but their symptoms and the fact they both were ill suggested a neurotoxin. Doctors at Salisbury General Hospital had been trained in chemical weapons treatment due to the proximity of Porton Down, an establishment that had made chemical weapons until they were banned by a series of treaties, and so could recognise the symptoms of a nerve agent. Doctors use the unappealing acronym SLUDGE – short for salivation, lacrimation, urination, diaphoresis, gastrointestinal discomfort and emesis – as shorthand for the "wet" symptoms of nerve agents. The pieces began to fall into place and by the next morning, a major incident had been declared. The Counter Terror Command of London's Metropolitan Police would take over the investigation, moving into the quiet town in force. Salisbury would never be the same.

Shortly after the Skripals had been admitted to intensive care, Detective Sergeant Nick Bailey tried to open the front door of their house in the suburbs of the city. His life would never be the same again. Smeared on the door in some form of gel was what was eventually identified as a Novichok, also known as A series Agents, in this case A-234. During an inspection of the house, the police officer would inadvertently spread the agent. He would then take it back to his police station and eventually to his own home, leaving traces of the toxin across the city, even on his children's toys. A day after visiting the Skripals's house and not finding anything untoward, Bailey would start to show symptoms of poisoning.

Eventually, six people would be poisoned with Novichok in Salisbury and one would die. On 30 June 2018, Charlie Rowley found a bottle of perfume and gave it to his girlfriend, Dawn Sturgess. She rubbed some on her wrists and not long afterwards began to experience symptoms. She died eight days later in Salisbury General Hospital. Rowley was also admitted for treatment but survived. Once again, places visited by the couple were sealed off, along with their house in Amesbury, eight miles outside Salisbury. The poison had been in a sealed bottle of brand-name perfume. Rowley said he had spilled some on his hands while opening it but washed it off. When Sturgess dabbed some on her wrists she was mystified by the fact that it had no aroma.

Nick Bailey suffered from long-term problems related to the poisoning while another police officer was contaminated without suffering from significant symptoms. Nerve agents may have enduring effects on many people, something noticed by the families of workers in factories producing the chemicals in the United States. There is also speculation that those with so-called Gulf War syndrome, a range of illnesses suffered by veterans of the conflict there in 1991, were sickened by the release of nerve agents when Iraqi weapons stocks were destroyed. What is clear is that these agents can have terrible effects on people even if they survive and that their use is a source of enduring fear.

THE DRY, CLINICAL descriptions do not capture the howling animal pain that took over the victim, his body contorting, his face dripping with sweat. A forty-four-year-old man, previously in good health, became confused and his face flushed ten minutes after the plane took off. He soon vomited until he briefly lost consciousness. The plane made an emergency landing in Omsk and the man was admitted for treatment about two hours after the initial symptoms. In a video taken by a fellow passenger, medics can be seen rushing to the back of the plane while a loud groaning is heard throughout the cabin. By the time he reached hospital he was in respiratory failure. His admission notes mentioned diaphoresis (sweating), myoclonic status (jerky involuntary movements), hyper-salivation and metabolic encephalopathy (confusion and poor motor skills). Two days later he was airlifted to the Charite-Universitatsmedizin in Berlin where he was diagnosed with severe poisoning with a cholinesterase inhibitor. A German military lab soon confirmed that the poison was one of a group of nerve agents known as Novichoks.

Alexei Navalny, leading opponent of Russian President Vladimir

Putin, fell ill on a domestic flight from the Siberian city of Tomsk to Moscow. The dissident had built his political movement on the back of social media, making popular YouTube videos that documented corruption among Kremlin leaders, and so all his moves ahead of the flight were captured by the unceasing eye of Instagram and Twitter. During his short working visit to the central Russian city, he had posted multiple selfies taken with volunteers and supporters. Just before boarding the plane, he was shown to be bright-eyed and lively, taking a photograph with a young man on the bus out to the aircraft. Not long after take-off, a colleague travelling with him said he started to lose focus. Soon the symptoms began and the pilots diverted the jet.

The quick thinking of the pilots and the medical staff that met the plane on the ground saved Navalny's life. In December 2020, while in Germany recovering from the poisoning, he prank-called one of the team of agents who had poisoned him and got the whole story, an almost unique case in which the victim learned how he was poisoned. The investigative team Bellingcat had identified some of those involved in the operation and Navalny managed to get one of them of the phone by pretending to be an aide to the secretary of the Security Council of the Russian Federation, a close ally of Putin.

Navalny managed to convince the FSB officer that he was gathering material for a report requested by his boss, who was concerned about the fallout from the assassination attempt. Although the officer, Konstantin Kudryavtsev, was only involved in the clean-up, he soon gave away the names of the poisoners and made it clear that this was an attempt to kill Navalny, not just to sicken or scare him. Kudryavtsev said that they had even upped the dose to try to ensure the dissident's death on the four-hour flight to Moscow but his life had been saved by the quick actions of the pilots and the medics on the ground in Omsk. Kudryavtsev had been sent to the Siberian city to make sure no evidence remained and had ensured that Navalny's clothes were cleaned. Navalny asked a trick question about some grey underwear, but Kudryavtsev confirmed that the poison had been applied to Navalny's blue underwear along the seams in the crotch.

Seven FSB officers, including a doctor and chemical weapons

experts, had been tailing Navalny during his travels around Russia since at least 2017. The opposition leader, a handsome and charismatic figure with a laconic and light-hearted style, was winning a growing following among the well-educated in cities who did not rely on state media. He was clearly becoming something of a personal threat to Putin, a sour, charmless man who had been selected to rise up the ladder of state security precisely because he lacked any charisma. While Navalny had an easy charm and humour, Putin has become increasingly eccentric and remote as he ages, a dead-eyed autocrat sitting alone at the end of a ridiculously long table.

Bellingcat tracked the travels of men from the FSB Counter-Extremism Unit and the Criminalistics Institute, showing that they were likely to have been involved in several other poisonings, including two attempts on the life of Vladimir Kara-Murza, an opposition figure sentenced to twenty-five years in prison in April 2023. In May 2015 and February 2017, the otherwise healthy Kara-Murza was afflicted with sudden illnesses that resulted in prolonged comas and the shut-down of vital organs. Kara-Murza, a journalist by profession, had pushed for the US Magnitsky Act that enabled the government in Washington DC to impose sanctions on Russian officials and institutions involved in human rights abuses.

Kara-Murza suffered severe medical problems after the poisonings and had such extensive nerve damage, he required a stick to walk. Russian doctors carried out toxicological tests but could never determine what chemical had been used against him. Eventually his wife got a blood sample to the FBI who carried out screenings. They refused initially to release the results and when they did, all they would say was that he was the victim of an intentional poisoning with a biotoxin. Some 275 pages of the report were withheld pending discussions with other agencies and ten pages were redacted entirely. That week, three senior Russian security officials were visiting Washington for high-level discussions on terrorism, perhaps explaining the agency's reticence over releasing the results.

Both Navalny and Kara-Murza survived poisonings carried out by the Russian government. Navalny returned to Moscow on 17 January

2021 and was immediately taken into custody for violating probation by leaving for emergency medical care in Germany after his poisoning. He then began a gruelling and cruel odyssey through the Russian penal system, finally ending up in the IK-6 prison camp, above the Arctic circle. The camp went by another name, "Polar Wolf", and was known for its extreme brutality. On 16 February 2024, the Russian Prison Service announced he had died after taking a walk and feeling unwell. He had been seen on screens at a court appearance the previous day looking gaunt but cheerful. A previously healthy, athletic and very active man was dead at the age of forty-seven from what the prison dismissively called "sudden-death syndrome".

NEITHER OF THESE two relentless opponents of Putin has been entirely silenced, but almost everyone else in Russia has been. The Russian president, dismissed by Navalny as "Vladimir, the poisoner of underpants", has had no compunction in killing his enemies. The investigative journalist Anna Politkovskaya was shot and killed entering her apartment building in 2006. The former deputy prime minister and opposition leader Boris Nemtsov was murdered on the street right by the Kremlin walls in 2015. Assassinations that could be pinned on hapless Chechens or low-level criminals have proved an effective way of ensuring there have been no serious challenges to Putin's rule. Guns have always been an easy way to kill, so why use poisons?

Vil Mirzayanov, a chemical weapons expert and author of a book on the Russian state poison programme, maintains that the aim was always to kill without being detected. This was the conclusion of the investigation into the death of Litvinenko which said that the poison would not have been detected by any normal postmortem examination. The State Research Institute of Organic Chemistry and Technology wanted to produce poisons that were tasteless, odourless and undetectable. The aim was to kill at a slight remove or to murder while provoking symptoms that could be mistaken for natural causes. However, at some stage, something changed. The use of polonium and a Novichok nerve agent meant that deaths were accompanied by dramatic symptoms. Litvinenko died a terrible death in University College Hospital, his cells killed off by alpha radiation. Novichok clearly causes pain and distress as the victims choke and endure violent muscle spasms. Both polonium and Novichoks are detectable, with difficulty, but now that they are known to have been used, tests would be likely in any suspicious case. They are no longer useful as surreptitious killers.

The purpose of certain killings or attempts at murder – particularly the poisonings of former intelligence officers in the United Kingdom in 2006 and 2018 – appear to be about signalling to others that if they betray Putin, they will die a painful death no matter where they might hide. The medium was the message. You will never be safe. Sending this message was vital for the leader of a mafia state, as any sign of leniency only signalled weakness to those around the leader. Transgressions had to be punished, something Putin even enshrined in law when in 2006 the Russian Duma allowed for people accused of terrorism or extremism to be assassinated by the state anywhere in the world. For those who betray the *slivoviki,* the secret world of the KGB and its successors that produced Putin and almost all those around him, death in a hail of bullets was too quick and easy. Putin clearly wanted something painful, theatrical and simultaneously visible and invisible. Skripal had become a double agent while serving as a military intelligence agent in Spain. For a decade he provided British intelligence with information about his colleagues in the GRU before being arrested and convicted in Russia in 2004. In 2010 he was allowed to go to the United Kingdom in a spy swap, living a low-key existence in Salisbury.

Assassination lies at the very heart of Russian politics and goes way back, long before the Revolution. Ivan the Terrible killed his son, the Tsarevich Ivan Ivanovich. Catherine the Great crowned herself in 1762 having had her husband murdered. Alexander II and his grandson Nicholas II were both assassinated. More recently murder has been commonplace in local politics. In 2002, Moscow's deputy mayor survived his second assassination attempt but the head of a local district outside the city was shot dead. The head of another district near the capital and his wife died in a car bombing. The Moscow education administrator, two deputy mayors, the head of the national sports foundation and even the country's chief notary were all killed in a short period.

Putin has continued a tradition of assassinating opponents that goes back to the early days of the Revolution. The Chekists who turned into the KGB, who became today's FSB, have long used poisons as a means of murder, but formerly with the aim of not being detected. Stepan Bandera, a right-wing Ukrainian and enemy of the Soviet Union, was killed with a cyanide gun in 1959. Another opponent, Lev Rebet, was killed by the same assassin in Germany. Their deaths were ascribed to heart attacks, the true cause only being revealed when the assassin defected.

To advance, or even to stay in place, it has been necessary for Russian rulers to maintain loyalty, mostly through lavish rewards but also by ensuring that any dissent is punished. In earlier days, those who supported Leon Trotsky against Stalin were hunted down and killed. Trotsky himself would die in exile in Mexico City in 1940, an ice pick embedded in his skull. In many cases, agents who betrayed Moscow mysteriously died of illness or suicide. One such figure left three suicide notes in his Washington DC hotel room in case anyone was in doubt that he had killed himself. Putin did not pioneer political murder in Russia – he has simply continued what has long been a pattern.

Like many sociopaths, Putin knows when to exploit the weaknesses of his opponents. The 2017 attacks in Salisbury came at a time when Britain was bogged down in its departure from the European Union and was led by a weak prime minister. Its secret services were focused on Islamic extremism rather than Russian agents, and its economy, still climbing out of the battering it took in 2008, was dependent on Russian money. Putin almost certainly calculated there would be little in the way of a response to the killing of Skripal or his other actions around this time, including interference in the US elections in favour of Donald Trump.

Poison might have been the choice in the London and Salisbury killings because both had some tactical advantages in allowing the killers to get away, while also being powerful in the messages it sent. The photographs of Litvinenko dying in hospital, bald and weak, only amplified the warning to any potential traitors. With Skripal, the British authorities tried to deny Putin the pleasure of broadcasting the pain of his victims by ensuring no photographs or graphic descriptions of their symptoms emerged. No information has emerged on their whereabouts or condition, and it is not known if they have suffered any long-term disability as the police officer who was poisoned entering their house has endured.

"IN A FEAR-RULED country you have to send signals to the population about what is acceptable and what is not." said the former CIA station chief John Sipher. The origins of the Novichok agent used to poison Russian dissidents are still murky, but the class of chemicals was probably first produced in 1975 in a special lab in a small town on the Volga called Shikhany. The word "Novichok" might be translated as "newcomer", or more colloquially "newbie", as they were the first new chemical weapons developed in some time. One of the Soviet scientists who worked on this, Vladimir Uglev, would later say the poisoning weighed on his conscience. He had been part of a group of scientists developing chemical weapons for military purposes in the belief that the United States was doing the same, although President Richard Nixon had unilaterally wound down the US programme shortly after coming to power in 1969. The Soviet scientists said they never expected it to be used for assassinations or terrorism.

Much of what is known about the Soviet development of nerve agents comes from the chemist Mirzayanov, who broke ranks and wrote about the secret programme. His book is a lengthy diatribe against the communist apparatchiks who made his life hell in a variety of top-secret weapons establishments. One of the reasons for the development of Novichoks was that the Soviet Union had built up a vast network of laboratories, testing centres and manufacturing plants employing tens of thousands of people and they all needed to be kept busy. Many occupied themselves with make-work projects to justify their costly special privileges. Others pressed forward with improving on the weapons in the Soviet arsenal, making them less dangerous to handle or more effective. The military establishment always had a desire for new chemical weapons, in part to be able to defeat NATO detectors and deny their troops the crucial few minutes to respond to an attack. At the State Research Institute of Organic Chemistry and Technology, which went by the unwieldy Russian acronym GosNIIOKhT, dozens of chemicals were analysed to test for their toxicity, persistence on the battlefield, resistance to cold, heat and humidity and other qualities.

Novichoks were a breakthrough in chemical weapons. They were a binary weapon – that is, they were made up of two relatively safe chemicals that had to be mixed before the substance became dangerous. They were stable, meaning they could be stored for long periods, and did not decay, dissolve or evaporate. They persisted in any environment, even in the cold of a Russian winter, making them potentially useful in tying up troops in lengthy decontamination efforts. A tank or truck sprayed with the agent would be out of use for some time. This quality also made them a massive threat to public safety if they leaked or were used, as they proved very hard to clean up. Although there is no official data on their toxicity, it was believed to be between five and eight times as deadly as VX, the most dangerous chemical weapon developed up to then. But Mirzayanov made a confession about the data he handed over: many Russian scientists were bored, lazy, underpaid and underappreciated by their managers. They told their superiors in the system what they wanted to hear, without checking the science, and they rarely used the most reliable methods. Their data often would not stand up to scrutiny. Several people have survived attempts on their lives with Novichok; it is possible that it is not as deadly as Russian scientists thought.

Novichoks are still very nasty substances. The nervous system operates by transmitting messages along special cells. To send a message to a muscle, the nerve has to transmit the message across a gap between cells. This is usually done with a chemical messenger called acetylcholine. When the message is sent and received, another chemical called acetylcholinesterase breaks down the chemical messenger so that the message is not repeated. If the acetylcholinesterase is not produced or is not working, the gap between the cells becomes flooded with the messenger and the signal is sent over and over. Saliva glands get repeated messages to release saliva; muscles will twitch as they receive repeated messages to contract; the throat contracts, making it hard to swallow; some muscles, such as those in the chest, freeze from the overload of signals. The victim froths at the mouth, unable to swallow and their breathing stops.

Nerve agents were discovered in Germany just before World War II, when scientists looking for effective insecticides discovered that the tiniest amounts of some chemicals shrank their pupils to pinpricks and left them breathless. Tests showed that these killed insects very effectively but were so toxic they could not be used commercially. They were quietly passed on to the German army's chemical weapons division and production of the first two nerve agents – known as tabun and sarin – was developed with the help of the chemical giant I. G. Farben. Production was highly secretive – the Allies only knew the full extent of German chemical weapons when the war was over and there was a frantic race with the Soviet Union to capture as many weapons and scientists as possible. The assumption from then on, on both sides, was that they were developing the German technology for their own use.

A post-war arms race began in the late 1940s to develop sufficient stockpiles of sarin. Making nerve agents is a massive challenge, even with the head-start provided by the German experience. Many of the precursor chemicals are toxic or corrosive in themselves. Special equipment, handmade at massive cost, is needed and often fails. What emerges from the chemical processes is often a mix of nerve agent, by-products and precursors, as well as some chemicals that are so intensively corrosive they would eat through shells or rocket casings. Leaks were common, as were explosions, which sometimes proved deadly. The early days of testing involved human subjects, mostly soldiers, in a manner that was profoundly unethical and dangerous. The legacy of the manufacture was equally troubling: very toxic chemicals were dumped at sea or pumped into the bedrock. Fortunately, the British and American stockpiles of sarin were never used in war.

"THIS NOVICHOK WHERE *is it?*" asked Vladimir Putin. A Novichok manufacturing plant was to be set up in Nukus – in what is now Uzbekistan – amid very tight security as Russia was playing a double game: while talking to the US about arms control, it was pushing ahead with biological and chemical weapons development, convinced that America was doing the same. Mirzyanov became increasingly concerned about this as the Soviet Union collapsed and exposed the programmes, which had been poorly designed and managed, contributing to the environmental blight that risked vast areas of the country. He was put on trial for disclosing state secrets, but the case fell apart as the government would not acknowledge the existence of Novichoks and if they did not exist, how could anyone reveal their secret?

The Soviets possibly made a few tonnes of Novichoks before their empire collapsed, but the chemical and biological programmes persisted under the new regime. It is unlikely that the Russians developed a capacity to manufacture the substances in sufficient quantities that they could be used in warfare. Regardless, as with all chemical weapons, it remains uncertain what tactical role they might play in battle, given their tendency to blow back, quite literally, on their users. Nevertheless, chemical weapons have their uses.

Navalny is only the latest of a line of well-known figures who have been poisoned in recent years. Most public was the killing of Kim Jong Nam, the brother of the North Korean dictator Kim Jong Un, who had his face smeared with a VX nerve agent by two women who thought they were filming a TV gag show at Kuala Lumpur airport. There are clearly other cases that are less known or were not fully investigated: the assassination attempt on a Bulgarian arms dealer and the suspicious poisonings of several Russian dissidents including the Pussy Riot activist Pyotr Verzilov. Novichok may have been used as a weapon of murder as early as 1995 when a businessman and his assistant died under suspicious circumstances. Rumours have persisted that some of the Russian mob acquired a small stockpile at a time of total anarchy across the country.

Part of the point of using poison is its very invisibility; it is impossible to know which car door handle is tainted, which meal is infused with a toxin. Anything could kill you. One of Putin's enemies, the oligarch Berezovsky, used to say of his security "they are not bodyguards, they are just witnesses", mordantly revealing he knew he was vulnerable anywhere, even with his massive resources. Paradoxically, poisons are invisible but also can be highly traceable. Everyone knows that Novichoks come from Russia, but the government there can maintain its usual pretence that it is innocent and demand evidence, something that states are often reluctant to make public as this would reveal intelligence methods and sources.

Not only can Russia infuse a certain world with fear but it can provoke a wave of conspiracy theories that sow doubts, muddy the waters, and steadily weaken trust in democratic institutions. The Kremlin's disinformation machine and its useful idiots began questioning whether these poisonings ever happened at all, suggesting they were simply part of the rampant Russophobia in the West. By early 2018, 90 per cent of Facebook posts mentioning Skripal had originated in Russia, most of them from pro-Kremlin media and disinformation bots. Distract, dismiss, distort and dismay – the 4D approach – has been consistently effective for Putin's criminal state. The Skripal attack was beneficial in many ways for Putin. It also created a distraction for Russians from their poorly performing economy and allowed him to complain about Russophobia. When Putin did finally make a public comment, it was to stir the disinformation pot. He had only read about the "tragedy" in the media but thought "that if it had been a military-grade nerve agent, the people would have died on the spot".

Meanwhile, the perpetrators of the Skripal attack are back in Russia, immune from justice, joking about their escapade that killed a British citizen with a nerve agent and severely harmed five other people. Kirill Kleimyonov, a news anchor on Russia's Channel One, sent a clear message to any disaffected potential defectors watching:

> Don't choose England as a place to live. Whatever the reasons, whether you are a professional traitor to the motherland or you just hate your country in your spare time, I repeat, no matter, don't move to England... Something is not right there. Maybe it's the climate? But in recent years there have been too many strange incidents with a grave outcome. People get hanged, poisoned, they die in helicopter crashes and fall out of windows in industrial quantities.

This is all part of the global game now. Assassinations or the use of poison gas in Syria are followed by information warfare, which breaks down the barriers of war and peace and ensures a constant global instability. It has been enormously successful for Putin, ensuring that there has been relatively little outrage over his activities in Syria. It is not a simple matter of countering the truth with lies, but ensuring that people give up even thinking the truth is a possibility.

Following the nerve agent attacks in Salisbury, it became fairly obvious to all but the most gullible that the Russian state, and almost certainly Putin himself, was behind the attacks. The nerve agent used was extremely rare and only ever made by the Soviets. The two men tracked on CCTV footage making highly unlikely trips to the suburbs of Salisbury on successive days were quickly shown to the agents of Russian military intelligence.

Almost immediately, however, the Russian disinformation machine sprang into action. A later study would count 130 narratives put out by the Russian disinformation machinery. Two outlets dominated the coverage by the state media – the television channel RT and the website Sputnik. Their stories were varied, contradictory and often seemingly at cross purposes but the mission was to churn up everything rather than clarify or inform. They succeeded at that mission.

Poison plays into this mistrust and fear; its ambiguity has always made it suitable for sowing doubt or inspiring a conspiracy theory. The attack in Salisbury was ripe for disinformation because of the proximity of Porton Down, the British weapon's testing establishment where work on vaccines and other countermeasures was done.

Among the stories that were dominant was that Russia had offered to cooperate in the investigation of the attack but that the United Kingdom had refused (true in itself but no state would have accepted the involvement of the main suspect in a criminal act). The story was about Russophobia, the go-to complaint in Moscow's media about any criticism of Putin. There was no evidence for Russian involvement and it was all a conspiracy aimed at damaging Moscow's image. When Prime Minister Theresa May announced that the poison was a Novichok, RT and Sputnik came out with a series of denials: the poison could not possibly be a Novichok, the Russian disinformation sites said, as these did not exist and if they did exist they were so deadly that nobody could have possibly survived. If it was a Novichok, it must have come from Porton Down, perhaps there was a leak or a rogue chemist poisoned the Skripals? It was not from Russia but if it was it might have come from a private source, as they could have bought the poison that was not made in Russia and did not exist anyway. Novichok was not even developed in Russia and probably came from Iran, or the United States. The British could not know if it was a Novichok, as they did not have stocks of the nerve agent with which to compare the samples. Britain was trying to get rid of its Novichok stocks so as to blame Russia. Most of all there was no evidence that Russia did it. When a Swiss lab confirmed the presence of a Novichok in samples from the Skripal attack, the Russian Foreign Minister Sergei Lavrov, notorious for his insistent, shameless dishonesty, said that tests showed the presence of a chemical agent that the Russians had never produced. (They had produced it and the chemical was not found in the Salisbury samples but in control samples that had been submitted to the lab.) And so on and so forth.

Churnalism is the neologism academics use to describe forms

of media that take a loosely sourced factoid, most often from some public relations statement, that gets recycled over and over as news in a variety of forms. Russian disinformation is very skilled at inserting its message into this; the aim is that it is endlessly churned around on Twitter and in new articles published on websites. Russia has never had a shortage of people willing to take their money to recycle conspiracy theories and indeed there are many who do it for free. The aim is to get as wide a circulation as possible for as many different stories, mostly by using troll farms and sometimes hapless, sometimes collusive, partners on the far left and far right of the political spectrum. Eventually, nobody can keep track of where a story originated.

Credibility has never been the issue with these tall tales, some of them utterly absurd. The point is to create confusion, a blizzard of alternatives that reduce the capacity of anyone to know what is true and what is false. It is a technique that has been used with much success by Putin, particularly in Syria where its machinery of hackers, paid embellishers, dupes and useful idiots whipped up scepticism about Assad's use of chemical weapons. Creating mistrust and anxiety has already succeeded for Putin in his elevation of the far right across Europe and the United States. He can never beat the West in any conflict or arms race, but he can weaken democracy by undermining it with lies.

19. GOT MILK?

*The FBI is warning police nationwide to be alert
for people carrying almanacs, cautioning that the
popular reference books covering everything from
abbreviations to weather trends could be used for
terrorist planning.*

Associated Press report, 29 December 2003

Each day in the United States, fifty-five million gallons of milk, enough to fill eighty-five Olympic swimming pools, moves from farms through dairies, cooperatives, processing centres, warehouses and supermarkets to millions of refrigerators. It is one of the most complex distributions of goods anywhere in the world, involving hundreds of thousands of people and billions of dollars. Developed over decades, it is one of the most cost-effective supply chains in the world, taking a highly perishable product and making it ubiquitous and cheap. It may also be the most efficient means yet devised to poison a large number of people.

Lawrence Wein and Yifan Liu, business professors at Stanford University, published a study of how to kill several hundred thousand people by putting the botulinum toxin in the milk supply. Crystalline botulinum toxin is one of the most toxic substance known to man. Just a gram of the poison theoretically could kill a million people if evenly distributed. It is seven thousand times as poisonous as cobra venom. The poison cuts up the protein that is essential for the release of the chemical acetylcholine into the gap between nerves and muscle cells. Transmission along nerves ceases, causing a creeping paralysis starting in the head and throat. Patients suffer what doctors call the 4Ds – diplopia, dysarthria, dysphonia, and dysphagia, or double vision, difficulty in speaking, deafness and problems swallowing – all while suffering an increasing symmetrical paralysis. Patients are lucid and suffer no fever, but gradually become more paralysed, eventually suffocating to death if they are not put on ventilators and treated with an anti-toxin.

Botulinum toxin was weaponised by both the United States and the Soviet Union. It is commonly used in a highly diluted form as Botox. This now common cosmetic product reduces wrinkling on the face by paralysing muscles temporarily. The toxin is fairly stable although it is denatured by heat. There is an anti-toxin, but anyone who falls ill from botulinum poisoning needs intensive medical care. Many cases at once would overwhelm even the best-prepared health system.

The toxin is produced by about two hundred variants of a bacteria that occurs widely in nature, living mostly in soil. Very few cases of natural poisoning occur each year – the United States usually reports fewer than two hundred. In the days of home canning, it was more common, as the bacteria thrived in the oxygen-free but non-sterile environment of poorly treated jars or cans. Death rates have declined markedly with better anti-toxins and improved ventilation treatment, but recovery from botulinum poisoning is still a long and arduous struggle. In the 1960s, about 60 per cent of those poisoned died. Today only about 6 per cent perish.

Wein and Liu studied the potential impact of a terrorist putting botulinum toxin into the milk supply in the early stages of the nine-step chain between cow and consumer. Heat treatment of milk would deactivate about 70 per cent of the poison but enough would remain to poison a vast number of people. In their study, if all the poisoned milk was consumed, 568,000 people would have taken in some of the toxin. If one gram – about the weight of a paperclip – was used, around 100,000 people could fall sick. If ten grams were used, then more than half a million could be affected, becoming ill within three to five days. The exact number depends on many factors: how much poison was used; how long it takes to detect the poison; how rapidly the milk is distributed, purchased and consumed; and at what stage in the processing the dose of poison is mixed in.

In reality, the problem would be noticed quite rapidly, the milk would be tested and people warned not to consume it. But even a small number of deaths could have an immense psychological impact as we saw with the anthrax killings of 2001 and earlier with the poisoning of consumer products such as the Tylenol murders in Chicago in 1982. A mass poisoning of this type would press all the buttons that make us fear such events. Manufactured disasters cause more anxiety than natural events, unfamiliar health threats in which we are unaware of the symptoms raise the level of concern further. The risk to children from poisoned milk would amplify the panic, sending every parent rushing to a hospital.

The costs would be enormous. If fifty thousand people fell ill, treatment costs could go as high as US$8.6 billion and death rates would probably be as high as 60 per cent due to the lack of treatment. In some cases of people being poisoned with unauthorised versions of Botox injected into their faces, treatment in the first two weeks was so intensive that it costs US$350,000. Treating all the victims of a mass poisoning might not even be possible, as hospitals would lack the necessary ventilators and staff. The milk industry – a twenty-billion-dollar-a-year business – would be devastated. The healthcare system would be pushed to the point of collapse by people presenting with what they imagine to be symptoms of poisoning. The fear would likely spread to other foods, destroying demand for fruit juices and fresh produce that might be subject to poisoning. Other countries would ban the export of US foods. A small envelope of poison could transform America.

Maimonides was among the first to recognise how difficult it is to poison someone.

> Whoever thinks he can prepare a deadly poison with no
> undue taste and smell about it, neither affecting the color

of the dish it is intended for nor its consistency, because of the small amount used but allegedly sufficient to kill a man, he who thinks so has not the slightest idea or knowledge of medicine because there are no such small quantities, the popular belief notwithstanding. The fact is that all deadly or harmful poisons, irrespective of their origins, deviate in taste and smell from the norm of their kind.

For Maimonides, the proper defence against poisoning was to only eat food prepared by those you trust.

Although we are biological and culturally programmed to assess the food we eat for its possible impact and although we have learned to enjoy acceptable risks, humans have shown themselves prone to misunderstanding very slight risks. People are poor at making judgements about low probability risks. When presented with these types of risk, people are simply not interested in the real data on probability, preferring to make gut choices. After the attacks of 9/11 many people decided to drive rather than fly because of the perceived risks of their planes being hijacked. Passenger numbers fell 18 per cent in the year following the attack. However, the reality is that it was sixty-five times more dangerous to drive than to fly. Media events, particularly those as all-consuming as 9/11, are a far more significant factor in the way people make decisions about risk than any statistical analysis can ever be. Risks are taken more seriously if they are presented as lifetime risks rather than annual or journey risks. Likewise, if a 1-in-100 risk is presented as a 10-in-1,000 risk or 100-in-10,000, it is more likely to be taken seriously.

We are not good at making judgements about being poisoned. As many attempts to poison people in modern times have shown, it is not that easy. Even with the advanced technologies available to governments, none have found poisonous weapons very effective beyond inculcating a climate of fear. The rareness and difficulty of poisoning someone have not really permeated our consciousness to the same depth that the fear of poisons has. Mass hysteria, notably over poisonings, has a long history, as does the social amplification of risks associated with poison. Before the first use of poison gas in World War I, it most often centred around isolated groups that had preoccupations with religion or witchcraft. Nuns were vulnerable to outbreaks of mass hysteria over poisonings. Many of the young women in convents were sent there against their will. They endured harsh discipline, restrictive and unvarying diets, and often appalling physical and psychological abuse. It appears that the more strict the discipline and the more cruel the regime, the more likely a convent was to succumb to mass sociogenic illness, which before the twentieth century mostly took the form of motor hysteria – an illness characterised by dissociation, histrionics, twitching, shaking and contractures that could last from weeks to months.

The most notorious cases involved nuns who were believed to have been possessed by the devil. They used foul language, exposed their genitalia and pretended to copulate with priests sent to exorcise them. In the fifteenth century there was an epidemic of biting nuns across Europe. It started in a German convent with one nun who repeatedly bit others, and it spread from there. Attacks of this sort of hysteria often descended into popular freak shows with people gathering to witness the shocking behaviour. At Loudon in western France, Father Urbain Grandier was accused of bewitching an entire convent in an episode that lasted from 1632 to 1634. Nuns bleated like sheep or meowed like cats, leading to the belief that they were possessed by animals. Grandier was tortured, tried in a court – leaving behind extensive records for historians who have produced more than a hundred books on the events – and finally burned at the stake.

Outbreaks died down as convent life became less brutal, but they still occur in similar settings. There was an outbreak of mass hysteria at an Islamic girls' school in the Malaysian state of Kedah in the 1980s, in which thirty-six girls were said to have been possessed and required exorcism by healers known as *bomohs*. The situation culminated with a group of girls taking hostages at knife point and a former prime minister being called in to negotiate. By the eighteenth century, mass motor hysteria shifted from convents to factories and schools, environments that could be as harsh and oppressive as convents. Episodes were recorded across Europe with the first in a work setting occurring at a mill in Lancashire in 1787, when one man and twenty-three women suffered from convulsions and the sensation of suffocation. Schoolchildren in Germany suffered from outbreaks of twitching hands, and girls at a London school developed an apparently infectious form of paralysis of the left arm.

The twentieth century saw a change not just in the locations of mass sociogenic illness but in the way it manifested. Rather than developing seizures, hysterical dancing or convulsions, outbreaks tended to be of the physical symptoms of extreme anxiety – headaches, vomiting, hyperventilation and fainting. Schools are still common locations for outbreaks, but groups of soldiers and other self-contained groups are also susceptible. After the horrors of the use of poison gas in World War I, attacks tended to focus around the fear of noxious substances and were mostly triggered by harmless but sometimes foul-smelling gases from such sources as pig farms, cleaning fluids, forest fires, sewers and even incense. In Virginia in the 1930s, there was a spate of supposed attacks by someone who was said to be releasing poison gas into homes. Across Botetourt County in December 1933, people reported that a mysterious intruder had sprayed their homes with a gas that left them nauseous and numb. Vigilante groups were set up and the legislature in Richmond passed a law allowing for sentences of twenty years for anyone releasing noxious gases. After much investigation, each case was found to have been caused by some innocuous problem such as a backed-up chimney flue or, in one incident, a severe case of flatulence. Fears of gas attacks were such that during the infamous Orson Wells *War of the Worlds* radio broadcast on Halloween in 1938, many people believed that the Martian attacks being described were in fact gas attacks by the Germans.

The outbreak of mass sociogenic illness in the school in Jenin described in Chapter Four is typical of the way in which a slight chemical odour can develop into mass sickness from a feared chemical attack. Mass illness of this type has been reported around the world from Japan to the United Arab Emirates, from Europe to

Canada and most recently across schools in Iran.

What the outbreaks tended to have in common was an underlying anxiety – there were several outbreaks during the first Gulf War when fears were widespread that Saddam Hussein might use his chemical weapons. During the first Scud missile attack on Israel by Iraq during that conflict, nearly half of those in the area where it landed reported breathing difficulties due to fears that it would be carrying a chemical warhead. Schoolgirls in Iran exhibited signs of a mass sociogenic event as they claimed they were being poisoned. The outbreak came after the government there had killed many people protesting the murder of a young woman by the religious police after she was detained for wearing her hijab improperly.

The twenty-first century has brought with it new forms of mass sociogenic illness with a much larger reach beyond confined communities or small groups living under intense stress. So-called TikTok Tics and new forms of Tourette's syndrome were likely sparked in adolescents by exposure to social media, while several studies have suggested that long Covid, the supposedly persistent symptoms suffered by some after the pandemic, is a sociogenic illness spread in the same way. In many cases, patients suffer from what is known as a "nocebo effect" that can be driven by negative stories on social media. This is the expectation of negative symptoms or side effects that brings on these very symptoms. If doctors are unsure of a diagnosis or warn of possible effects of a drug, nocebo effects are more likely to occur.

It is not clear why certain people suffer during these episodes. There is no personality type that is susceptible. There is some evidence that those with below average intelligence are more likely to suffer, but equally studies have shown that it is those who score higher than average on IQ tests who succumb. There has also been a backlash against the idea, with researchers claiming that people really had been poisoned, often by what is known as "sick building syndrome" or by chemicals that have been released into the atmosphere and then dispersed rapidly so they were not picked up by subsequent tests. An outbreak of what was thought to be a sociogenic illness at a school in Britain in 1990 turned out to have been a genuine case of poisoning caused by pesticides on cucumbers served at lunch.

ON 5 JANUARY 2003, the Metropolitan Police raided a flat above a chemist shop in Wood Green in North London. Four men from North Africa were arrested and a close search of the flat turned up a recipe for ricin and twenty-two castor beans. A crew of crime scene investigators dressed in white biohazard suits began a detailed examination of the building under the glare of television lights. Two days later press reports had confirmed that traces of ricin had been found and government officials were expressing their deep concern that an Al Qaeda ring may have been developing chemical weapons in the heart of London. Prime Minister Tony Blair said that the threat was "present and real and with us now and its potential is huge".

The media was instantly obsessed with the horrors of ricin. Just one milligram of the poison – the tiniest dot – could kill someone, inducing fever, abdominal pains and eventually severe dehydration. Ricin is a type II ribosome inactivating protein, a chemical that inhibits protein synthesis in cells. It leads to rapid failure of most organs and the immune system. It is six thousand times more toxic than cyanide and twice as poisonous as cobra venom. When castor oil is produced from beans, 5 per cent of the mash left behind contains this toxin, which produces different effects depending on how it enters the body. Inhaled, ricin inflames the airways and the lungs fill with fluid, drowning a person in their own blood. Ingested, it results in severe nausea, diarrhoea and intensive bleeding in the intestinal tract. Injected, it kills cells and results in massive internal bleeding that leads rapidly to a failure of all organs. The poison is stable, can be used as an aerosol or spread through food. Weaponisation is not difficult, the relatively primitive Iraqi programme managed to produce ricin, but it is a difficult toxin to spread among large numbers of people.

After the Wood Green raid, most reports mentioned that ricin was easy to make. Almost every article raised the case of Georgi Markov, the only person known to have died of deliberate ricin poisoning.

Markov, a forty-nine-year-old Bulgarian dissident and broadcaster, was waiting for a bus near Waterloo Bridge when a man stabbed him in the thigh with an umbrella. He died three days later. During his autopsy a tiny perforated metal ball was found in the wound on his leg. The ball, made from platinum and iridium, was believed to have contained a tiny dose of ricin. No ricin was ever detected. The diagnosis was based on the tiny dose in the metal ball and the symptoms Markov showed before his death. A Danish antiques dealer was said to have been hired for the job by the Bulgarian secret police who, in turn, had obtained the poison and the metal ball from the Soviet KGB.

The return of ricin to the streets of London in 2003 prompted a frenzy in the media, which only worsened when another suspect in what was quickly dubbed an "Al Qaeda Poison Ring" stabbed a policeman to death while his flat was being raided. On 7 February, Secretary of State Colin Powell mentioned the ring in his now infamous speech to the United Nations Security Council, linking it to Al Qaeda and Saddam Hussein. Blair took up the call to arms in the weeks ahead of the invasion of Iraq, calling the Wood Green arrests "powerful evidence of the continuing terrorist threat". Recipes for ricin are available on the internet and one did appear in an Al Qaeda training manual that was discovered in 2000. But recipes have circulated for years on how to extract the poison from castor beans. *The Poisoner's Handbook* and now many internet sites have included a recipe using lye and acetone that is not an effective way of producing a useful amount of ricin. The same training manual also contained the bizarre idea that it was possible to assassinate someone with three cigarettes. A full reading of the manual suggests the authors might be evil, but they are not geniuses. Another manual, known as the *Mujahideen Poisoner's Handbook* and supposedly written by one Abdul Aziz, is even less useful. One of its chemical weapons recipes produced hydrogen sulphide, a gas that smells of rotten eggs but does not live up the handbook's claim that it could "kill in thirty seconds". Kamel Bourgass was convicted of "public nuisance" and sentenced to seventeen years in prison for the Wood Green plot.

He had already been given a life sentence for killing a policeman during the raid in which he was arrested. Bourgass and four other defendants were cleared of terrorism charges involving ricin. The Algerian did have some faulty recipes for ricin culled from the internet but there was no ricin at all. There were no clear links to Al Qaeda and the others arrested along with Bourgass apparently knew nothing of his poisonous ambitions. His recipe would have produced only the tiniest quantity of the poison, perhaps enough to kill one person if they were injected with it.

This dastardly terrorist plot to smear the deadly toxin on door handles across London turned out to be nothing of the sort, despite the spin put on it by the British government before and after the failed trial. Deputy Metropolitan Police Commissioner Peter Clarke said after Bourgass's conviction that "a real and deadly threat" had been thwarted and it was hard to over-estimate "the fear and disruption this plot could have caused across the country". It appears the police were being driven by intelligence reports from Algeria that probably resulted from the torture of a low-level Islamic militant who did have links to Al Qaeda.

Analysis of the "Ricin Ring" trial by Doctor George Smith, a fellow at a think-tank in Washington DC who was a consultant for the defence, showed how much of it was based on hype. There was no risk of mass poisoning. The various normal household chemicals that doubtless were found in the flat in Wood Green were more dangerous than the handful of castor beans that were the sum total of the "ricin death factory" that was whipped into being by an enthusiastic British press. Even the collapse of the case did not stop the police and the press from whipping it all up. The day after the trial, the *Evening Standard* quoted police as saying they were concerned that another Al Qaeda cell might have the missing ricin that was never found at the apartment.

Since Wood Green, ricin scares have become commonplace, with any stray white powder being taken out of a sealed building by a crew in full biohazard gear featuring on the evening news. In February 2006, ricin was said to have been discovered in a roll of quarters at a dormitory laundry at the University of Texas in Austin. It turned out to be nothing harmful – white powder in a college laundry room could be any number of substances, but is unlikely to be ricin. The incident was reported across the United States, producing hours of anxiety-inducing television coverage of men in biohazard suits evacuating buildings. Senate Majority Leader Bill Frist received some ricin in the mail, causing a scare that closed down three Senate office buildings. The White House also received a capsule containing the poison in the mail. A chronology of events linked to ricin put together in 2004 is a catalogue of angry white men engaged in attempts to scare officials or relatives. Of the nineteen incidents, only ten involved ricin, mostly in tiny quantities. There has only been one death that is known of – that of Georgi Markov.

THE STANFORD AUTHORS of the toxic milk study assume that terrorists would be able to get their hands on large quantities of botulinum toxin. But it is questionable how easy that is. Ed Schantz, a scientist who worked at the Food Research Institute of the University of Wisconsin at Madison, was one of the leading experts on botulinum production. It took him about three weeks to manufacture the toxin but it involved decades of work to perfect the technique. Schantz began his research into toxins at Fort Detrick during World War II. He later went on to develop forms of the toxin for use in treating muscle spasms, crossed-eyes and other tics. The use of botulinum for those medical reasons led to the discovery that by paralysing the muscles of the face, wrinkles were reduced. An even milder solution was developed into the cosmetic drug Botox.

Botulinum toxin is made by placing *Clostridium botulinum* into a blend of brewer's yeast extract, digested milk protein and dextrose. The bacteria develop rapidly in this nutrient mix, turning it a muddy brown. When the nutrients are exhausted, the cell walls of the bacteria break down and the mixture clears. The mixture now includes botulinum toxin and the waste from the bacteria. To separate out the toxin, it is precipitated with the addition of sulfuric acid. This creates a mud at the bottom of a flask, which is repeatedly redissolved in a salt solution and then precipitated with acid and alcohol at varying temperatures. The process is not impossible for any reasonably well-trained chemist but it requires a deft hand and experience.

The experiments carried out by Aum Shinrikyo show that even well-trained scientists found it remarkably difficult to produce effective botulinum toxin. Separating the toxin from the cells that produce it is complex and many batches end up with only a low level of toxin. For botulinum toxin to become a truly mass casualty weapon, it needs to be purified to 95 per cent, something that the US weapons programme in the 1960s struggled to achieve. The toxin also loses its effectiveness when exposed to sunlight, degrading by about 4 per cent each minute. It is such a poor weapon that the US abandoned plans to make it.

Wein and Liu's paper cites no real scientific analysis of how easy it would be for terrorists to produce the toxin. Instead it draws on conclusions by Richard Danzig, a corporate lawyer who was a former US Secretary of the Navy, and from Judith Miller, a political writer for the *New York Times* whose credulous reporting on weapons of mass destruction in Iraq were profoundly misleading. The authors did not use a scientific assessment of the ease with which terrorists might be able to produce botulinum toxin in sufficient quantities to kill thousands.

Botulinum toxin is a very large molecule, about eighty times the size of insulin. This makes it a delicate flower; it is one large poisonous protein accompanied by a group of other proteins that act to stabilise it. It is vulnerable to heat and does not remain potent if exposed to air for a long time. It can be added to the food supply but it would be vulnerable to pasteurisation of milk or other products. US milk processors raised the temperature of pasteurisation to ensure that a protein such as botulinum toxin would be denatured and left harmless, significantly reducing the risk from such an attack.

The paper also supposed that a jihadi manual available on the internet provides all the information necessary to make botulinum.

But the manual does not explain how to obtain a toxin strain of *Clostridium botulinum* in the first place. Many strains of the bacteria produce next to no toxin and finding the right one among the seven hundred variants is extremely difficult. It is unlikely that a jihadist would be able to obtain a sample from a recognised culture centre, as security has been tightened. In their critique of the paper, Milton Leitenberg and George Smith, two experts on bioterrorism, said that variations in three key assumptions made by Wein and Liu could reduce the outcome by nine orders of magnitude, that is, reducing the risk to a billionth of the results that the Stanford scholars found. In other words the jihadi scenario of poisoned milk is a lot less frightening than these researchers would have us imagine.

Bioweapons and poisons are the perfect weapon of terrorism because they cause far more fear than death. The anthrax scare that followed the 9/11 attacks engendered a massive wave of anxiety across the United States despite killing only six people. In the first two weeks of October 2001 there were 2,500 anthrax false alarms in the United States. The Illinois Department of Health received nearly 1,500 samples from anthrax scares in the following three months, although not one of them proved to be the germ. Elsewhere across America there was a near panic about every possible release of white powder. Airlines grounded planes because Sweet'n Low artificial sweetener was found on tray tables. Mail piled up in warehouses undelivered. Some guacamole spilled from a burrito caused a bioterrorism scare in Chicago.

The reality is that despite being quite finely tuned biologically to avoid poison, we have become culturally unmoored from an ability to judge when we are at risk. We are so bombarded with information and so inclined in Western cultures to feel that there must be an answer to every problem that our capacity to make judgements on issues such as terrorism is always hindered. We live in a world in which, for many of us, anxieties about scarcity have been replaced with a desire to eliminate risk entirely. Few of us have any direct experience with issues such as terrorism, environmental hazards or dangerous consumer products, so we rely on the media to bring us information on these issues. The media almost always inflates risks. Evening news shows are not inclined to show reports about how safe your children are in a world of airbags, car seats, non-toxic plastics, rubber playgrounds and fenced swimming pools. The media can take a relatively minor or contained problem and explode it to extraordinary proportions. We live in an increasingly media-saturated world, something that plays into the hands of terrorists. They have already become extraordinarily astute at media manipulation, providing images that terrify and provoke over-reactions by people and governments.

In 1987 two men in search of scrap metal found a cylinder in an abandoned clinic in the Brazilian city of Goiânia. They sold it to a junk dealer. One of his employees broke it open and found within a shiny blue powder that he took home. Children played with the 'carnival glitter' and smeared it on their bodies and faces. It was caesium 137, a radioactive element used in medical treatments. Four people died, twenty-one required hospitalisation and one person had to have a limb amputated. Scientists identified seven workplaces and forty-two homes that had been contaminated. Most of those contaminated received very low doses of radiation and their lives were not in danger. However, it was the most serious radiological incident in the Western Hemisphere up to that time.

There was little attention for two weeks after the accident, but then the story was picked up by a tabloid TV show in São Paolo. The impact was devastating. More than 100,000 people lined up to be tested for contamination. The value of foods produced in the state of Goiás fell by half. Hotel occupancy plummeted, even in tourist destinations more than an hour away from Goiânia. Pilots refused to fly passengers from the city, cars with license plates from Goiás were stoned. The city was permanently tagged with a radioactive reputation. A bioterrorist attack today would have even greater amplification of risk than this event. Terrorists know that a small attack using weapons that provoke such dread will have an impact far beyond the immediate deaths and illness.

Unfortunately, the media and politicians will play into the hands of any attackers by amplifying the effects. We expect political leaders to make those judgements for us and we expect academics to have solid information behind their writing. We also expect journalists to gather information with a measure of scepticism. But Bill Clinton launched a bioterrorism project after reading a scare-mongering novel *The Cobra Event* that has all the scientific depth and accuracy of a sixteenth-century pamphlet on Jews poisoning wells. CNN has repeatedly shown videos seized from an Al Qaeda base in Afghanistan that show dogs being poisoned, but rarely mentions that what this reveals is that their technological level is rudimentary. These chemical warfare experiments barely go beyond the cruel pranks of unpleasant schoolchildren.

Fear-mongering about chemical and biological attacks has become so normal that it is rarely challenged, even by scientists who should know better. Chemical weapons are difficult and dangerous to produce. They are even more difficult to produce and disseminate in quantities that would kill more than a handful of people. There is considerable art in both the production of biological and chemical toxins. There may well be jihadi guidebooks on the internet but they are neither new nor useful. Recipes for many poisons have been available for decades in most scientific libraries. In the New York Public Library it is easy to find recipes for ricin and botulinum toxin. Books like *The Poisoner's Handbook*, and Maynard Campbell's *Catalogue of Silent Tools of Justice, Biology for Aryans* and *Silent Death* have all been available for many years, long before the internet made such texts available to anyone. But having a recipe does not guarantee the cake will rise.

George Tenet, the former CIA chief who declared the presence of weapons of mass destruction in Iraq "a slam dunk", has long been one of the most prominent sources of disinformation on this front. This is in the nature of the job of modern intelligence agencies. Rather than get caught out by ever being wrong, it is politically safer to give the worst-case scenario of terrorist or other attacks. Groupthink is immensely prevalent within these agencies, as it is within any bureaucracy. Intelligence agencies suffer from it because they are sealed off from scrutiny. "Terrorist interest in chemical and biological weapons is not surprising, given the relative ease with which some of these weapons can be produced in simple laboratories," Tenet told a congressional committee in 1997. Danzig said a year earlier that "biological weapons are inexpensive and accessible. A small pharmaceutical industry or even moderately sophisticated university or medical research laboratory can generate a significant offensive capability." But these positions were contradicted by one of the most knowledgeable people in this field. Doctor David Franz, the former commander of the US Army Medical Institute for Infectious Diseases, said, "An effective, mass casualty producing attack on our citizens would require either a fairly large, very technically competent, well-funded terrorist program or state sponsorship." The bathtub weapons-lab scenario is greatly overstated.

Chemical weapons are so unlikely to cause mass casualties that their inclusion in the category of "weapons of mass destruction" is somewhat misleading, unless, as John and Karl Mueller have written, the category of weapons of mass destruction is stretched to include such weapons as machetes, which were used to kill hundreds of thousands of people in Rwanda. Machetes killed far more people in that conflict than have ever been killed in chemical or biological warfare. Even when states have developed and used chemical weapons, these have not proved the most effective way to kill people. Only 1 to 2 per cent of those gassed on the Western Front in World War I died. Conventional weapons killed a quarter of people exposed to them. It took one tonne of gas to kill a single person on average and gas accounted for less than 1 per cent of total battlefield casualties. The AK-47 rifle has much more of a history of mass destruction than any chemical or biological weapon but it produces few expressions of anxiety from political leaders or think-tank experts.

A study by the US Congressional Office of Technology Assessment said it would take a ton of sarin dispersed in ideal conditions against unprotected people in a densely populated area to kill between three thousand and eight thousand people. If the conditions changed – if for example there was a light wind or it was a sunny day – the death tolls would fall to a tenth of those figures. No terrorist group is currently known to have the ability to manufacture a ton of sarin, and the likelihood of such a group coming together is relatively small.

The rogue Russian scientist has become something of stock character in movies and airport novels, but very few put themselves on the market in this way. A study by the RAND Corporation think-tank found considerable obstacles to former Soviet scientists selling their expertise to terrorists. For a start, there are not that many who have the critical information on turning microbes into weapons. Leitenberg asserts that of the sixty thousand people who worked in the Biopreparat system that prepared biological weapons, probably only around 5 per cent were senior scientists and only a hundred knew how to carry out all the steps to create bioweapons. Scientists were mostly still living in isolated and once-closed cities, and so finding a Russian bioweapons engineer with all the necessary expertise would not be simple. Widely divergent skills are needed to produce effective bioweapons and assembling a team would be challenging. Likewise, their access to passports and travel documents are most likely limited. The Russian FSB has also done a better than expected job of monitoring weapons centres. Patriotism, pride in their work and the fact that they still received better than average salaries have also kept Russian scientists at home.

Some risks exist in sales of such weapons from stocks held by governments around the world. North Korea is believed to have significant stockpiles of chemical and biological weapons. If the country collapsed into political and economic chaos, it is easy to imagine a situation in which rogue military units or commanders might decide to sell off some of these stocks. Otherwise, the countries that are believed to have programmes to make chemical or biological weapons – Iran, Egypt, China, India, Israel, Syria, Russia and possibly Taiwan – are not likely to pass them on to terrorist groups, even those that they have backed in the past, because their use would likely have catastrophic results for these governments.

Bioweapons may not prove as terrifying as is often made out. The most commonly presented scenario is the terrorist with a tiny vial of smallpox that he drops in Times Square. It would be enormously difficult for a terrorist to get hold of such a vial and it is not clear that it would sweep through the population. In 1971, smallpox was released accidentally from a Soviet bioweapons lab in Aralsk in what is now Kazakhstan. There was no effective public health system in the area and life expectancy was extremely low at just forty years. Three people died in what was a fairly ideal situation for a runaway epidemic to break out. Vaccination rates were higher then than now, as the Soviet Union was playing a leading role in eliminating the disease. The government vaccinated fifty thousand people almost immediately, stopping the virus in its tracks. An attack in the US might have worse results due to the fact that people are no longer routinely vaccinated against smallpox, but the likelihood of such an attack is still not great. About half the US population still has some immunity to smallpox. The disease is not airborne and therefore quarantine and vaccination could bring it under control. A smallpox outbreak would be a terrible, terrifying event, requiring a massive mobilisation of resources and leading to vast disruption, but it would not end society as we know it. It is no wonder that many scientists regard bioweapons as weapons of mass disruption rather than destruction. We may not even notice a bioweapons attack. Authorities in Oregon took more than a year to find out that the Bhagwan Shree Rajneesh group had deliberately spread salmonella through salad bars. Until a confession by a member of the group, investigators had thought it was just another periodic outbreak that occurs each year in the United States. Japanese authorities did not even know about the five bioweapons attacks by Aum Shinrikyo until one of the cult members confessed. Likewise, natural outbreaks of disease have raised concerns that they might be a bioweapons attack. When the West Nile virus started to spread in the United States, there were fears that the disease, which does not normally occur in the Western Hemisphere, had been spread deliberately. The Cuban government accused the United States of a bioweapons attack after an outbreak of dengue fever in 1981 although there was no proof of this and it fits a pattern of baseless accusations from Havana. Of more concern to some is an attack on agriculture. Spreading a disease that affects crops or animals could create more disruption and be easier than causing an anthrax outbreak in a city. However, there are obstacles to this: getting hold of a pathogen is not easy; crop surveillance is sophisticated and agricultural production is widely dispersed.

There are, however, worrying trends. Forty of the fifty-six criminal uses of bioweapons occurred in the 1990s. Nineteen of twenty-seven of terrorist cases occurred in the same decade. But this needs to be put into context – they include cases in which terrorist groups and criminals expressed an interest in biological weapons, not necessarily cases where they acquired or used them. Although the trend is troubling, there are some figures that need to remain at the front of everyone's minds: From 1900 to 2000, ten people died in criminal and terrorist bioweapons attacks. From 2000 until 2006, six more people had joined the list. The deaths of sixteen people over more than a century must make bioweapons attacks one of the least likely ways to die.

The risks of large numbers of people dying in a chemical or biological attack are small but they remain the perfect weapons of terror because of the ways in which we regard risks. We ignore routine hazards such as smoking or fatty foods, but we greatly fear the rare and apocalyptic. We tend to exaggerate risks when we are surrounded by the media storm that accompanies terrorism. A combination of innate fear of poisoning, a horror of disease, anxieties about science and the failure of scientists to find ways to protect us contribute to the dread around chemical and biological weapons that have a profound impact on the way we respond to these problems. Poisoning arouses intense fears because of the sense of contamination and defilement, as well as the impossibility of knowing immediately if one has escaped from the threat. A lack of knowledge about a poison makes the situation worse – the less people are told, the more inclined they are to fear contamination. A horror of disease plays into this. A fear of contamination and infecting others makes the fear even greater. The less well known the disease, the greater the anxiety. Very rare viral illnesses such as Marburg and Ebola are the subject of terrifying books and movies. Ebola has killed around sixteen thousand people since it was first identified in 1976 and fewer than five hundred people are known to have died from the Marburg virus. Malaria, by contrast, kills more than a million people a year but features in no Hollywood blockbusters. If there is a terrorist attack using chemical or biological weapons, there is a likelihood of significant panic. The treatment of mental health problems and panic will be as important as the treatment of those physically injured in any attack. This would require a well-organised public health system, an effective mental health network, politicians who will calm rather than exaggerate and a media that does not jump to conclusions. There is also a need to be cautious about responding to every false alarm and hoax; a study of one such case showed that the sight of emergency workers in biohazard gear dealing with a threat provoked considerable anxiety and psychological trauma even though no attack took place. Assessing the real risks is vital; only 2.3 per cent of reports of suspicious substances provoked a full-scale response by British police. In the United States 15 per cent led to a response and in some European countries it was as high as 100 per cent.

Bioterrorism has become a "dreaded risk", something that evokes such a deep emotional response that calm, measured responses have become almost impossible. Just as people were once willing to consume human flesh, whale vomit or gemstones to evade poisoning, modern societies have become willing to accept a whole array of policies that may be more harmful than the threat they are designed to combat. An example is the desire among many officials to classify critical genetics research that might be used to develop bioweapons. Recently scientists have been able to produce the polio virus from scratch using a published genome and DNA ordered from labs. Others have changed the DNA of viruses to make them resistant to inoculations or the human immune system, making them vastly more dangerous. The increasing ease with which this research can be carried out has alarmed security experts, but scientists contend that closing off the free flow of information would reduce the pace of scientific development, slow efforts to produce countermeasures to diseases that could be spread deliberately and leave the world much more vulnerable in the case of natural outbreaks of diseases such as avian flu. A study by the Department of Defense in 1982 concluded that US security was enhanced by ensuring an open flow of scientific knowledge. This was borne out by the situation in the Soviet Union where scientific exchanges were tightly controlled and a great many problems, from covert bioweapons projects to terrible environmental disasters, were kept from public and scientific scrutiny. Policing of scientists by other scientists is essential.

The most effective way to deal with the threat of bioterrorism is not to impose draconian limits on research and science, but to improve public health systems around the world, many of which had been severely cut back before the Covid-19 pandemic. Surveillance of disease, early efforts to inoculate people around outbreaks, close scrutiny of the spread of such illnesses as the drug-resistant tuberculosis spreading in Russian prisons and of diseases such as coronaviruses from birds and animals will all be vital. Public health systems need to be effective, global and well-funded.

The Covid-19 pandemic ramped up all the anxieties about scientists and their behaviour, ranging from stories that the virus had been deliberately developed and released, that it was the result of "gain of function" experiments in a Chinese lab, to theories that it had been whipped up by Bill Gates to expand his wealth and surveillance over humanity. The stories were multitudinous and varied, spurred by the genuine fears of illness and the power of social media to ride roughshod over our cognitive processes. The endless repetition and ubiquity of such stories meant they found an ever expanding audience and gained credibility, even as governments and others struggled to tell the public what they knew.

HOW LIKELY IS an attack using chemical or biological weapons? How many will die? No one can answer these questions with any certainty but we can look at what sort of role fear plays in the scenarios in the minds of terrorists and those who have to respond to the risks. Mass poisonings have become a staple of modern fiction and movies. A whole string of movies since the 1990s – *The Rock, Outbreak, Executive Decision, 12 Monkeys* and *I Am Legend* –

centred around such events. *The Cobra Event*, a novel by Richard Preston, and Stephen King's *The Stand* all captured not just the wider public imagination but the fears of President Clinton as well. Television series such as *Chicago Hope, The X Files, Seven Days, Outer Limits, The Burning Zone, Spooks* and *Millennium* all featured chemical or biological weapons.

Many of the ideas presented as possible scenarios are much more difficult to carry out than is often suggested. The Soviet bioweapons expert Ken Alibek wrote that the real challenge was not so much the bacteria or virus, but the ability to weaponise it. "The most virulent culture in a test tube is useless as an offensive weapon until it has been put through a process that gives it stability and predictability. The manufacturing technique is, in a sense, the real weapon, and it is harder to develop than individual agents." Poisoning a water supply would be extremely difficult – the dilution of most poisons and the chlorination process would make it necessary to use an enormous volume of most toxins. Poisoning the water supply in a building is a more likely scenario, as a water tank could be contaminated. However, this is likely to be noticed fairly rapidly and contained.

After the break-up of the Soviet Union, scientists revealed a massive bioweapons programme carried out by the Biopreparat organisation, which was supposed to be involved in drug research. Among the many innovations they developed was the ability to insert plasmids, small circular sections of DNA that exist outside of chromosomes, into bacteria to make them more virulent. What this meant was that they could develop binary biological weapons. The bacteria would be significantly less virulent until mixed with the plasmids, meaning that bacteria production and transportation would be far safer and easier. The virulence of certain bacteria can also be enhanced through gain-of-function modifications, making much more deadly, infectious or drug-resistant forms of diseases.

Gene mapping, the ability to know the DNA sequence of any living thing, is already a readily available and affordable technology. What may become more common is the ability to build living things using DNA fragments inserted together in the correct manner. Smallpox may not exist outside the two high-security repositories in the United States and Russia, but it may become increasingly easy for a scientist to splice together the genetic material to make a new version of the virus. Scientists have reconstructed the virus that caused the 1918 Spanish Flu pandemic, drawing on RNA found in cadavers of victims. More recently, a lab created a synthetic Ebola virus to have studies to sample. While building a new virus, it would also potentially be possible to make it more virulent or resistant to existing treatments by inserting additional genes, a now common process known as "gain of function". It is believed that Soviet scientists did some on work creating "chimeras", bioweapons that, for example, attempted to combine the stability of smallpox with the deadliness of Ebola.

From 1970 to 2000, computers went from being enormous, expensive and difficult to operate to being ubiquitous and vastly adaptable. In the next thirty years, biotechnology, and notably synthetic biology that involves creating new combinations of genetic material, may take a similar trajectory. Just as a computer can be used to damage other computers and systems, these biotechnologies are likely to have a similar capacity in the wrong hands. Gene splicing may make it easy to get a humble *E. coli* bacterium to produce botulinum toxin or to produce everyday food plants laced with poisons. Genes could be inserted into populations stealthily and then be activated. Synthetic biology also has the potential for better monitoring of possible diseases and, as seen during the Covid-19 pandemic, the rapid creation of new vaccines. Much depends on how governments respond to the emerging risks.

Several nations may still have bioweapons programmes and some capacity to cause serious problems: Russia is the most likely, given its evasions of treaty obligations, but North Korea and Syria are suspects, too. The increasing commercial manipulation of DNA outside highly specialised laboratories, and the likelihood of both greater automation of these processes and the use of artificial intelligence in developing novel molecules, does raise a sinister spectre of groups other than states being able to master new bioweapons rather than relying on the limited number of natural pathogens that can be weaponised.

One possible threat is the weaponisation of one of the five million species of fungi that live around us. Fungi produce several highly dangerous toxins and infection is always difficult to treat, in part because scientists have found it a challenge to isolate differences in the metabolism of mushrooms that could be targeted. Nearly two million people a year die from fungal illnesses but they attract little research funding. Another concern that has not been well addressed is the toxic potential of nanotechnology, the realm of microscopic processes and products much smaller than the smallest organisms. Nanotechnology is already used in many products including pharmaceuticals, sun screen, sports gear and clothing. While it offers a vast array of potential benefits – self-cleaning clothes, new approaches to medicine – materials at this scale can have very different properties than when found in larger forms. Nanoparticles can easily enter the body and their impact is little understood even as their uses proliferate. For example, nanoparticles of silver used to provide antibacterial effects in food packaging allowing products to last longer, can enter the blood stream and can be shown to be toxic in lab studies. Speculative works by academics suggest nanobots the size of the tiniest flies might be unleashed armed with ricin or botulinum toxin. New lasers developed with nanomaterials could produce fission in small amounts of nuclear material, creating a nuclear hand grenade. Nanotoxins could be released to kill people in vast numbers after a certain time, although it is hard to imagine any combatant not suffering from some blowback from such an act, as the particles would likely persist and poison those who deployed them.

Nanotechnology is already used in the manufacture of weapons, for example making bombs more powerful by enhancing explosives, but to date nothing has emerged suggesting death flies are about to be unleashed on us all. Science fiction soon becomes reality in a world when computer power advances by leaps and bounds. It is now possible to synthesise a new bacterium but it remains vastly complex and expensive. That may not be the case for much longer. Many of the fantasies of medieval poisoners – delayed effects and undetectable toxins – might all come to pass.

POSTSCRIPT: THE BASILISK

*What availed, / Murrus, the lance by which thou
didst transfix / A Basilisk? Swift through the
weapon ran / The poison to his hand: he draws his
sword / And severs arm and shoulder at a blow: /
Then gazed secure upon his severed hand / Which
perished as he looked. So had'st thou died, / And
such had been thy fate!*
                    Lucan, *Pharsalia*

J. K. Rowling brought the basilisk back to
widespread attention when a dark wizard, Herpo
the Foul, bred one by hatching a chicken's egg
beneath a toad. Being a Parselmouth, Herpo was
able to talk to the giant green snake, which could
grow to a length of more than ten metres. The
basilisk's glance could kill and nobody, not even
Harry Potter, was immune to its deadly powers.
Basilisks were much feared, with the Ministry of
Magic banning their breeding, even inspecting
chicken coops to ensure no toads were hatching
them.

The animal plays an important role in *Harry
Potter and the Chamber of Secrets*, when Salazar
Slytherin hides a basilisk in a chamber at Hogwarts.
The giant snake was able to petrify people even if

> they looked into its eyes in a mirror rather than
> directly. The lives and powers of basilisks became
> the subject of much debate in the forums run by
> fans of the books. Much eye-rolling and derision
> was provoked by a Lego video game in which the
> obviously female basilisk was portrayed as a male
> with a red coxcomb.

T he history of poisoners is a history of echoes. Aum Shinrikyo and the Peoples Temple recall medieval apocalyptic movements in which the rootless poor, often stressed by famine or violence, gathered around a leader in a spiritual mission that first offered redemption but turned to violence. The Brethren of the Free Spirit, a millenarian sect that started in the thirteenth century and persisted for several hundred years in various guises, encouraged a breaking of all the regulations of society. They were pantheistic, believing that everyone and everything was God and therefore nothing was sinful. Rape, theft and murder were encouraged among the chosen people.

A later group, the Taborites, appeared in Bohemia in the fifteenth century. They started out as pacifists but violence took over the group. The "elect", as they called themselves, wanted to kill to purify society in preparation for the return of Christ. At one stage they were even led by a blind general, Jan Zizka, who promised them freedom from a materialistic society and purity through murder. Jim Jones created a closed, paranoid world in which he constantly reiterated that outsiders were out to get him. He even went as far as to stage attacks and kidnappings to convince his followers that they were genuinely threatened. The paranoia was fuelled by rigorous discipline and even violence against dissenters within the group, as it had been in the Peoples Temple. Many Aum Shinrikyo followers were so convinced that the group was under threat that they actually believed the subway attack was aimed at them.

Al Qaeda's justification for violence often included the idea that it would purify the world and take it back to a better time – in this case, the time when the Prophet Mohammed walked the earth. Islamic State offered a similar vision of a purer time. Poisons have an appeal in the idea that they can cleanse the world, an idea that stretches back to the basilisk brought in to rid the Temple of Apollo of all living things. Several James Bond movies took on this idea with a deranged visionary imagining the purging of the planet and the creation of a better world.

There is certainly a small risk from a chemical or biological attack by an extremist group. But even if no incident occurs, the fear of one may have taken hold in our society – something that may cause more harm than a real attack. Our responses to the fear of poison are starting to raise echoes of their own. We have returned to an age of anxiety about the invisible.

Perhaps more dangerous are those states that have used poisons with impunity, most recently Russia and Syria. Their attacks have killed hundreds but serve little real tactical ends. They are all about inducing fear and uncertainty, as well as rattling the West. Although both nations are signatories of the Chemical Weapons Convention, they have ignored international law because they know they can act with impunity. Whenever they have used poisonous weapons, either for assassinations or against large numbers of their enemies, they have blamed everyone else, denied using them, turned the tables on their opponents and created confusion through their media and their useful idiots. Every use of poison is said to be a false flag attack on them, an effort to undermine them and large numbers of people are willing to go along with these lies. The disinformation spreads rapidly on social media but the dry reports with their evidence of who did carry out the attacks are barely noticed. It all ramps up the information warfare that has always been part of poisonings, tapping into our deep fears.

These fears are as contagious as any bioweapon and more difficult to clean up than any chemical toxin. A terrorist attack in the future that uses chemical or biological weapons will only worsen these fears, creating an array of unpredictable responses. It is likely that far more people have died in panicked over-reactions to fears of poisons than from mass poisonings themselves. History is littered with the bodies of those killed for poisoning or being seen as poisonous in themselves. Pogroms against Jews, the burning of witches and the murder of enslaved people are all extreme expressions of the fear that came in part from the lack of understanding of the dreaded risks of poison. But today we understand much more about the real risks of being poisoned. We have chemical sensors that can pick up minute traces of toxins, tests for almost all deadly substances, an understanding of the mechanisms by which poisons kill and effective treatments for a great many of them. New biotechnologies offer as many or more opportunities as they create threats.

Our fears endure, undiminished, and those fears will always be open to exploitation. History shows us that claims of terrible deadly threats – new, invisible, unknown – almost always exceed the risk. There is always someone out there urging us to be afraid of the basilisk.

ACKNOWLEDGEMENTS

My thanks are owed to many people, too many to mention here. Above all I am most grateful to my agents Deborah Rogers and David Miller who sadly did not live to see this book published. They were enormously important friends, advisers and supporters, the sort all authors need, and their loss is deeply felt. Abby Robinson and Kate Bourne read early drafts without rolling their eyes too much. Minh Bui Jones both improved the book immeasurably and has taken the risk of publishing it. The staff at the British Library, the Wellcome Institute Library, the New York Public Library and at Cafelunya on Ronda de Sant Pere in Barcelona are owed many thanks, as are all my friends and family.

ENDNOTES

Introduction: The Empty Sand

Details of Fred Lewis at: Fagan, Kevin. Haunted by Memories of Hell. 12
    November. *SF Gate*. https://www.sfgate.com/news/article/haunted-by-
    memories-of-hell-2979612.php (1998).

This chapter contains quotes from the last speech recorded by Jim Jones,
    known as the "Death Tape". A vast trove of records and writings are
    available at the Jonestown Project at the Library of the San Diego State
    University at Jonestown.sdsu.edu. FBI Transcript of the "Death Tape"
    13 June 1979.
    Available at: https://jonestown.sdsu.edu/wp-content/uploads/2019/03/
    Serial-2303.pdf

Some details come from Templer, Robert. Jonestown. *The Richmond Review*.
    1998. Available at https://www.richmondreview.co.uk/temple02/

Ann Moore's letters have been published in several books:

Background of Jim Jones in Indiana is from: Carpenter, Dan. Countdown to
    Armageddon: The Reverend Jim Jones and Indiana. *Traces of Indiana
    and Midwestern History*. Indiana Historical Society. 2018. Available at:
    https://jonestown.sdsu.edu/?page_id=80779

The New West article that precipitated the crisis within the Peoples Temple
    can be found at: Kidruff, Marshall and Tracy, Phil. Inside Peoples
    Temple. *New West*. 1 August (1977) https://jonestown.sdsu.edu/?page_
    id=14025

The chapter draws from the many books written on Jonestown as well as these
    articles:

On cults and new religions: Olson, Paul J. "The Public Perception of 'cults'
    and 'New Religious Movements'." *Journal for the Scientific Study of
    Religion*. 45.1 (2006). Chidester, David. 'Rituals of Exclusion and the
    Jonestown Dead'. *Journal of the Academy of Religion* (1988). Guinn,
    Jeff. *The Road to Jonestown. Jim Jones and the Peoples Temple*. Simon
    and Schuster. (2017)

On the legacies of the Temple: Kutulas, Judy. *After Aquarius Dawned. How the Revolutions of the Sixties Became the Popular Culture of the Seventies.* University of North Carolina Press 2017. Feltmate, David. People's Temple. A Lost Legacy for the Current Moment. *Nova Religio: The Journal of Alternative and Emergent Religions*, Volume 22, Issue 2, Crockford, Susannah 'How do you know when you are in a Cult?" *Nova Religio: The Journal of Alternative and Emergent Religions*, Volume 22, Issue 2. (2018) Moore, Rebecca. Jonestown Forty Years On. *Nova Religio: The Journal of Alternative and Emergent Religions,* Volume 22, Issue 2, (2018). Moore, Rebecca. The Stigmatized Deaths at Jonestown. *Death Studies*, 35: 42–58, (2011). Moore, Rebecca. *In Defense of People's Temple... And Other Essays.* Edward Mellen Press. (1988). Wendy Edmonds. *intoxicating FOLLOWERSHIP: in the Jonestown Massacre.* Emerald Publishing (2021).

Other sources include: The Jonestown Memorial Project: https://web.archive. org/web/20201203091505/https://www.jones-town.org/ Higgins, Chris.. Stop Saying 'Drink the KoolAid.' *The Atlantic.* 8 November (2012). https://www.theatlantic.com/health/archive/2012/11/stop-saying-drink-the-kool-aid/264957/ Guyana Coroner's Report. 13th December 1978. Available at https://jonestown.sdsu.edu/wp-content/uploads/2013/10/GuyanaInquest.pdf Judge, John *The Black Hole of Jonestown.* (1985). Available at: https://ratical.org/ratville/JFK/JohnJudge/Jonestown. html. FBI Audiotape Collection from Jonestown. Available at https:// jonestown.sdsu.edu/?page_id=27280

Part I

1. The Soul-Hunting Fog

This chapter draws from the extensive work of Adrienne Mayor on ancient warfare and the uses of poisons. Mayor, Adrienne. Greek Fire, Poison Arrows, and Scorpion Bombs. Princeton University Press. (2003). Matyszak, Philip. Ancient Magic: A Practitioner's Guide to the Supernatural in Greece and Rome. Thames and Hudson. (2019). Thorndike, Lynn. History of Magic and Experimental Science. Six Volumes. Columbia University Press. (1943). Golden, Cheryl. The Role of Poison in Roman Society. Dissertation. University of North Carolina at Chapel Hill. (2005).

On Homer's Odyssey and Iliad, it draws from the translations by Emily Watson. Watson, Emily: The Iliad. (2023) W.W. Norton and Company and The Odyssey. (2017) W.W. Norton and Company.

"Ever Living Gods..." Homer. *The Odyssey*, Book I.

Arrows were not always fatal: Christine Salazar. 'The Treatment of War Wounds in Graeco-Roman Antiquity'. *Studies in Ancient Medicine.* (2000). The recipe for Scythion is described in: Claudius Aelianus, *De Natura Animalium*, 9.15

On Scythians from Renate Rolle. *The World of the Scythians.* B.T Batsford. (1989).

On Herodotus: *The Histories,* Book 1, Chapter 120.

On the Battle of Kirrha: Mayor. *Greek Fire.*

On Strabo: Strabo. *Geography.* Book XI Chapter 2. Section 19.

On "Writing on Against a Stepmother:" See Gagarin, Michael *Antiphon: The Speeches.* Cambridge: Cambridge University Press. (1997). Freeman, Kathleen. *'Against a Stepmother'.* In *The Murder of Herodes and Other Trials from the Athenian Law Courts* (1963). Also, see Davidson, James. *Courtesans and Fishcakes* (1998). Carey, Christopher *Trials from Classical Athens.* London: Routledge. (1997)

On Jacques Derrida's ideas on *pharmakon* see: Derrida, Jacques. "Plato's Pharmacy" In: *Dissemination*, translated by Barbara Johnson, Chicago, University of Chicago Press. (1981).

On the Chinese character *du* see Liu, Yan. 'Poisonous Medicine in Ancient China'. In *History of Toxicology and Environmental Health.* Elsevier. (2015).

On Hindu taboos on poisoned weapons see: *The Laws of Manu*, Chapter VII. Line 90. Translated by G. Buhler (1886).

On Nicander see. Theriaca. Translated by.Gow, A and Scholfield, A (1953) and available at http://www.attalus.org/poetry/theriaca.html. In 1771, Albrecht von Haller dismissed Nicander's poem in his *Biblioteca Botanica.*

On Greek deities and women healers see: Achterberg, Jeanne. *Woman as Healer* (1991).

On Greek and Byzantine pharmacological history in Scarborough, John. Early Byzantine Pharmacology. *Dumbarton Oaks Papers* (1984). On Love Philtres see: Faraone, Christopher. *Ancient Greek Love Magic.* Harvard University Press. (1999)

On Medea see: Dutta, Shomit and Goldhill, Simon. Eds. *Greek Tragedy.* Penguin Classics. (2004).

On the Nessus Shirt see: Mayor, Adrienne. The Nessus Shirt in the New World: Smallpox Blankets in History and Legend. *Journal of American Folklore* 108:427:54 (1995). Bennett, Gillian. Bodies: *Sex, Violence, Disease and Death in Contemporary Legend* University Press of Mississippi. (2005).

2. An Evil Notoriety

The words "Like the basilisk softly whistling in its cave..." is Machiavelli's description of Cesare Borgia. From Machiavelli, Nicolo. *The Prince.* Independently Published. (2020).

For background on poisonings in the Ancient World see: Matyszak, Philip. Ancient Magic: A Practitioner's Guide to the Supernatural in Greece and Rome. Thames and Hudson (2019). Southon Emma. A Fatal Thing Happened on the Way to the Forum: Murder in Ancient Rome. Abrams Press. (2021). Southon, Emma. Agrippina: The Most Extraordinary Woman of the Roman World. Pegasus Books. (2019). Barrett. Anthony A. Livia: First Lady of Imperial Rome. Yale University Press. (2021). Beard. Mary Twelve Caesars: Images of Power from the Ancient World to the Modern. Princeton University Press. (2021) Beard, Mary. Emperor of Rome. Liveright (2023). Mayor. Adrienne. Poison King: *The Life and Legend of Mithridates,* Rome's Greatest Enemy. Princeton University Press. (2011) Golden. Cheryl. The Role of Poison in Roman Society. PhD Dissertation. Chapel Hill. (2005).

Pliny the Elder describes the basilisk in *Natural History* Book VIII. Chapter 33. https://www.perseus.tufts.edu/hopper/text?doc=Perseus:text:1999.02. 0137:book=8:chapter=33

On the events of 331 BC in Livy. *The History of Rome*, book viii. 18. Also see: Tenney, Frank. 'The Bacchanalian Cult of 186 BC'. *Classical Quarterly* (1927).

Gyara was an island in the Aegean on which prisoners were confined.

On drugs, medicines and poisons: Cilliers, Louise and Retief, F. P.. 'Poisons, Poisoning and the Drug Trade in Ancient Rome'. *Akroterion* (2014). Horstmanshoff, Manfred. 'Ancient Medicine Between Hope and Fear: Medicament, Magic and Poison in the Roman Empire'. *European Review* (1999).

Tacitus does not mention this detail but it is described by Cassius Dio and others. Tacitus. *The Annals*, book xii. 66.

On deaths among emperors: Grimm-Samuel, Veronika 'On the Mushroom that Deified the Emperor Claudius'. *Classical Quarterly* (1991). Cilliers Louise and Retief F. P. Causes of Death among the Caesars (27 BC–AD 476). *Acta Theologica* (2006). Scheidel. Walter. Emperors, Aristocrats and the Grim Reaper: Towards a Demographic Profile of the Roman Elite. *Classical Quarterly* (1999). Wolf, Greg. *Rome: An Empire's Story*. Oxford University Press. (2021).

On the issue of lead in the Ancient World: Cilliers Louise and Retief F.P. Lead Poisoning in Ancient Rome. *Acta Theologica* (2010). Emsley J. Ancient

World Was Poisoned by Lead. *New Scientist* (1994). Makra Lazlo. Anthropogenic Air Pollution in Ancient Times. In: *History of Toxicology and Environmental Health.* (2015). Scarborough J. The Myth of Lead Poisoning among the Romans: An Essay Review. *Journal of the History of Medicine.* (1984). Needleman, Lionel and Diane. Lead Poisoning and the Decline of the Roman Aristocracy. *Classical Views* (1985). Cilliers, Louise and Retief F.P. Lead Poisoning and the Downfall of Rome: Myth or Reality? *In History of Toxicology and Environmental Health.* Elsevier. (2014). Gilfillan, S.C. Lead Poisoning in Ancient Rome. *Acta Theologica Supplementum* 7 (2005).

On the decline of the Roman Empire see: Harper, Kyle. *The Fate of Rome: Climate, Disease and the End of an Empire* (2017). Meier, Mischa. The "Justinianic Plague": The Economic Consequences of the Pandemic in the Eastern Roman Empire and its Cultural and Religious Effects. *Early Medieval Europe* (2016).

On laws on magic in Rome: Pharr, Clyde. The Interdiction of Magic in Roman Law. *Transactions and Proceedings of the American Philological Association* (1932).

On the death of Germanicus: Tacitus. *The Annals,* book ii. 69-73.

On Martina killing herself with poison hidden in her hair: Tacitus. *The Annals*, book iii. 7.

On the blurring of medicines and poisons in Rome: Horstmanhoff. 'Ancient Medicine'.

On Pliny on poisons and empire see: Jones-Lewis, Molly Ayn. Poison: Nature's Argument for the Roman Empire in Pliny the Elder's Naturalis Historia. *Classical World* Vol 106. No. 1 (2012).

On Tacitus and The Annals in shaping our understanding of the Julio-Claudian Dynasty and its concentration of power in the hands of the emperor see: Rach, JuliAnne. *Practicing Magic: An Evaluation of Magic, Gender and Power in Tacitus*. Dissertation at University of Arizona. (2022).

Alain Touwaide. Pietro d'Abano, De venenis: Reintroducing Greek Toxicology into Late Medieval Medicine in *Toxicology in the Middle Ages and Renaissance*, 2017, p.43-52

Collard, Franck. *The Crime of Poison in the Middle Ages* Deborah Nelson-Campbell (Translator) Wiley. (2003) Gibbs, Frederick. *Poison, Medicine, and Disease in Late Medieval and Early Modern Europe* Routledge (2016).

Kaye, J.M. The Early History of Murder and Manslaughter. *Law Quarterly Review* (1967).

3. What a Splendid Result of Reason

On poisoning in the Middle Ages: Collard, Franck. The Crime of Poison in the Middle Ages Deborah Nelson-Campbell (Translator) Wiley. (2003) Gibbs, Frederick. Poison, Medicine, and Disease in Late Medieval and Early Modern Europe Routledge (2016).

On Chaucer see: Margaret Hallissy. Poison: Imagery and Theme in Chaucer's Canterbury Tales. Dissertation at Fordham University. (1974). Margaret Hallissy. Poison and Infection in Chaucer's Knight's and Canon Yeoman's Tales. Essays in Arts and Sciences, 1981-05, Vol.10 (1), p.31. Richard Ireland. Chaucer's Toxicology. The Chaucer Review. Vol 29. No. 1 (1994).

On Peter of Abano. Touwaide, Alain. Pietro d'Abano, De venenis: Reintroducing Greek Toxicology into Late Medjeval Medicine in Toxicology in the Middle Ages and Renaissance, (2017).

On crimes in Medieval England see Hanawalt, Barbara. Crime and Conflict in English Communities, 1300–1348 Harvard University Press. (1979).

First law in England on poisoning: Richard Ireland. Medicine, Necromancy and the Law: Aspects of Medieval Poisoning. *Cambrian Law Review* (1987).

On Piers Plowman: Rosanne Gasse. The Practice of Medicine in "Piers Plowman". *The Chaucer Review* (2004).

Franck Collard writes extensively on poisoning in the medieval world in his works: Collard, Franck. Le salut par la couleur. Mutations chromatiques et détection du poison dans les Giftschriften du Moyen Âge latin. Pallas, Vol 117. (2021). Collard, Franck. Les écrits sur les poisons. Typologie des Sources du Moyen Âge Occidental, 88. Collard, Franck. *The Crime of Poison in the Middle Ages.* Praeger, (2008). Collard, Franck and Hanna, Ralph. Malachy the Irishman, On Poison: A Study and an Edition, *Cahiers de civilisation médiévale*, 262 | (2023). Collard, Franck. Le poison et le sang dans la culture médiévale. *Médiévales*, Vol.60. (2012).

Deaths among Byzantine leaders: Lascaratos, John. *The Poisoning of the Byzantine Emperor Romanos III Argyros* (1995). Padover, Saul. 'Patterns of Assassination in Occupied Territory'. *Public Opinion Quarterly* (1943).

On Islamic toxicology see: Urdang, George. Pharmacopeias as Witnesses of World History. *Journal of the History of Medicine and Allied Sciences* (1946). Levey, Martin. Medieval Arabic Toxicology; the Book on Poisons of Ibn Wahshiya and its relation to early Indian and Greek Texts. *Transactions of the American Philosophical Society* v. 56, pt. 7.

(1966). Levey, Martin. Ibn al-Waḥshīya's "Book of Poisons," "Kitāb al-Sumūm": Studies in the History of Arabic Pharmacology II. *Journal of the History of Medicine and Allied Sciences,* Vol.18 (4) (1963). Levey, Martin. Chemistry in the "Kitab Al-Sumum" ("Book of Poisons") by "Ibn Al-Wahshiya" *Chymia*, Vol.9. (1964). Mozhgan Ardestani, Roja Rahimi, Mohammad Esfahani, Omar Habbal, Mohammad Abdollahi. The Golden Age of Medieval Islamic Toxicology. In *Toxicology in the Middle Ages and Renaissance. History of Toxicology and Environmental Health.* Elsevier. (2017). Frederick Gibbs. *Poison, Medicine, and Disease in Late Medieval and Early Modern Europe.* Routledge. (2019).

On Avicenna: Wickens, G.M.. *Avicenna, Scientist and Philosopher.* Luzac. (1952). Gaynes, Robert. Avicenna, a Thousand Years Ahead of His Time. *Germ Theory.* 2011. Ebrahim Nasiri, Jamal Orimi, Mohammad Hashemimehr, Zahre Aghabeiglooei, Maedeh Rezghi, Mohammad Amrollahi-Sharifabadai. Avicenna's clinical toxicology approach and beneficial materia medica against oral poisoning. *Archives of Toxicology*, Vol.97 (4). (2023-04).. M. Heydari, M.H. Hashempur, A. Zargaran. Medicinal aspects of opium as described in Avicenna's Canon of Medicine. *Acta Med Hist Adriat* 11. (2013). Samaneh Soleymani and Arman Zargaran. From Food to Drug: Avicenna's Perspective, a Brief Review. *Research Journal of Pharmacognosy*, Vol.5 (2), (2018).

On Maimonides: Suessman Muntner, Suessman. '*Maimonides' Book for Al-Fadil'*. Isis (1944). Manguel, Alberto. *Maimonides: Faith in Reason.* Yale University Press (2023). Ferrario, Gabriele. Maimonides' Book on Poisons and the Protection Against Lethal Drugs in *Toxicology in the Middle Ages and Renaissance.* Elsevier. (2017).

4. Poisoning Wells

For more on the development of Medieval antisemitism see: Trachtenberg, Joshua. *The Devil and the Jews.* Meridian Books. 1961.

On the blood libel against Jews see Hannah Johnson. *Blood Libel. The Ritual Murder Accusation at the Limit of Jewish History.* University of Michigan Press. (2012). Gregory X Letter on the blood libel at The Jewish Virtual Library: https://www.jewishvirtuallibrary.org/gregory-x-letter-about-jews-against-the-blood-libel?utm_content=cmp-true

On the *foetor judaicus:* Lanfranchi, Pierluigi *Foetor judaicus*: archéologie d'un préjugé. *Pallas.* Vol 104 No. 2 (2017).

On the imagery around Jews in the Middle Ages: Strickland, Debra Higgs. *Saracens, Demons, and Jews: Making Monsters in Medieval Art.* Princeton University Press. (2003).

Angolo di Tura. Cronaca Senese. In Rerum Italicarum scriptores (1931–39).

There is considerable debate over whether the Black Death and other epidemics were the plague now known to be caused by the bacterium Yersinia pestis or some other disease or mix of diseases. There are many inconsistencies. Modern plague does not seem nearly as virulent but it might have changed over time and more people may have some resistance. Few plague accounts mention the expected mass death of rats before the disease breaks out in people. It seems to have spread faster than expected, given the limited roaming range of urban rats. People also described a range of symptoms other than the classic buboes; indeed, some descriptions point more clearly to anthrax. Malnutrition may have led to higher death rates than modern plagues because of the damage it does to the immune system. It may also have been spread by the human flea rather than the rat flea. Yersinia pestis may have simply been much more virulent in earlier forms or people may have better developed immunity that was built up over centuries of outbreaks. The debates will likely continue for some time, particularly as the Black Death remains a popular area for historical study. There are at least eight *Yersinia pestis* strains and four biovars, and all have emerged within the last 5000 to 20,000 years

On plague and well poisonings: Bennett, Gillian. Bodies: Sex Violence, Disease and Death in Contemporary Legend University Press of Mississippi. (2005). Carmichael, Ann. Universal and Particular: The Language of Plague, 1348–1500. *Medical History Supplement*. 27. (2008). Ginzburg, Carlo. *Ecstasies: Deciphering the Witches' Sabbath.* University of Chicago Press. (1990). Azzolini, Monica. 'Plagues, Poisons and Potions: Plague-Spreading Conspiracies in the Western Alps, c. 1530-1640 *Bulletin of the History of Medicine* (2004). Finley, Theresa and Koyama, Mark. Plague, Politics, and Pogroms: The Black Death, the Rule of Law, and the Persecution of Jews in the Holy Roman Empire. *The Journal of Law and Economics*, Vol.61 (2) (2018). Voigtländer, Nico and Voth, Hans-Joachim. Persecution Perpetuated: The Medieval Origins of Anti-Semitic Violence in Nazi Germany. *Quarterly Journal of Economics* 127. (2012). Aberth, John. *From the Brink of the Apocalypse: Confronting Famine, War, Plague, and Death in the Later Middle Ages*. 2d ed. Routledge. (2010) Breuer, Mordechai. The "Black Death" and Antisemitism. In *Antisemitism through the Ages*, edited by Shmuel Almog. Pergamon Press. (1988). Barkai, Ron. 'Jewish Treatises on the Black Death (1350–1500)'. *In Medicine from the Black*

Death to the French Disease (1998). Girard, Rene. *The Scapegoat.* Johns Hopkins University Press. (1986). Moore, Robert. *The Formation of a Persecuting Society.* Wiley-Blackwell. (2007). d'Agramont, Jacme. *Regimen de Perservacio de Pestilencia.* (1348). de Cordoba. Alfonso. *Epistola et Regimen de Pestilencia,* written in Montpellier in 1349. Naphy, William. *Plagues, poisons and potions Plague-spreading conspiracies in the Western Alps, c. 1530–1640* Manchester University Press (2002). Winslow, C. and Duran-Reynals, M. Jacme d'Agramont and the First of the Plague Tractates. *Bulletin of the History of Medicine.* Vol 22 No. 6. (1948). Barzilay, Tzafir. Early Accusations of Well Poisoning against Jews: Medieval Reality or Historiographical Fiction? *Medieval Encounters: Jewish, Christian, and Muslim Culture in Confluence and Dialogue,* Vol.22 (5). (2016). Barzilay, Tzafir. *Poisoned Wells. Accusations, Persecutions and Minorities in Medieval Europe 1321-1422.* University of Pennsylvania Press. (2022).

On school poisonings in the Middle East: Silver, Eric. 300 West Bank Girls 'Poisoned. *The Guardian.* 28 March 1983. Albert Hefez. The Role of the Press and the Medical Community in the Epidemic of "Mysterious Gas Poisoning" in the Jordan West Bank. *American Journal of Psychiatry* (1985). See CDC summary of the report at http://www.cdc.gov/mmwr/preview/mmwrhtml/00000068.htm. Bartholomew, Robert and Sirois, Francois. 'Occupational Mass Sociogenic Illness: A Transcultural Overview'. *Transcultural Psychiatry* (2000). When Poison Is Not Poison'. *Journal of Palestine Studies* Vol 13. No. (1983). Bartholomew, Robert and Wessely, Simon. Protean Nature of Mass Sociogenic Illness. *British Journal of Psychiatry.* 180. (2002). Tirosh, B., Modan, M., and Weissenberg, E. et al. The Arjenyattah Epidemic: A Mass Phenomena: Spread and Triggering Factors *Lancet,* 1472-4. (1983).

5. Plants Die, Shrubs Wither, Dogs Run Mad
On Dena Thompson: Cowan, Rosie. 'Black widow' jailed for life for killing husband. *The Guardian.* !6 December 2003. Bird, Steve. "This woman is every man's nightmare . . . They can sleep safe tonight knowing she has been taken off the streets'. *The Times.* 16 December 2003. Dena Thompson. The Black Widow and the Poisoned vindaloo. *Courtnewsuk. co.uk.* 13 December 2003. Gatton, Adrian. *Black Widow: The True Story of How Dena Thomson Lured Men into a Twisted Web of Sex, Lies and Murder.* John Blake. (2010) Thompson, Flora. Black Widow Killer to be released from jail: Parole Board. *Evening Standard.* 23 May (2022).
On cosmetics: Morag Sarah Martin. Consuming Beauty: The Commerce of

Cosmetics in France 1750–1800' Dissertation. University of California, Irvine. (1999). Neville Williams. Powder and Paint: A History of the Englishwoman's Toilet, Elizabeth I–Elizabeth II. Longman, Green and Co. (1957).

On Machiavelli's *The Mandrake*: Giannetti, Laura and Ruggiero, Guido. *Five Comedies from the Italian Renaissance* Johns Hopkins University Press. (2003).

On *Secretum Secretorum:* Williams, Steven. *The Secret of Secrets: The Scholarly Career of a Pseudo-Aristotelian Text in the Latin Middle Ages.* University of Michigan Press. (2003).

On Kautilya: Boesche, Roger. Kautilya's *Arthasastra* on War and Diplomacy in Ancient India. *The Journal of Military History.* Vol 67. No 1. (2003).

On poisonous women. Hallissy, Margaret *Venomous Woman: Fear of the Female in Literature.* Greenwood Press (1987). Muir, Edward. *Sex and Gender in Historical Perspective.* Johns Hopkins University Press. (1990). Helen Rodnite Lemay, trans. *Women's Secrets: A Translation of Pseudo-Albertus Magnus' 'De Secretis Mulierum' with Commentaries.* New York University Press. (1992). Cohen, Monte. 'The Diabolical and Diverse Dangerously Devastating Drugs and Dead Poisons and Potions of the Demonic and Devilish Dr Fu Manchu'. *Mithridata* (2006).

6. More Wine, More Excitement

On Alexander VI: Burchard, John. *Pope Alexander VI and his court—extracts from the Latin Diary of John Burchard,* Perennial Press, (2015). Cobb, Cathy. The Chemistry of Lucrezia Borgia et al. *ACS Symposium Series,* (20130. Bradford, Sarah. *Lucrezia Borgia: Life, Love, and Death in Renaissance Italy.* Penguin, (2004).Cobb, Cathy. The Case Against the Borgias: Motive, Opportunity, and Means. *Toxicology in the Middle Ages and Renaissance,* 2017. Ferrara, Orestes. *The Borgia Pope: Alexander VI* (1942), 398. Frederick Baron Corvo. Chronicles of the House of Borgia (1901). Ferdinand Gregorovius. Lucrezia Borgia (1904). Paolo Giovio Histories. II 47. Mittag, Martina. These Pale Alchemies: Lucretia Borgia in Nineteenth-Century Literature. 103 Simon, C. and Arco, M. The Death of Alexander VI: A Medical-Historical Analysis. *Medizinische Monatsschrift,* Vol.12 (10) (1958). Burchard, Johann. *At the Court of the Borgia: Being an Account of the Reign of Alexander VI.* Folio Society. (1963).

On the historiography and image of the Borgias. Phillips, Mark. *Francesco Guicciardini: The Historian's Craft* (1977). J. N. Hillgarth. 'The Image of Alexander VI and Cesare Borgia in the 16th and 17th Centuries'.

Journal of the Warburg and Courtauld Institutes (1996).

On the denunciations of the Borgias: John Bale. *The Pageant of Popes* (1574). The Latin original, published in Basel in 1558, is even racier in its denunciation of papal behaviour.

On the Borgias in English drama: Barnabe Barnes. *The Devil's Charter: A Tragedy Containing the Life and Death of Pope Alexander VI* (1999). Somerset, Anne. Unnatural Murder: Poison. Machiavelli. The Discourses. Playwright. *Archaeologia Aeliana* xxiv. 1946. Lucanie, Ralph. Papal Poisoning: Preferred Pastime of the Pontiffs. Mithridata. (1998).

On the de Medicis: Leonie Frieda. Catherine de Medici (2003). Letter from Cosimo De Medici on poisoning Piero Strozzi. See online Medici archives at http://www.medici.org/news/dom/dom999.html.

Quote from Niccolo Machiaveli. *The Discourses.* Conspiracies. III.6.

On the death of Francesco de' Medici. Mari F. Potettini A. Lippi D. Bertol E. The mysterious death of Francesco I de' Medici and Bianca Cappello: An arsenic murder? *British Medical Journal.* 2006

7. Gazing at Piss

On the basilisk: Krzyszczuk, Łukasz and Morta, Krzysztof Basilisk – The History of the Legend. *Alea.* Vol. 25/1 (2023). Bondeson, Jan. *The Feejee Mermaid and Other Essays in Natural and Unnatural History*. Ithaca, NY: Cornell University Press, (1999). Breiner, Laurence. The Career of the Cockatrice. *Isis.* A Journal of the History of Science Society, v. 70, no. 1, (1979). Cohen, Meredith. The Bestiary beyond the Book. In: Morrison, E. and Grollemond, L. (eds.). *Book of Beasts. The Bestiary in the Medieval World.* The J. Paul Getty Museum, (2019).

On Brunetto Latini: Holloway, Julia. *Brunetto Latini : An Analytic Bibliography.* Grant and Cutler. (1986)

On Isidore of Seville. Barney, Stephen et al. *Isidore. The Etymologies of Isidore of Seville.* Cambridge University Press, (2006). Bednarski, Steven. *A Poisoned Past: The Life and Times of Margarida de Portu, a Fourteenth-Century Accused Poisoner.* University of Toronto Press. (2014).

On Paracelsus: Ball, Philip. *The Devil's Doctor: Paracelsus and the World of Renaissance Magic and Science.* Farrar, Strauss and Giroux. (2006). Daniel, Dane Thor and Gunnoe, Charles. *Paracelsus.* Oxford University Press (2023). Webster, Charles. *From Paracelsus to Newton: Magic and the Making of Modern Science.* Cambridge, UK: Cambridge University Press, 1982. Thorndike, Lynn. *A History of Magic and Experimental Science.* 8 vols. Macmillan, (1923–1958.) Grandjean, Philippe.

Paracelsus Revisited: The Dose Concept in a Complex World. *Basic and Clinical Pharmacology and Toxicology*, Vol.119 (2) (2016). Moran, Bruce. Paracelus. An Alchemical Life. Reaktion Books. (2019). Gibbs, F. W. Specific Form and Poisonous Properties: Understanding Poison in the Fifteenth Century. *Preternature*, Vol.2 (1) (2013).

On Italians, the English and ideas of poison: George Parks. The First Italianate Englishmen. *Studies in the Renaissance* (1961). McKerrow, R.B. ed. 'Pierce Penilesse'. In *The Works of Thomas Nashe*. Fox, Alistair. *Identity and Representation in Elizabethan England.* Blackwell (1997).

On poison and drama: Alexander, Nigel, Poison, Play and Duel: A Study in Hamlet, Routledge (1971). Price, Hereward. 'The Function of Imagery in Webster'. *Publications of the Modern Language Association* (1955). Pollard, Tanya. *Drugs and Theatre in Early Modern England.* Oxford University Press. (2005). Klippel, Heike ; Wahrig, Bettina ; Zechner. Anke. *Poison and Poisoning in Science, Fiction and Cinema: Precarious Identities*. Springer. (2017).

On Paré. Hamby, Wallace. *Ambroise Paré: Surgeon of the Renaissance.* Warren Green. (1967).

Donne, John. *The Sermons of John Donne*. Edited by George Potter and Evelyn Simpson University of California Press. (1962).

On medicine and magic: Thomas, Keith Thomas. *Religion and the Decline of Magic Studies in Popular Beliefs in 16th and 17th Century England* (2003), Porter, Roy. 'Medicine and the Decline of Magic'. *Journal of the History of the Behavioral and Social Sciences* (1988). Cole, Richard. In Search of a New Mentality: The Interface of Academic and Popular Medicine in the 16th Century. *Journal of Popular Culture* (1993). Volumen Medicinae Paramirum in Ball. *Devil's Doctor.* 239. Multhauf, Robert. The Origins of Chemistry cited in Ball. Devil's Doctor. 228. Derrida, Jacques. "Plato's Pharmacy" In: *Dissemination*, translated by Barbara Johnson, Chicago, University of Chicago Press. (1981).

8. Within Half a Degree of Poison

On the basilisk: Krzyszczuk, Łukasz and Morta, Krzysztof Basilisk – The History of the Legend. *Alea.* Vol. 25/1 (2023). Bondeson, Jan. *The Feejee Mermaid and Other Essays in Natural and Unnatural History.* Cornell University Press, 1999. Breiner, Laurence. The Career of the Cockatrice. Isis. *A Journal of the History of Science Society*, v. 70, no. 1, (1979). Cohen, Meredith. The Bestiary beyond the Book. In: Morrison, E. and Grollemond, L. (eds.). *Book of Beasts. The Bestiary in the Medieval World.* The J. Paul Getty Museum, (2019). Conrad Gessner's *Historia*

Animalium was a massive compendium of all animals as well as sources on natural history. It was published in Zurich from 1551

On drama and Jews: James Shapiro. *Shakespeare and the Jews.* Columbia University Press. (2016).

On imagery of Jews: Gil Harris, Jonathan. Foreign Bodies and Body Politics. Discourse of Social Pathology in Early Modern England. Cambridge University Press (1998).

On Maimonides: Nuland, Sherwin. *Maimonides.* Knopf Doubleday. (2005).

On the Lateran Councils: Canons 67-70 of the Fourth Lateran Council in 1215 set various limits on Jewish life in Europe. Beckham, Linda Ray. *The Fourth Lateran Council of 1215: Church reform, exclusivity, and the Jews.* Dissertation. The University of Kentucky. (2005).

Harris, Robert. Paxman, Jeremy. *A Higher Form of Killing: The Secret History of Chemical and Biological Weapons.* Random House. (2002).

On the Holocaust: Friedlander, Henry. *The Origins of Nazi Genocide: From Euthanasia to the Final Solution.* University of North Carolina Press. (1995).Arad, Yitzhak. *Belzec, Sobibor, Treblinka: The Operation Reinhard Death Camps.* Indiana University Press (1987) Lewy, Guenter. *The Nazi Persecution of the Gypsies.* Oxford University Press (2000). Rees, Laurence. *Auschwitz: A New History.* Public Affairs. (2006) Kogon, Eugen. *Nazi Mass Murder: A Documentary History of the Use of Poison Gas.* Yale University Press (1994). *Schmidt. Ulf. Secret Science: A Century of Poison Warfare and Human Experiments.* Oxford University Press. (2015). Kogon, Eugen; Langbein, Hermann and Ruckerl, Adalbert. Eds. *The Theory and Practice and Hell: the German Concentration Camps and the System Behind Them.* Farrar, Strauss and Giroux. (2006).

9. Feste, Farine, Forche

On Mozart: Paul Kolecki. 'The Poison Theory and Mozart's Death'. *Mithridata* (2001). Borowitz, Albert 'Salieri and the "Murder" of Mozart'. *The Musical Quarterly* Vol LIX no. 2 (1973).

On witches: Oster, Emily, 'Witchcraft, Weather and Economic Growth in Renaissance Europe'. *Journal of Economic Perspectives* (2004). Roper, Lyndal. *Witch Craze.* (2004). Kramer, Heinrich and Sprenger, James. *The Malleus Maleficarum* (1971). Nowadays Sprenger is regarded as a hanger-on rather than a true co-author and the work is seen as by Kramer's hand. See also Broedel, Hand Peter. To Preserve the Manly Form from so Vile a Crime: Ecclesiastical Anti-Sodomitic Rhetoric and the Gendering of Witchcraft in the Malleus Maleficarum. *Essays in Medieval Studies* (2002). Ankarloo, Bengt and Clark, Stuart, eds. *Witchcraft and Magic in Europe: The Eighteenth and Nineteenth Centuries.* University of Pennsylvania Press. (1999).

On syphilis: Carol, Anne. Une brève historiographie de la syphilis. In Pouget, Benoit, Ardagna Yann (eds). *La syphilis. Itinéraires croisées en Méditerranée et au-delà XVIe-XXIe siècles*, PUP. Stanislaw Andreski. *Syphilis, Puritanism and Witch-hunts.* Macmillan. (1989). Rothschild, Bruce. History of Syphilis. *Clinical Infectious Diseases,* Vol.40. No. 10. (2005) Quétel, Claude, *The History of Syphilis.* Johns Hopkins University Press (1990). Grzybowski, Andrzej ; Pawlikowska-Łagód, Katarzyna. Some lesser-known facts on the early history of syphilis in Europe. *Clinics in Dermatology.* (2023). Maatouk, Ismael ; Moutran, Roy. History of Syphilis: Between Poetry and Medicine. *Journal of Sexual Medicine.* Vol.11 (1), (2010). Heymann, Warren R. The History of Syphilis. *Journal of the American Academy of Dermatology.* Vol.54 (2) (2006).

On Guilia Tofana and women poisoners in Italy: Pere Labat. Travels in Spain and Italy, Book 2, 24. Giovanna Fiume. 'The Old Vinegar Lady or the Judicial Modernization of the Crime of Witchcraft'. *In History from Crime* (1994).

Part II

10. Excellent Against All Venome

On the Arthasastra: Weber, Max. 'Politics as a Vocation'. In *Weber: Selections in Translation* Cambridge University Press. (1978). Jolly, Julius and Schmidt, Richard. *The Arthasastra of Kautilya* (1923), Volume 1, 25. Charpentier, Jarl. 'Poison Detecting Birds'. *Bulletin of the School of Oriental Studies.* (1929).

On invisibility of poisons and their detection: Thomas Middleton. John Webster. *Five Jacobean Tragedies.* Wordsworth Editions. (2001). Round, J.H. The King's Pantler. Archaeological Journal. Vol 60. 1 (1903) Collard, Franck. Le salut par la couleur. Mutations chromatiques et détection du poison dans les Giftschriften du Moyen Âge latin. Pallas 12. (2021) Pollard, Tanya. *Drugs and Theater in Early Modern England. PhD Dissertation.* (2005)

On food tasters: Golden, Cheryl. *The Role of Poison in Roman Society.* University of North Carolina at Chapel Hill. (2005). Xenophon, *Cyropedia.* George Bell. (1898). Mishan, Ligaya, What if the Powerful (and Paranoid) Starting Using Food Tasters Again. *New York Times.* 31

October (2018).

On theriacs: Watson, G. *Theriac and Mithridatium: A Study in Therapeutics.* The Wellcome Historical Medical Library. (1966). Heberden, William. *Antithēriaka: An Essay on Mithridatium and Theriaca.* Gale Echo (2010).

On Unicorns. They are mentioned in the Bible in Numbers 23:22 and Numbers 24:8. *The Unicorn by* Lise Gotfredsen (1999). Duffin, Christopher J. 'Fish', fossil and fake: medicinal unicorn horn. *Geological Society special publication,* Vol.452 (1) (2017). Curley, M.J. *Physiologus. A Medieval Book of Nature* Lore. Chicago University Press, Chicago, IL. (1979). Weitbrecht, Julia. "Thou hast heard me from the horns of the unicorns:" The Biblical Unicorn in Late Medieval Religious Interpretation. *Interfaces* (Milano), 12 (5) (2018). Ariew, R. Leibniz on the Unicorn and various other curiosities. *Early Science and Medicine,* 3, (1998). Fotheringham, B. The unicorn and its influence on pharmacy and medicine. *Pharmacy History,* 10, 3–7. (2000). Findlen, Paula, *Possessing Nature: Museums, Collecting, and Scientific Culture in Early Modern Italy.* University of California Press (1996).

On theriacs and antidotes: Mari, Francesco; Polettini, Aldo; Lippi Donatella and Bertol, Elisabetta. The Mysterious Death of Francesco I de' Medici and Bianca Cappello: An Arsenic Murder? *British Medical Journal* (2006). Homer. *The Odyssey,* book x. Mayor, Adrienne. *Greek Fire, Poison Arrows, and Scorpion* Bombs (2003). Plutarch. *The Fall of the Roman Republic* Penguin Classics. (1958).

On the Renaissance pharmacy: Benedicenti, Alberico. Malati-Medici e Farmacisti ii (1925). Meier Reeds, Karen. 'Renaissance Humanism and Botany'. *Annals of Science* (1976). Palmer, Richard. Pharmacy in the Republic of Venice in the Sixteenth Century. In *The Medical Renaissance of the Sixteenth Century.* (1985).

On the end of theriacs: Renehan, Robert. An Eighteenth Century Carved Piece with Medical Associations. *Journal of the Warburg and Courtauld Institutes* (1963).

11. Emeralds and Hairballs

On antidotes: Gibbs. F.W. *Specific Form and Poisonous Properties: Understanding Poison in the Fifteenth Century. Preternature,* Vol.2 1 (2013). Wexler, P. (Ed) *Toxicology in the Middle Ages and Renaissance.* Elsevier/Academic Press; (2017)

On Terra Sigilata: Dannenfeldt, K.H. The Introduction of a new sixteenth-century drug: Terra Silesiaca. *Medical History,* 28, 1984. Barroso,

M.D. Bezoar Stones, Magic, Science and Art. In: Duffin, C.J., Moody, R.T.J. & Gardner- Thorpe, C. (eds) *A History of Geology and Medicine.* Geological Society, London, Special Publications, 375. (2013) Mayor, Adrienne.. *Greek Fire, Poison Arrows, and Scorpion Bombs* (2003).

On Medieval Poisons and the Leech Book of Bald: This Anglo-Saxon work is one of the earliest medical texts from England. The title is in old English and does not refer to leeches but to medicine. There is an original copy in the British Library. *Thomas Oswald Cockayne (ed. & transl.) (1865).* "Chronicles and Memorials of Great Britain and Ireland - The Middle Ages - Volume II: Leechdoms, wortcunning, and starcraft of early England, books i, ii, and iii". Internet Archive. London: Longman, Green, Longman, Roberts, and Green. Digitised online, original and translation side-by-side. Meaney, A. L. 'Variant Versions of Old English Medical Remedies and the Compilation of Bald's Leechbook, *Anglo-Saxon England* 13 (1984)

On Bede. *History of the English Church and People*. Cited in 'Medicine, Necromancy and the Law: Aspects of Medieval Poisoning'. *Cambrian Law Review* 18/52 (1987).

On bezoars: The full title of the book is "*Joyfull Newes Out of the Newe Founde World: A Booke Which Treateth of Two Medicines Excellent Against all Venome" Which are the Bezaar stone and the Herbe Escuronera. Wein Are Declared Their Marvellous Errectes and Great Virtues With the Manner How to Cure The Sayd Venome and The Order Which is To Be Used to be Preserved From Them. Where Shall Be Seen Greate Secrets in Medicine and Many Experiences.* Oliver Impey and Arthur MacGregor, eds. *The Origins of Museums* (2001). Elias, N. and Ross, E.D. in their *History of the Moghuls of Central Asia.* (1898). Slare, Frederick. *Observations Upon Bezoar Stones Which Prove Them to be No Use in Physick* (1715). D. A. Hutton. 'The Goa Stone'. *Pharmacology Journal* (1980). On bezoars: Duffin, C.J.. Lapis de Goa: the 'cordial stone'. *Pharmaceutical Historian,* 40, 22–32. (2010). Eamon, W. *Science and the Secrets of Nature: Books of Secrets in Medieval and Early Modern Culture.* Princeton University Press. (1994).

On ambergris. 'Medieval Magic and Science in the Seventeenth Century' by Lynn Thorndike. *Speculum* (1953). *The Sixth Voyage of Es-Sindibad of the Sea: Stories from 1001 Nights* (1909) Camporesi, Piero. *Bread of Dreams: Food and Fantasy in Early Modern Europe* (1996). Schroeder, Johann. *Pharmacopeia Medico-Chymica: Sive Thesaurus Pharmacologicus* (1646).

On mumia: Roach, Mary. *Stiff: The Curious Life of Human Cadavers* Penguin.

(2004). Dannenfeldt, Karl. 'Egyptian Mumia: The Sixteenth Century Experience and Debate'. *The Sixteenth Century Journal* Vol 16. No. 2 (1985).

On the origins of experimentation: Keynes, Geoffrey ed. *The Apologie and Treatise of Ambroise Paré* Dover Publications/Peter Smith (1952). See also Paré, Ambroise. *Oeuvres Completes 3 vols. (1840–1841).* Tribby, Jay. Cooking (with) Clio and Cleo: Eloquence and Experiment in Seventeenth-Century Florence. *Journal of the History of Ideas* 52.3 (1991).

On antidotes. Robert Flanagan, Alison Jones, Robert Maynard. *Antidotes: Principles and Clinical Applications.* Taylor and Francis. (2001).

On hormesis: Jocelyn Kaiser. Sipping from a Poisoned Chalice. *Science* (2003), 376. Axelrod, Deborah and Burns, Kathy et al. 'Hormesis': An Inappropriate Extrapolation from the Specific to the Universal'. *International Journal of Occupational Environmental Health* (2004).

12. Arsenic

On arsenic: Parascandola, John. *The King of Poisons: A History of Arsenic.* Potomac Books. (2012). Christisson, Robert. *A Dispensatory, or Commentary on the Pharmacopoeias of Great Britain : comprising the natural History, Description, Chemistry, Pharmacy, Actions, Uses, and Doses of the Articles of the Materia Medica.* 2nd Ed. Lea & Blanchard, (1848). Digital edition by the University and State Library Düsseldorf. Meharg, Andrew. *Venomous Earth: How Arsenic Caused the World's Worst Mass Poisoning.* (2005). Mead, M. Nathaniel. Arsenic: In Search of an Antidote to a Global Poison. *Environmental Health Perspectives,* Vol.113 (6), (2005) Cullen, William. Is arsenic an aphrodisiac? The Socio-chemistry of an Element. *Royal Society of Chemistry.* (2008). Christisson, Robert, A treatise on poisons, in relation to medical jurisprudence, physiology, and the practice of physic (1836). Bertomeu-Sánchez, José Ramón and Nieto-Galan, Agustí (eds). *Chemistry, Medicine, and Crime. Mateu J.B. Orfila (1787–1853) and His Times.* Science History Publications (2006). Orfila, Mathieu. *A General System of Toxicology.* Legare Street Press. (2023).

On criminal uses of arsenic: Watson, Katherine. *Poisoned Lives: English Poisoners and Their Victims* (2004).

On the plague: Herring, Francis. *Certaine Rules for This Time of Pestilentiall Contagion: With a Caveat to Those that Wear Impoisoned Amulets as a Preservative from the Plague.* London. 1625, originally published 1603. 'The Opinion of Peter Turner Doct: in *Physicke, Concerning Amulets or*

Plague Cakes Whereof Perhaps Some Holde Too Much, and Some Too Little '. London. (1603).

On accidental poisoning by arsenic: Hogg, Jabez. Arsenical Poisoning by Wall-papers and Other Manufactured Articles. *Journal of Science* (1885). Carr, Henry. Our Domestic Poisons. *Scientific American* 9/220 (1880). Harwood, John. *Our Best Bedroom*. Chambers Journal. 13 September. (1862) Whorton, James. *The Arsenic Century: How Victorian Britain Was Poisoned at Home, Work and Play*. Oxford University Press. (2010)

On Napoleon and arsenic poisoning: Keynes, Milo. The Medical Health of Napoleon Bonaparte. *Journal of Medical Biography.4.2* (1996). Hindmarsh, J. Thomas and Corso, Philip. The Death of Napoleon Bonaparte: A Critical Review of the Case. *Journal of the History of Medicine* 53.3. (1998).

On arsenic and the plague: Kari Konkola. More Than a Coincidence? The Arrival of Arsenic and the Disappearance of Plaque in Early Modern Europe. *Journal of the History of Medicine and Allied Sciences* 47.2 (1992).

13. This All Blasting Tree

On the upas tree: Beekman, E.M. ed. and trans. *The Poison Tree: Selected Writings of Rumphius on the Natural History of the Indies* (1981). Foersch. J.N. *The London Magazine* LII (1783), 512–17. Bastin. John 'New Light on J.N. Foersch and the Celebrated Poison Tree of Java'. *Journal of the Malaysian Branch of the Royal Asiatic Society* (1985). Beekman, E.M. *The Ambonese Curiosity Cabinet*. G.E. Rumphius. Yale University Press. (1999).

On colonialism and poison: Ondaatje, Michael. *Running in the Family*. Picador. (1982). Leonard Cole. The Poison Weapons Taboo: Biology, Culture and Policy. *Politics and the Life Sciences* (1998). Banerjee, Pompa. Hard to Swallow: Women, Poison and Hindu Widow Burning, 1500-1700. *Continuity and Change* (2000). Grémont, Johann. Le complot des Empoisonneurs de juin 1908, Terrain, 77 (2022). Fioravanti, Marco. "Domestic Enemy: Poisoning and Resistance to the Slave Order in the 19th Century French Antilles." Historia Constitucional, 14, (2013). Brekke, Torkel. The Ethics of War and the Concept of War in India and Europe. Numen, Vol. 52, Fasc. 1, *Religion and Violence* (2005).

On arrow poisons: Knoefel, Peter. 'Felice Fontana on Poison'. *Clio Medica* 15/ ½ (1980).

On poisonings in Hong Kong: Eitel, E.J. *Europe in China* (1895). Sayer.

Geoffrey Robley. *Hong Kong, 1841-1862*. Hong Kong University Press. (1980). Griffin. J.P. 'Famous Names: The Esing Bakery, Hong Kong'. *Adverse Drug Reactions Toxicological Review* (1997). Laurence Oliphant. *Travels with Lord Elgin.* Hong Kong: Odyssey Press. (1997). Lowe, Kate and McLaughlin, Eugene. 'Caution! The Bread is Poisoned': The Hong Kong Mass Poisoning of January 1857, *The Journal of Imperial and Commonwealth History*, 43:2, (2015). Larkin, Thomas. *A Life of Suspicion and Distrust. Sino-American Relations, Racialized Anxieties, and Anglo Panic in Nineteenth-Century Hong Kong.* University of California Press. (2023).

On poison in the 19th century: Burney, Ian. 'A Poisoning of No Substance: The Trials of Medico-Legal Proof in Mid-Victorian England'. *Journal of British Studies* (1999). Forbes, Thomas. Deadly Parents: Child Homicide in Eighteenth and Nineteenth Century England. *Journal of the History of Medicine and Allied Sciences* (1986). Essig. Mark Regan. *Science and Sensation: Poison Murder and Forensic Medicine in Nineteenth Century America.* Dissertation. Cornell University.(2000). Watson, Katherine. *Poisoned Lives: English Poisoners and Their Victims.* Hambledon. (2004). Wolfgang, Marvin. *Patterns in Criminal Homicide.* University of Pennsylvania Press. (1958).

Women and poisonings: Pollak, Otto. The Criminality of Women. A.S. Barnes. (1961). Robb, George. 'Circe in Crinoline: Domestic Poisoning in Victorian England'. *Journal of Family History.* Vol 22. No. 2 (1997). Farell. A. Lethal ladies. Revisiting What We Know About Female Serial Murderers." *Homicide Studies* 15(3). (2011). Nagy, Victoria M. "Narratives in the Courtroom: Female Poisoners in Mid-Nineteenth Century England." *European Journal of Criminology* 11(2). (2014). Mansel, *Philip. King of the World.* The Life of Louis XIV. University of Chicago Press. (2020). Somerset. Anne. *Unnatural Murder. Poison at the Court of James I.* George Weidenfeld and Nicholson. (1997) Somerset, Anne. *The Affair of the Poisons. Murder Infanticide and Satanism at the Court of Louis XIV.* George Weidenfeld and Nicholson. (2004).

On murder: Orwell, George. *Decline of the English Murder and Other Essays (*1978). Lane, Roger. Murder in America: A History (1997), Essig. Science and Sensation. Adelson, Lester. "Homicidal Poisoning: A Dying Modality of Lethal Violence?" The *American Journal of Forensic. Medicine and* Pathology 8(3). (1987). Trestrail III, John. H. *Criminal Poisoning: Investigational Guide for Law Enforcement, Toxicologists, Forensic Scientists, and Attorneys.* 2nd ed. Humana Press. (2007) Shepherd, Greene and Ferslew, Brian.. Homicidal

Poisoning Deaths in the United States 1999–2005. *Clinical Toxicology* 47 (2009). Zaitsu, Wataru."Homicidal Poisoning in Japan: Offender and Crime Characteristics." *International Journal of Police Science and Management* 12(4). (2010) Emsley, John. *The Elements of Murder.* Oxford University Press. (2006) Emsley, John. *Molecules of Murder: Criminal Molecules and Classic Cases.* Royal Society of Chemistry (2008)

14. To Traffic in Souls

On medical ethics in Ancient Greece: Ratzen, Richard and Ferngren, Gary. A Greek Progymnasma on the Physician-Poisoner. *Journal of the History of Medicine and Allied Sciences* (1993). Kudlein, Rudolf. 'Medical Ethics and Popular Ethics in Greece and Rome'. *Clio Medica* (1970).

On Agatha Christie: Gerald, Michael. *The Poisonous Pen of Agatha Christie. University of Texas Press.* (1993). Sollmann, Torald. A Manual of Pharmacology and Its Applications to Therapeutics and Toxicology. *Indian Medical Gazette.* (1948).

On doctors in sixteenth century England: Brewste, Paul. 'Physician and Surgeon as Depicted in 16th and 17th Century English Literature'. *Osiris* 14 (1962). Green, Dominic. *The Double Life of Doctor Lopez: Spies, Shakespeare and the Plot to Poison Elizabeth I* Century. (2004).

On Harold Shipman: See documents at http://www.the-shipman-inquiry. org.uk. Home Secretary, Secretary of State for Health. Learning from Tragedy, Keeping Patients Safe. Overview of the Government's Action Programme in Response to the Recommendations of the Shipman Inquiry. *The Stationery Office.* (2007). Available at: https://assets.publishing. service.gov.uk/government/uploads/system/uploads/attachment_data/ file/228886/7014.pdf.

On healthcare murders: Yorker, B.C., Kizer, K.W.et al. Serial Murder by Healthcare Professionals. *Journal of Forensic Science* 51(6) (2006). Kizer, K.W. and Yorker, B.C. Health Care Serial Murder: A Patient Safety Orphan. *Journal of Quality of Patient Safety.* 36(4). (2010). Eddy, M. German nurse convicted of killing 85 patients. *The New York Times.* 6 June 6 (2019). Available from: https://www.nytimes.com/2019/06/06/ world/europe/germany-nurse-killed-patients.html. Yardley E, and Wilson D. In Search of the 'Angels of Death': Conceptualising the Contemporary Nurse Healthcare Serial Killer. *Journal of Investigative Psychological and Offender Profiling* 2016;13(1):39-55.

Part III

15. A Cold-Blooded Calculation

On the taboo on the use of poisons: Weekly, Terry. 'Proliferation of Chemical Warfare: Challenge to Traditional Restraints'. *Parameters* (1989). William of Malmesbury. *Gesta Regnum Anglorum or The History of the Norman Kings.* Translated by Joseph Stevenson (1991). See Grotius, Hugo. *The Law of War and Peace.* Translated by Francis Kelsey (1925), Book 3, Chapters 15 and 16. Roberts, Adam and Guelff, Richard. *Documents on the Laws of War* (1982), 29. Roberts. A. *Poison in Warfare* (1915), 52–57. Price, Richard 'A Genealogy of the Chemical Weapons Taboo'. *International Organization* (1995). Doniger, Wendy and Smith, Brian, trans. *The Laws of Manu* Penguin Classics. (1991). Tezcür, Güneş Murat ; Horschig, Doreen. A conditional norm: chemical warfare from colonialism to contemporary civil wars. *Third World Quarterly.* Vol 42 (2) (2021).

On the fear of snakes: Kellert, Stephen R; Wilson, Edward O; Doughty, Robin W. The Biophilia Hypothesis. *Annals of the Association of American Geographers.* Vol. 86, Iss. 4. (1996)

On the history of poison gases: West, Clarence. 'The History of Poison Gases'. *Science* (1919). Whiteside, Thomas. 'Annals of the Cold War'. *The New Yorker.* 11 February 1991. Bundt, Thomas. Gas, Mud, and Blood at Ypres. The Painful Lessons of Chemical Warfare. *Military Review.* Vol 84 No 4. (2004). Florian Schmaltz. *One Hundred Years of Chemical Warfare: Research, Deployment, Consequences.* Springer. (2017).

On the development of German nerve agents: Tucker, Jonathan. *War of Nerves: Chemical Warfare from World War I to Al-Qaeda.* Anchor. 2005). van Courtland Moon, John Ellis. 'Controlling Chemical and Biological Weapons Through World War II'. In *Encyclopedia of Arms Control and Disarmament* (1993). Dunikowska, Magda and Turo, Ludwig. Fritz Haber: The Damned Scientist. *Angewante Chemie International Edition.* 50. (2011)

Churchill Memo dated 6 July 1944, quoted in Robert Harris and Jeremy Paxman. *A Higher Form of Killing* (2002).

On the use of chemical weapons in World War II: van Cortland Moon. 'Chemical Weapons and Deterrence: The World War II Experience'. *International Security* (1984). Allen, Thomas 'Gassing Japan'. *MHQ: The Quarterly Journal of Military History* (1997). van Courtland Moon, John Ellis. The development of the norm against the use of poison: What literature tells us. *Politics and the Life Sciences,* Vol.27 (1) (2008). van Bergen, Leo. The poison gas debate in the inter-war years, *Medicine, Conflict and Survival,* 24:3. (2008). Bernstein, Barton J. Why We Didn't Use Poison Gas in World War II. *American Heritage.* Vol 36 (5). (1985).

Wolf, René. Judgement in the Grey Zone: The Third Auschwitz (Kapo) Trial in Frankfurt 1968, *Journal of Genocide Research*, 9:4. (2007).

16. The Song of Sarin the Magician

On Kurdistan and Halabja: Hiltermann, Joost. *A Poisonous Affair: America, Iraq, and the Gassing of Halabja. Camridge University Press.* (2007). Kelly, Michael. *Ghosts of Halabja. Saddam Hussein and the Kurdish Genocide.* Praeger Security International. (2006). Palkki, David and Rubin, Lawrence. Saddam Hussein's role in the gassing of Halabja, *The Nonproliferation Review*, 28:1-3, (2021).

On Aum Shinrikyo: Metraux, Daniel. 'Religious Terrorism in Japan: The Fatal Appeal of Aum Shinrikyo'. *Asian Survey* (1995). Takahashi, Hiroshi et al. 'Bacillus anthracis Bioterrorism Incident, Kameido, Tokyo 1993'. *Emerging Infectious Diseases* (2004). Lifton, Robert Jay. *Destroying the World to Save It* (1999). Tucker, Jonathan. *War of Nerves: Chemical Warfare from World War I to Al-Qaeda* (2005). Olson, Kyle. 'Aum Shinrikyo: Once and Future Threat?' *Emerging Infectious Diseases* (1999). Gardiner, Richard. 'Lost in the Cosmos and the Need to Know'. *Monumenta Nipponica* (1999). Aum Shinrikyo: Olson, K.B. Aum Shinrikyo: once and future threat? *Emerging infectious diseases,* Vol.5 (4), (1999) Tanimoto, Tetsuya ; Ozaki, Akihiko ; Harada, Kayo ; Tani, Yuta ; Kami, Masahiro Involvement of doctors in Aum Shinrikyo *The Lancet*, Vol.392 (10153), (2018) Lifton, Robert Jay. Aum Shinrikyo: The Threshold Crossed. *Journal of Aggression, Maltreatment & Trauma*, Vol.9 (1-2) (2004) Murders with VX: Aum Shinrikyo in Japan and the assassination of Kim Jong-Nam in Malaysia. Nakagawa, Tomomasa ; Tu, Anthony T. *Forensic Toxicology,* Vol.36 (2) (2018). Reader, Ian. Religious violence in contemporary Japan : the case of Aum Shinrikyô. *Nordic Institute of Asian Studies Monograph Series.* No. 82. (2013). Reader, Ian. *A Poisonous Cocktail? Aum Shinrikyo's Path to Violence* (1996). Ushiyama, Rin. *Aum Shinrikyo and religious terrorism in Japanese collective memory* (British Academy Monographs) (2023). Murakami, Haruki. *Underground: The Tokyo Gas Attack and the Japanese Psyche.* Vintage International. (2010). Kaplan, David and Marshall, Andrew. *The Cult at the End of the World: The Terrifying Story of the Aum Doomsday Cult, from the Subways of Tokyo to the Nuclear Arsenals of Russia.* (1996). Rosenau, William. Aum Shinrikyo's Biological Weapons Program: Why Did It Fail? *Studies in Conflict and Terrorism* (2001).

17. Men in Black

Pushkin quote from: Pushkin, Alexander. Mozart and Salieri. Translated by

Shaw, Alan. *Russian Language Journal* (1984).

On the killing of Litvinenko: Owen, Robert. The Litvinenko Inquiry. *The National Archives*. January (2016) Available at https://webarchive. nationalarchives.gov.uk/ukgwa/20160613090324/https://www. litvinenkoinquiry.org/report Hugh Pennington. Lethal Specks. *London Review of Books*. Vol 28 No. 14. (2006). Harrison, John. The Polonium Poisoning of Mr. Alexander Litvinenko. *Journal of Radiological Protection*. Vol 37. No. 1 (2017). Harrison, John. Collateral Contamination to the Polonium-210 Poisoning of Mr. Alexander Litvinenko. *Journal of Radiological Protection* Vol 37. No. 4. (2017). Nathwani, Amit. Polonium-210: A Firsthand Account. *The Lancet* Vol 388. Issue 10049. (2016). Also, see: On Russia and assassinations below.

On Arafat: Kapeliouk, Amnon. Yasser Arafat. 'A-t-il ete Assassine'. *Le Monde Diplomatique* November. (2005). Froidevaux, Pascal et al. Improving Forensic Investigation for Polonium Poisoning. *The Lancet*. Vol 382. No. 9900. (2013). Froidevaux, Pascal. 210-Po Poisoning as a Possible Cause of Death: Forensic Investigations and Toxicological Analysis of the Remains of Yasser Arafat. *Forensic Science International*. Vol 259 (2016).

Attempted assassination of Khaled Meshal. Report of the Commission Concerning the Events in Jordan September 1997. *Jerusalem Government Press Office* (17 February 1998). Bergman, Ronen, *Rise and Kill First*. Random House. (2018)

On Callas and others: Charles, Jeremy. Callas Was Killed in Poison Plot, Claims Director Franco Zefferelli. *The Scotsman* (24 November 2004). Shane, Scott. Poison's Use as Political Tool: Ukraine Is No Exception. *New York Times* (15 December 2004). Chivers C. J. A Dinner in Ukraine Made for Agatha Christie. *New York Times* (20 December 2004). Volkogonov, Izvestiia Dmitrii. Stalin's Plan to Assassinate Tito. *History and Public Policy Program*. Wilson Center (1993). Thomas. D.M. *Alexander Solzhenitsyn: A Century in His Life* (1998). Andrew, Christopher and Mitrokhin, Vasili. *The Mitrokhin Archive II: The KGB and the World* (2005). Violet, Bernard. 'Les Services Secrets francais tentent d'empoisonner le Rais'. *Historia* 598 (1996). Human Rights Watch. Indonesia: *No Justice Two Years After Munir's Death. Investigation Stalled, Killers of Leading Rights Activist Remain Free.* (2006)

On Castro: 'Interview with Samuel Halpern'. *The National Security Archives*. 29 November 1998. Many of the details of assassination attempts against Castro were revealed in testimony to the Church Commission, a US Senate investigation into the activities of the CIA. Mayor, Adrienne.

'The Nessus Shirt in the New World'. *Journal of American Folklore* (1995).

18. The Medium is the Message

On the poisoning of the Skripals: Gioe, David, Goodman, Michael, and Frey, David. Unforgiven: Russian intelligence vengeance as political theatre and strategic messaging, Intelligence and National Security, 34:4, (2019). Organisation for the Prohibition of Chemical Weapons. "Incident in Salisbury." 2018. https://www.opcw.org/special-sections/ salisbury-incident/ Rimmer, Abi. Salisbury novichok attack: five minutes with . . . Christine Blanshard. BMJ (Online), Vol. 366. (2019) Counter Terrorism Policing. Salisbury & Amesbury investigation update. https://www.counterterrorism.police.uk/salisbury/. (2018) Kanygin P. "Novichok" is too much for one Skripal "Novaya Gaz. https:// novayagazeta.ru/articles/2018/03/23/75908-demon strativnoe-ubiystvo-mozhno-bylo-organizovat-prosche. (2018) Ridley D. The Inquest touching upon the death of Dawn Kelly Sturgess. In: Wiltshire Council. https://www.wiltshire.gov.uk/media/3904/Scope-ruling-Sturgess-20-December-2019/pdf/Scope-ruling-sturgess-20-december-20191.pdf. (2019) Salisbury, Novichok and the OPCW Loyola, Benjamin Ruiz LOJ Pharmacology and Clinical Research, 2019-08, Vol.1 (4)

On Russia and assassinations: Carrère d'Encausse, H. *The Russia Syndrome. One Thousand Years of Political Murder*. New York: Holmes & Meier, 1993. Lisitsina, D. 2018. Andrew, C. *The Sword and the Shield: The Mitrokhin Archive and the Secret History of the KGB*. Basic Books (1999) Bellingcat Investigation Team. The Dreadful Eight: GRU's Unit 29155 and the 2015 Poisoning of Emilian Gebrev. In: *Bellingcat.* https:// www.bellingcat.com/news/uk-and-europe/2019/11/23/the- dreadful-eight-grus-unit-29155-and-the-2015-poisoning-of-emili an-gebrev/ (2019) Dzutsati V. Poisonings of Activists in the North Caucasus: A Low Threshold for Chemical Weapons Use Inside Russia? *Eurasia Daily Monitor* https://jamestown.org/program/poisonings- of-activists-in-the-north-caucasus-a-low-threshold-for-chemical- weapons-use-inside-russia/. (2021) Stanley A. Moscow Journal; To the Business Risks in Russia, Add Poisoning. *New York Times.* https://www.nytimes. com/1995/08/09/world/moscow-journal-to-the-business-risks-in-russia-add-poisoning.html?searchResultPosition=1 (1995). Blake, Heidi. *From Russia with Blood. The Kremlin's Ruthless Assassination Program and Vladimir Putin's Secret War on the West.* Mulholland Books. (2019). Cowell, Alan. *The Terminal Spy: A True Story of Espionage, Betrayal* and Murder. Crown. (2008). Kaszeta, Dan. *A History of Nerve Agents from Nazi Germany to Putin's Russia.* Oxford University Press. (2021). Knight, Amy. Orders to Kill. The Putin Regime and Political Murder. Thomas Dunne Books. (2017). Sweeney, John. *Killer in the Kremlin.* Transworld Digital. (2022). Urban, Mark. *The Skripal Files.* Macmillan. (2018). Volodarsky, Boris. *The KGB's Poison Factory From Lenin to Litvinenko.* Frontline Books. (2013). Volodarsky, Boris. *Assassins. The KGB's Poison Factory 10 Years On.* Frontline Books. (2020). Spiers, Filippa. *Agents of War: A History of Chemical and Biological Weapons.* Reaktion Books. (2021). Farrell, Michael. *Criminology of Poisoning Contexts.* Palgrave MacMillan. (2020). Harding, Luke. *A Very Expensive Poison: The Assassination of Alexander Litvinenko and Putin's War with the West.* Vintage. (2017). Kinzer. Steven. *Poisoner-in-Chief. Sidney Gottleib and the CIA Search for Mind Control* Henry Holt and Co. (2019). Walton, Calder. *Spies: The Epic Intelligence War Between East and West.* (2023) Simon and Schuster.

On Novichoks: Mirzayanov, Vil. *State Secrets. An Insider's Chronicle of the Russian Chemical Weapon's Program.* Outskirts Press. (2008). Steindl, David ; Boehmerle, Wolfgang et al. Novichok Nerve Agent Poisoning. The Lancet (British edition), Vol.397 (10270) (2021) Bolt, Hermann M. ; Hengstler, Jan G. Recent research on Novichok. Archives of Toxicology Vol.96 (5), (2022). Vale, J. Allister ; Marrs, Timothy C. ; Maynard, Robert L. Novichok: a murderous nerve agent attack in the UK. Clinical toxicology (Philadelphia, Pa.), Vol.56 (11) (2018). Voiţă-Mekereş, Florica ; Delcea, Cristian ; Buhaş, Camelia Liana ; Ciocan, Veronica. Novichok Toxicology: A Review Study. Archives of Pharmacy Practice, Vol.14 (3) (2023) Franca, Tanos C C and Kitagawa, Daniel A S et al. Novichoks: The Dangerous Fourth Generation of Chemical Weapons. International journal of molecular sciences, Vol.20 (5), (2019). Blom, Tess L ; Wingelaar, Thijs T. Current Perspectives on the Management of Patients Poisoned With Novichok: A Scoping Review. *Military Medicine,* (2023) Cockburn H. Soviet-era scientists contradict Moscow's claims Russia never made Novichok nerve agent. *The Independent.* https://www.independent.co.uk/news/world/europe/novichok-nerve-agent-poison-soviet-russia-salisbury-attack-leonid-rink- vladimir-uglev-the-bell-a8265626.html. (2018) Roth A and McCarthy T. 'It's got me': the lonely death of the Soviet scientist poisoned by novichok. In: The Guardian. https://www. theguardian.com/world/2018/mar/22/andrei-zheleznyakov- soviet-scientist-poisoned-novichok. (2018). Stone R. How German military scientists likely identified the nerve agent used to attack Alexei Navalny. *Sci AAAS.* (2021) Von Hippel F Russian whistleblower faces jail. *Bulletin of Atomic Scientists.* (1993) Noga, Maciej; Jurowski, Kamil What do we currently know about Novichoks? The State of the Art. *Archives of Toxicology*, Vol.97 (3) 2023. Tucker JB *War of nerves: chemical warfare from world war I to Al-Qaeda*, 1st edn. Pantheon Books, New York (2006)

On Putin and the media: Putin's Media Strategy? 'A Free Jazz Orchestra'. *Public Radio International,* May 18. https://www.pri.org/ stories/2018-05-18/putin-s-media-strategy-free-jazz-orchestra Warrick, J., and A. Troianovski. 2018. "Agents of Doubt. How a Powerful Russian Propaganda Machine Chips Away at Western Notions of Truth." *Washington Post*, December 10. https://www.washington post.com/ graphics/2018/world/national-security/russian-propaganda-skripal-salisbury/?utm_term=.f14e4fb5de40 Wright, R. Putin, a Little Man Still Trying to Prove His Bigness. *The New Yorker*, March 20. (2018)https:// www.newyorker.com/news/news-desk/putin-a-little-man-still-trying-to-prove-his-bigness Barojan, D.. "#Putinatwar: Kremlin Narratives on Skripal Continue to Grow Online." *Atlantic Council Digital Forensic Research Lab*, July 5. (2018) https://medium.com/dfrlab/putinatwar-kremlin-narratives- on-skripal-continue-to-grow-online-c5495da731db . *Myers, Steven Lee. The New Tsar : The Rise and Reign of Vladimir Putin.* Alfred A. Knopf. (2015)

19. Got Milk?

On Botulinum poisoning: Wein, Lawrence and Liu, Yifan. 'Analyzing a Bioterror Attack on the Food Supply: The Case of Botulinum Toxin in Milk'. *Proceedings of the National Academy of Sciences* (2005). Arnon, Stephen, Schechter Robert, and Inglesby, Thomas et al. 'Botulinum Toxin as a Biological Weapon'. *Journal of the American Medical Association* (2001).

On assessment of risk. Sivak, Michael and Flanagan, Michael. 'Flying and Driving after the September 11 Attacks'. American Scientist (2003). Kunreuther, Howard. 'Risk Analysis and Risk Management in an Uncertain World'. The Wharton Financial Institutions Center Working Paper December. (2001). 22 Beck, Ulrich. Risk Society: Toward a New Modernity (1992). Also Kasperson, Roger and Kasperson, Jeanne. 'The Social Amplification and Attenuation of Risk'. The Annals of the American Academy of Political and Social Science (1996). Carus, W. Seth. Bioterrorism and Biocrimes: The Illicit Use of Biological Agents Since 1900. Fredonia Books.(2002). Gleick, Peter H. 2006. "Water and

Terrorism." *Water Policy* 8:481–503.

On mass sociogenic illness: Markush R.E. Mental Epidemics: A Review of the Old to Prepare for the New. *Public Health Reviews* (1973). Bartholomew, Robert and Wessely, Simon. 'Protean Nature of Mass Sociogenic Illness'. *British Journal of Psychiatry* (2002). Boss, Leslie. Epidemic Hysteria: A Review of the Published Literature. *Epidemiological Reviews* (1997). Bartholomew, Robert and Wessely, Simon. Epidemic Hysteria in Virginia: The Case of the Phantom Gasser of 1933–1934. *Southern Medical Journal* 92/8 (1999). Amin, Yousreya, Hamdi, Emad, and Eapen, Valsamma. Mass Hysteria in Arab Culture. *The International Journal of Social Psychiatry* (1997). Faasse, Kate and Petrie, Keith. The Nocebo Effect: Patient Expectations and Medication Side Effects. *Postgraduate Medical Journal* (2013). Joffe, Ari and Elliot, April. Long Covid as a Functional Somatic Symptom Disorder Caused by Abnormally Precise Prior Expectations During Bayesian Perceptual Processing: A New Hypothesi s and Implications for Pandemic Response. August 24. *Sage Open Medicine* (2023).

On ricin and on Georgi Markov: Nehring, C. Umbrella or Pen? The Murder of Georgi Markov: New Facts and Old Questions. *Journal of Intelligence History* 16, no. 1 (2017) BBC News. Terror Police Find Deadly Poison. BBC News. 7 January 2003. Knight, Bernard. Ricin—a Potent Homicidal Poison. *British Medical Journal* (1979). Smith, George. 'The Recipe for Ricin. Part II. *GlobalSecurity.org.* Burke, Jason. An Informer Told Algerian interrogators that Britain Is Facing Al Qaeda Threat. *The Observer* 17 April (2005). Chronology of Incidents Involving Ricin. *James Martin Center for Nonproliferation Studies.* Waters, Tim The Fine Art of Making Poison. *Discover* (August 1992). Archer, Lawrence and Bawdon, Fiona. *Ricin! The Inside Story of the Terror Plot that Never Was.* Pluto Press. (2010)

On the challenges of making weapons of mass destruction: Rosenau, William 'Aum Shinrikyo's Biological Weapons Program: Why Did It Fail?' *Studies in Conflict and Terrorism* (2001). Mueller, John and Mueller, Karl. 'Sanctions of Mass Destruction'. *Foreign Affairs* (1999). Parachini, John et al. Diversion of Nuclear, Biological and Chemical Weapons Expertise from the Former Soviet Union. Understanding an Evolving Problem. *RAND* (2005). Danzig, Richard Aum Shinrikyo: Insights into How Terrorists Develop Biological and Chemical Weapons. *Center for a New American Security.* (2011)

On risk of bioweapons: 'The Invisible Enemy'. *The Economist* (16 November 2000). Easterbrook, Gregg. 'The Meaningless of Term Limits'. *The*

New Republic (7 October 2002). Stern, Jessica. 'Dreaded Risks and the Control of Biological Weapons'. *International Security* (2002/2003). Jason Pate and Gavin Cameron. *'Covert Biological Weapons Attacks against Agricultural Targets: Assessing the Impact against US Agriculture'* (2001). Carus, Seth. 'Bioterrorism and Biocrimes'. *Center for Counterproliferation Research.* National Defense University. Washington DC. (1998, revised February 2001). Diamond, Dickson, Pastor, Larry, McIntosh, Roger. 'Medical Management of Terrorism-related Behavioural Syndromes'. *Psychiatric Annals* (2004). Coignard. B. 'Bioterrorism Preparedness and Response in European Public Health Institutes'. *Eurosurveillance* (2001). 'Poll: Majority Reject Evolution'. CBS News (23 October 2005). Ken Alibek. *Biohazard: The Chilling True Story of the Largest Covert Biological Weapons Program in the World* (1999).

On next generation weapons: Ainscough, Michael. 'Next Generation Bioweapons: Genetic Engineering and Biological Warfare'. In *The Gathering Biological Warfare Storm* (2004). There have been three critical defections from the Soviet/Russian programme: Vladimir Pasechnik, a former director of the Institute for Ultra Pure Biological Preparations; a lower-level scientist known by his code "Temple Fortune"; and Kanatjan Alibekov, now known as Ken Alibek, the author of *Biohazard*, a personal account of his work for the bioweapons programme. E rik Frinking, et al. The Increasing Threat of Biological Weapons: Handle with Sufficient and Proportionate Care. *Hague Centre for Strategic Studies* (2016). Casadevall A. and Imperiale, M. Risks and Benefits of Gain-of-function Experiments with Pathogens of Pandemic Potential, Such as Influenza Virus: A Call for a Science-based Discussion. *mBio* (2014). Lentzos, Filippa. and Rose, Nikolas. 'Governing Insecurity: Contingency Planning, Protection, Resilience'. *Economy and Society* (2009). Lentzos, Filippa. *Biological Threats in the 21st Century: The Politics, People, Science and Historical Roots.* Imperial College Press. (2016). Lentzos. Filippa 'Dual-use Biology: Building Trust and Managing Perceptions of Intent'. *The Nonproliferation Review* (2020). Lentzos. Filippa. 'How to Protect the World from Ultra-targeted biological Weapons'. *Bulletin of the Atomic Scientists* (2020). Rees. M. 'Denial of Catastrophic Risks'. *Science* (2013). Ananthan, Reuben and Dass Santhana. 'Bioterrorism: Lessons from the COVID-19 Pandemic'. *Counter Terrorist Trends and Analyses* (2021). Arturo Casadevall. 'Global Catastrophic Threats from the Fungal Kingdom'. In *Global Catastrophic Biological Risks* (2019). Zariati, Parisa et al. 'A Comprehensive Review: Toxicity of Nanotechnology in the Food Industry'. *Journal of Medical Discovery* (2018). Del Monte, Louis. *Nanoweapons: A Growing Threat to Humanity*

Potomac Books. (2017). Kuhn, Jens and Leitenberg, Milton. 'The Soviet Biological Warfare Program'. In *Biological Threats in the 21st Century* (2016), 79–102. And Wilton Park. *Powerful Actor, High Impact Bio-Threats – Initial Report*. West Sussex: UK Foreign and Commonwealth Office. 2018.

Postscript: The Basilisk

On Harry Potter and the basilisk. Details from Harry Potter Wiki entry on the Basilisk: https://harrypotter.fandom.com/wiki/Basilisk and Villains Wiki at https://villains.fandom.com/wiki/Basilisk_(Harry_Potter) and Simon Reisinger. The Basilisk in the Wizarding World. Undated at https://simonreisinger.com/the-basilisk/

On Heretical Movements: Robert Lerner. *The Heresy of the Free Spirit in the Later Middle Ages.* University of Notre Dame Press. (1972).

On fear: Dozier Jr., V. *Fear Itself. The Origin and Nature of the Powerful Emotion That Shapes Our Lives and Our World.* St. Martin's Griffin. (2015).

BIBLIOGRAPHY

Aberth, John. From the Brink of the Apocalypse: Confronting Famine, War, Plague, and Death in the Later Middle Ages. 2d ed. Routledge. (2010)

Achterberg, Jeanne. Woman as Healer. Boulder, CO: Shambhala. (1991).

Ainscough, Michael. 'Next Generation Bioweapons: Genetic Engineering and Biological Warfare'. In *The Gathering Biological Warfare Storm*. Edited by Jim Davis and Barry Schneider. Connecticut: Praeger. (2004).

Alexander, Nigel. *Poison, Play and Duel: A Study in Hamlet*. London: Routledge and Kegan. (1971).

Alibek, Ken. *Biohazard: The Chilling True Story of the Largest Covert Biological Weapons Program in the World*. New York: Random House. (1999).

Allen, Thomas. Gassing Japan. *MHQ: The Quarterly Journal of Military History* (1997).

Amin, Yousreya. Hamdi, Emad. Eapen, Valsamma. 'Mass Hysteria in Arab Culture'. *The International Journal of Social Psychiatry* 43 (1997).

Andrew, Christopher. Mitrokhin, Vasili. *The Mitrokhin Archive II: The KGB and the World*. London: Allen Lane. (2005).

Ankarloo, Bengt. Clark, Stuart, eds. *Witchcraft and Magic in Europe: The Eighteenth and Nineteenth Centuries*. Philadelphia: University of Philadelphia Press. (1999).

Arad, Yitzhak. *Belzec, Sobibor, Treblinka: The Operation Reinhard Death Camps*. Indiana University Press (1987)

Ardestani, Mozhgan et al. The Golden Age of Medieval Islamic Toxicology. In *Toxicology in the Middle Ages and Renaissance. History of Toxicology and Environmental Health*. Elsevier. (2017).

Arnon, Stephen. Schechter, Robert. Inglesby, Thomas, et al. 'Botulinum Toxin as a Biological Weapon'. *JAMA* 258 (2001).

Axelrod, Deborah. Burns, Kathy. Davis, Devra. von Larebeke, Nicolas. '"Hormesis": An Inappropriate Extrapolation from the Specific to the Universal'. *International Journal of Occupational Environmental Health* 10/3 (2004).

Azzolini, Monica. Plagues, Poisons and Potions: Plague-Spreading Conspiracies in the Western Alps, c. 1530-1640 (review)'. *Bulletin of the*

History of Medicine 78/1 (2004).

Bale, John. *The Pageant of Popes*. London. (1574).

Ball, Philip. *The Devil's Doctor: Paracelsus and the World of Renaissance Magic and Science*. New York: Farrar, Straus and Giroux. (2006).

Banerjee, Pompa. Hard to Swallow: Women, Poison and Hindu Widow burning, 1500-1700. *Continuity and Change* 15/2 (2000).

Barkai, Ron. 'Jewish Treatises on the Black Death (1350–1500)'. In *Medicine from the Black Death to the French Disease*. London: Routledge. (1998).

Barnes, Barnabe. *The Devil's Charter: A Tragedy Containing the Life and Death of Pope Alexander VI*. London: Globe Quartos. (1999).

Baron Corvo, Frederick. *Chronicles of the House of Borgia*. New York: E. P. Dutton and Co. (1901).

Barrett. Anthony A. *Livia: First Lady of Imperial Rome*. Yale University Press. (2004)

Bartholomew, Robert and Sirois, Francois. Occupational Mass Sociogenic Illness: A Transcultural Overview. *Transcultural Psychiatry* 37/4 (2000).

Bartholomew, Robert and Wessely, Simon. Epidemic Hysteria in Virginia: The Case of the Phantom Gasser of 1933–1934. *Southern Medical Journal* 92/8 (1999).

Bartholomew, Robert and Wessely, Simon. Protean Nature of Mass Sociogenic Illness. *British Journal of Psychiatry* 180/4 (2002).

Barzilay, Tzafir. Early Accusations of Well Poisoning against Jews: Medieval Reality or Historiographical Fiction? Medieval Encounters: Jewish, Christian, and Muslim Culture in Confluence and Dialogue, Vol.22 (5). (2016).

Barzilay, Tzafir. Poisoned Wells. Accusations, Persecutions and Minorities in Medieval Europe 1321-1422. University of Pennsylvania Press. (2022).

Bastin, John. New Light on J.N. Foersch and the Celebrated Poison Tree of Java. *Journal of the Malaysian Branch of the Royal Asiatic Society* 58/1 (1985).

Beard, Mary. Twelve Caesars: Images of Power from the Ancient World to the Modern. Princeton University Press. (2021).

Beard. Mary *Emperor of Rome*. Liveright (2023).

Beck, Ulrich. *Risk Society: Toward a New Modernity*. Thousand Oaks CA: Sage. (1992).

Beckham, Linda Ray. *The Fourth Lateran Council of 1215: Church reform, exclusivity, and the Jews*. Dissertation. The University of Kentucky. (2005).

Beekman, E. M., ed. and trans. *The Poison Tree: Selected Writings of Rumphius on the Natural History of the Indies*. University of Massachusetts Press. (1981).

Bellingcat Investigation Team. The Dreadful Eight: GRU's Unit 29155 and the 2015 Poisoning of Emilian Gebrev. In: *Bellingcat*. https://www.bellingcat.com/news/uk-and-europe/2019/11/23/the-dreadful-eight-grus-unit-29155-and-the-2015-poisoning-of-emili an-gebrev/ (2019).

Bennett, Gillian. *Bodies: Sex Violence, Disease and Death in Contemporary Legend.* University Press of Mississippi. (2005).

Bergman, Ronen. *Rise and Kill First.* Random House. (2018)

Bernstein, Barton J. Why We Didn't Use Poison Gas in World War II. *American Heritage.* Vol 36 (5). (1985).

Bertomeu-Sánchez, José Ramón and Nieto-Galan, Agustí (eds). *Chemistry, Medicine, and Crime. Mateu J.B. Orfila (1787–1853) and His Times.* Science History Publications (2006).

Blake, Heidi. *From Russia with Blood. The Kremlin's Ruthless Assassination Program and Vladimir Putin's Secret War on the West.* Mulholland Books. (2019).

Blom, Tess and Wingelaar, Thijs T. Current Perspectives on the Management of Patients Poisoned With Novichok: A Scoping Review. *Military Medicine,* (2023)

Boesche, Roger. 'Kautilya's *Arthasastra* on War and Diplomacy in Ancient India'. *The Journal of Military History* 67/1 (2003).

Borowitz, Albert. 'Salieri and the "Murder" of Mozart'. *The Musical Quarterly* 59/2 (1973).

Boss, Leslie. 'Epidemic Hysteria: A Review of the Published Literature'. *Epidemiological Reviews* 19/2 (1997).

Breiner, Laurence. The Career of the Cockatrice. *Isis. A Journal of the History of Science Society*, v. 70, no. 1, (1979).

Breuer, Mordechai. The "Black Death" and Antisemitism. In *Antisemitism through the Ages*, edited by Shmuel Almog. Pergamon Press. (1988).

Brewster, Paul. 'Physician and Surgeon as Depicted in 16th and 17th Century English Literature'. *Osiris* 14 (1962).

Broedel, Hans Peter. To Preserve the Manly Form from so Vile a Crime: Ecclesiastical Anti-Sodomitic Rhetoric and the Gendering of Witchcraft in the *Malleus Maleficarum' Essays in Medieval Studies* 19 (2002).

Burke, Jason. 'An Informer Told Algerian interrogators that Britain Is Facing Al Qaeda Threat'. *The Observer.* 17 April. (2005).

Burney, Ian. 'A Poisoning of No Substance: The Trials of Medico-Legal Proof in Mid-Victorian England'. *Journal of British Studies* 38 (1999).

Camporesi, Piero. *Bread of Dreams: Food and Fantasy in Early Modern Europe.* Hoboken, NJ: Wiley. (1996).

Carol, Anne. Une brève historiographie de la syphilis. In Pouget, Benoit, Ardagna Yann (eds). *La syphilis. Itinéraires croisées en Méditerranée et au-delà XVIe-XXIe siècles*, PUP. (2017)

Carr, Henry. 'Our Domestic Poisons'. *Scientific American* 9/220 (1880).

Carrère d'Encausse, H. *The Russia Syndrome. One Thousand Years of Political Murder.* New York: Holmes & Meier. (1993)

Carus, W. Seth. 'Bioterrorism and Biocrimes'. Center for Counterproliferation Research. National Defense University. Washington DC. February

(2001).

Casadevall, A. Imperiale, M. 'Risks and Benefits of Gain-of-function Experiments with Pathogens of Pandemic Potential, Such as Influenza Virus: A Call for a Science-based Discussion'. *mBio* 5/4 (2014).

Casadevall, Arturo. 'Global Catastrophic Threats from the Fungal Kingdom'. In *Global Catastrophic Biological Risks*. Edited by Thomas Inglesby and Amesh Adalja. (2019).

Charles, Jeremy. 'Callas Was Killed in Poison Plot, Claims Director Franco Zefferelli'. *The Scotsman*. 24 November (2004).

Charpentier, Jarl. 'Poison Detecting Birds'. *Bulletin of the School of Oriental Studies* 5/2 (1929).

Chidester, David. 'Rituals of Exclusion and the Jonestown Dead'. *Journal of the Academy of Religion* 56/4 (Winter 1988).

Chivers, C. J. 'A Dinner in Ukraine Made for Agatha Christie'. *New York Times*. 20 December (2004).

Christison, Robert. *A Dispensatory, or Commentary on the Pharmacopoeias of Great Britain: comprising the natural History, Description, Chemistry, Pharmacy, Actions, Uses, and Doses of the Articles of the Materia Medica.* 2nd Ed. Lea & Blanchard, (1848)

Cilliers, Louise. Retief, F. P. 'Poisons, Poisoning and the Drug Trade in Ancient Rome'. *Akroterion* 45/2000 (2014).

Cilliers, Louise and Retief, F.P. Lead Poisoning in Ancient Rome'. *Acta Theologica* (2010).

Cilliers, Louise and Retief F.P. Lead Poisoning and the Downfall of Rome: Myth or Reality? *In History of Toxicology and Environmental Health.* Elsevier. (2014).

Cobb, Cathy. The Chemistry of Lucrezia Borgia et al. ACS Symposium Series, (2013)

Cobb. Cathy. The Case Against the Borgias: Motive, Opportunity, and Means. Toxicology in the Middle Ages and Renaissance, (2017).

Cockburn H. Soviet-era scientists contradict Moscow's claims Russia never made Novichok nerve agent. *The Independent.* https://www.independent.co.uk/news/world/europe/novichok-nerve-agent-poison-soviet-russia-salisbury-attack-leonid-rink-vladimir-uglev-the-bell-a8265626.html. (2018).

Cohen, Meredith. The Bestiary beyond the Book. In: Morrison, E. and Grollemond, L. (eds.). *Book of Beasts. The Bestiary in the Medieval World*. The J. Paul Getty Museum, (2019).

Cohen, Monte. 'The Diabolical and Diverse Dangerously Devastating Drugs and Dead Poisons and Potions of the Demonic and Devilish Dr Fu Manchu'. *Mithridata* 16/ 2 (2006).

Coignard, B. 'Bioterrorism Preparedness and Response in European Public Health Institutes'. *Eurosurveillance* 6/11 (2001).

Collard, Franck. The Crime of Poison in the Middle Ages Deborah Nelson-

Campbell (Translator) Wiley. (2003).

Collard, Franck. Le salut par la couleur. Mutations chromatiques et détection du poison dans les Giftschriften du Moyen Âge latin. *Pallas*, Vol 117. (2021).

Collard, Franck. Les écrits sur les poisons. Typologie des Sources du Moyen Âge Occidental, 88. Collard, Franck. *The Crime of Poison in the Middle Ages.* Praeger, (2008).

Collard, Franck and Hanna, Ralph. Malachy the Irishman, On Poison: A Study and an Edition, *Cahiers de civilisation médiévale*, 262 | (2023).

Collard, Franck. Le poison et le sang dans la culture médiévale. *Médiévales*, Vol.60. (2012).

Cole, Leonard. 'The Poison Weapons Taboo: Biology, Culture and Policy'. *Politics and the Life Sciences* 17/2 (1998).

Cole, Richard. In Search of a New Mentality: The Interface of Academic and Popular Medicine in the 16th Century. *Journal of Popular Culture* 26/4 (1993).

Counter Terrorism Policing. Salisbury & Amesbury investigation update. https://www.counterterrorism.police.uk/salisbury/. (2018).

Cowell, Alan. *The Terminal Spy: A True Story of Espionage, Betrayal and Murder.* Crown. (2008).

Crockford, Susannah. How do you know when you are in a Cult?" *Nova Religio: The Journal of Alternative and Emergent Religions*, Volume 22, Issue 2 (2018).

Dannenfeldt, Karl. 'Egyptian Mumia: The Sixteenth Century Experience and Debate'. *The Sixteenth Century Journal* 16/2 (1985).

Dass, Reuben Ananthan Santhana. 'Bioterrorism: Lessons from the COVID-19 Pandemic'. *Counter Terrorist Trends and Analyses* 13/2 (2021).

Davidson, James. *Courtesans and Fishcakes*. New York: St Martin's Press. (1998).

Del Monte, Louis. *Nanoweapons: A Growing Threat to Humanity*. Lincoln, NE: University of Nebraska Press. (2017).

Derrida, Jacques. "Plato's Pharmacy" In: *Dissemination*, translated by Barbara Johnson, Chicago, University of Chicago Press, (1981).

Di Tura, Agnolo. 'Cronaca Senese'. In *Rerum Italicarum Scriptores.* Bologna. (1931).

Diamon, Dickson. Pastor, Larry. McIntosh, Roger. 'Medical Management of Terrorism-related Behavioural Syndromes'. *Psychiatric Annals* 34/9 (2004).

Doniger, Wendy. Smith, Brian, trans. *The Laws of Manu.* New York: Penguin. (1991).

Donne, John. *The Sermons of John Donne.* Edited by George Potter and Evelyn Simpson. Berkeley, CA: University of California Press. (1962).

Dozier Jr., V. *Fear Itself. The Origin and Nature of the Powerful Emotion That Shapes Our Lives and Our World.* St. Martin's Griffin. (2015).

Dunikowska, Magda and Turo, Ludwig. Fritz Haber: The Damned Scientist. *Angewante Chemie International Edition.* 50. (2011).

Dutta, Shomit, ed. *Greek Tragedy.* London: Penguin. (2004).

Easterbrook, Gregg. 'The Meaningless of Term Limits'. *The New Republic.* 7 October (2002).

Dzutsati V. Poisonings of Activists in the North Caucasus: A Low Threshold for Chemical Weapons Use Inside Russia? *Eurasia Daily Monitor* https://jamestown.org/program/poisonings-of-activists-in-the-north-caucasus-a-low-threshold-for-chemical- weapons-use-inside-russia/. (2021)

Edmonds. Wendy. *intoxicating FOLLOWERSHIP: in the Jonestown Massacre.* Emerald Publishing. (2021).

Eitel, E. J. *Europe in China.* London: Luzac and Co. (1895).

Emsley, J. 'Ancient World Was Poisoned by Lead'. *New Scientist* 143 (1994).

Essig, Mark Regan. *Science and Sensation: Poison Murder and Forensic Medicine in Nineteenth Century America.* PhD dissertation. Cornell University. (2000).

Faasse, Kate. Petrie, Keith. 'The Nocebo Effect: Patient Expectations and Medication Side Effects'. *Postgraduate Medical Journal* 89/1055 (2013).

Fagan, Kevin. Haunted by Memories of Hell. 12 November 1998. SF Gate. https://www.sfgate.com/news/article/haunted-by-memories-of-hell-2979612.php (1998)

Faraone, Christopher. *Ancient Greek Love Magic.* Harvard University Press. (1999)

Farrell, Michael. *Criminology of Poisoning Contexts.* Palgrave MacMillan. (2020).

Feltmate, David. People's Temple. A Lost Legacy for the Current Moment. *Nova Religio: The Journal of Alternative and Emergent Religions,* Volume 22, Issue 2, (2018).

Ferrara, Orestes. *The Borgia Pope: Alexander VI.* Translated by F. J. Sheed. London: Sheed and Ward. (1942).

Ferrario, Gabriele. Maimonides' Book on Poisons and the Protection Against Lethal Drugs in *Toxicology in the Middle Ages and Renaissance.* Elsevier. (2017).

Finley, Theresa and Koyama, Mark. Plague, Politics, and Pogroms: The Black Death, the Rule of Law, and the Persecution of Jews in the Holy Roman Empire. *The Journal of Law & Economics,* Vol.61 (2) (2018).

Fioravanti, Marco. "Domestic Enemy: Poisoning and Resistance to the Slave Order in the 19th Century French Antilles." *Historia Constitucional,* 14, (2013).

Fiume, Giovanna. 'The Old Vinegar Lady or the Judicial Modernization of the Crime of Witchcraft'. In *History from Crime.* Translated by Margaret Gallucci. Baltimore, MD: Johns Hopkins University Press. (1994).

Flanagan, Robert. Jones, Alison. Maynard, Robert. *Antidotes: Principles and*

Clinical Applications. London: CRC Press.(2001).

Foersch, J. N. *The London Magazine* LII (1783).

Forbes, Thomas. 'Deadly Parents: Child Homicide in Eighteenth and Nineteenth Century England'. *Journal of the History of Medicine and Allied Sciences* 41 (1986).

Fox, Alistair. *Identity and Representation in Elizabethan England.* Hoboken, NJ: Blackwell. (1997).

Franca, Tanos C C and Kitagawa, Daniel A S et al. Novichoks: The Dangerous Fourth Generation of Chemical Weapons. International journal of molecular sciences, Vol.20 (5), (2019).

Freeman, Kathleen. 'Against a Stepmother'. In *The Murder of Herodes and Other Trials from the Athenian Law Courts.* New York: Norton Library. (1963).

Frieda, Leonie. *Catherine de Medici.* London: Wiedenfeld and Nicholson. (2003).

Frinking, Erik, et al. 'The Increasing Threat of Biological Weapons: Handle with Sufficient and Proportionate Care'. Hague Centre for Strategic Studies. (2016).

Froidevaux, Pascal et al. Improving Forensic Investigation for Polonium Poisoning. *The Lancet.* Vol 382. No. 9900. (2013).

Froidevaux, Pascal. 210-Po Poisoning as a Possible Cause of Death: Forensic Investigations and Toxicological Analysis of the Remains of Yasser Arafat. *Forensic Science International.* Vol 259 (2016).

Gasse, Rosanne. 'The Practice of Medicine in "Piers Plowman"'. *The Chaucer Review* 39/2 (2004).

Gatton, Adrian. *Black Widow: The True Story of How Dena Thomson Lured Men into a Twisted Web of Sex, Lies and Murder*. John Blake. (2010).

Gaynes, Robert. Avicenna, a Thousand Years Ahead of His Time. *Germ Theory.* 2011.Gardiner, Richard. 'Lost in the Cosmos and the Need to Know'. *Monumenta Nipponica* 54/2 (1999).

Gerald, Michael. *The Poisonous Pen of Agatha Christie.* Austin, TX: University of Texas Press. (1993).

Giannetti, Laura and Ruggiero, Guido. *Five Comedies from the Italian Renaissance.* Baltimore, MD: Johns Hopkins University Press. (2003).

Gibbs, Frederick. Poison, Medicine, and Disease in Late Medieval and Early Modern Europe Routledge (2016).

Gilfillan, S.C. Lead Poisoning in Ancient Rome. *Acta Theologica Supplementum* 7 (2005).

Gil Harris, Jonathan. Foreign Bodies and Body Politics. Discourse of Social Pathology in Early Modern England. Cambridge University Press (1998).

Gioe, David, Goodman, Michael, and Frey, David. Unforgiven: Russian intelligence vengeance as political theater and strategic messaging, *Intelligence and National Security*, 34:4, (2019).

Ginzburg, Carlo. *Ecstasies: Deciphering the Witches' Sabbath.* London:

Hutchinson. (1990).

Girard, Rene. *The Scapegoat*. Baltimore, MD: Johns Hopkins University Press. (1986).

Gleick, Peter H. Water and Terrorism. Water Policy 8. (2006).

Gotfredsen, Lise. *The Unicorn*. London: The Harvill Press. (1999).

Grandjean, Philippe. Paracelsus Revisited: The Dose Concept in a Complex World. *Basic and Clinical Pharmacology and Toxicology*, Vol.119 (2) (2016).

Green, Dominic. *The Double Life of Doctor Lopez: Spies, Shakespeare and the Plot to Poison Elizabeth I*. Century Hutchinson. (2003).

Gregorovius, Ferdinand. *Lucrezia Borgia*. New York: Benjamin Blom. (1904).

Grémont, Johann. Le complot des Empoisonneurs de juin 1908, *Terrain*, 77 (2022).

Griffin, J. P. 'Famous Names: The Esing Bakery, Hong Kong'. *Adverse Drug Reactions Toxicological Review* 16/2 (1997).

Grimm-Samuel, Veronika. 'On the Mushroom that Deified the Emperor Claudius'. *Classical Quarterly* 41/1 (1991).

Grotius, Hugo. *The Law of War and Peace*. Translated by Francis Kelsey. Indianapolis: Bobbs-Merrill. (1925).

Grunwald, Lisa and Adler, Stephen, *Letters of the Century: America 1900-1999*, The Dial Press (1999).

Grzybowski, Andrzej ; Pawlikowska-Łagód, Katarzyna. Some lesser-known facts on the early history of syphilis in Europe. *Clinics in Dermatology.* (2023).

Hallissy, Margaret. 'Poison: Imagery and Theme in Chaucer's Canterbury Tales'. PhD Thesis. Fordham University. New York. (1974).

Hallissy, Margaret. *Poison and Infection in Chaucer's Knight's and Canon Yeoman's Tales. Essays in Arts and Sciences*, Vol.10 (1). (1981).

Hallissy, Margaret. *Venomous Woman: Fear of the Female in Literature*. London: Bloomsbury. (1987).

Hamby, Wallace. *Ambroise Paré: Surgeon of the Renaissance*. St Louis, MO: Warren H. Green. (1967).

Hanawalt, Barbara. *Crime and Conflict in English Communities, 1300–1348*. Harvard, MA: Harvard University Press. (1979)

Harding, Luke. *A Very Expensive Poison: The Assassination of Alexander Litvinenko and Putin's War with the West.* Vintage. (2017).

Harper, Kyle. *The Fate of Rome: Climate, Disease and the End of an Empire*. Princeton, NJ: Princeton University Press. (2017).

Harris, Robert. Paxman, Jeremy. *A Higher Form of Killing: The Secret History of Chemical and Biological Weapons*. New York: Random House. (2002).

Harrison, John. The Polonium Poisoning of Mr. Alexander Litvinenko. *Journal of Radiological Protection.* Vol 37. No. 1 (2017).

Harrison, John. Collateral Contamination to the Polonium-210 Poisoning of Mr.

Alexander Litvinenko.. *Journal of Radiological Protection* Vol 37. No. 4. (2017).

Harwood, John. Our Best Bedroom. *Chambers Journal.* 13 September. (1862).

Heberden, William. *Antithēriaka: An Essay on Mithridatium and Theriaca.* Gale Echo (2010).

Hefez, Albert. 'The Role of the Press and the Medical Community in the Epidemic of "Mysterious Gas Poisoning" in the Jordan West Bank'. *American Journal of Psychiatry* 142/7 (1985).

Higgins, Chris.. Stop Saying 'Drink the KoolAid.' *The Atlantic.* 8 November 2012. https://www.theatlantic.com/health/archive/2012/11/stop-saying-drink-the-kool-aid/264957/

Hillgarth, J. N. 'The Image of Alexander VI and Cesare Borgia in the 16th and 17th Centuries'. *Journal of the Warburg and Courtauld Institutes* 59 (1996).

Hiltermann, Joost. *A Poisonous Affair: America, Iraq, and the Gassing of Halabja.* Cambridge: Cambridge University Press. (2007).

Hindmarsh, J. Thomas. Corso, Philip. 'The Death of Napoleon Bonaparte: A Critical Review of the Case'. *Journal of the History of Medicine* 53 (1998).

Hogg, Jabez. 'Arsenical Poisoning by Wall-papers and Other Manufactured Articles'. *Journal of Science* 3/vii (1885).

Homer. *Odyssey and Iliad*, Translations by Emily Watson. Watson, Emily: *The Iliad.* (2023) W.W. Norton and Company and *The Odyssey.* (2017) W.W. Norton and Company.

Horstmanshoff, Manfred. 'Ancient Medicine Between Hope and Fear: Medicament, Magic and Poison in the Roman Empire'. *European Review* 7/1 (1999).

Human Rights Watch. Indonesia: *No Justice Two Years After Munir's Death. Investigation Stalled, Killers of Leading Rights Activist Remain Free.* (2006).

Impey, Oliver. MacGregor, Arthur, eds. *The Origins of Museums.* London: House of Stratus. 2001.

Ireland, Richard. 'Medicine, Necromancy and the Law: Aspects of Medieval Poisoning'. *Cambrian Law Review* 18/52 (1987).

Joffe, Ari. Elliot, April. 'Long Covid as a Functional Somatic Symptom Disorder Caused by Abnormally Precise Prior Expectations During Bayesian Perceptual Processing: A New Hypothesis and Implications for Pandemic Response'. *Sage Open Medicine* 24/11 (2023).

Johnson, Hannah. *Blood Libel. The Ritual Murder Accusation at the Limit of Jewish History.* University of Michigan Press. (2012).

Jolly, Julius. Schmidt, Richard. *The Arthasastra of Kautilya.* Lahore: Punjab Sanskrit Series, (1923).

Jones-Lewis, Molly Ayn. Poison: Nature's Argument for the Roman Empire in Pliny the Elder's Naturalis Historia. *Classical World* Vol 106. No. 1

(2012).

Judge, John *The Black Hole of Jonestown*. (1985). Available at: https://ratical.org/ratville/JFK/JohnJudge/Jonestown.html.

Kaiser, Jocelyn. 'Sipping from a Poisoned Chalice'. *Science* 302/5644 (2003).

Kapeliouk, Amnon. Arafat, Yasser. 'A-t-il ete Assassine'. *Le Monde Diplomatique*. November (2005).

Kaplan, David and Marshall, Andrew. *The Cult at the End of the World: The Terrifying Story of the Aum Doomsday Cult, from the Subways of Tokyo to the Nuclear Arsenals of Russia.* (1996).

Kasperson, Roger. Kasperson, Jeanne. 'The Social Amplification and Attenuation of Risk'. *The Annals of the American Academy of Political and Social Science* 545 (1996).

Kaszeta, Dan. *A History of Nerve Agents from Nazi Germany to Putin's Russia.* Oxford University Press. (2021).

Kaye, J. M. 'The Early History of Murder and Manslaughter'. *Law Quarterly Review* 83/3 (1967).

Kellert, Stephen R; Wilson, Edward O; Doughty, Robin W. The Biophilia Hypothesis. *Annals of the Association of American Geographers.* Vol. 86, Iss. 4. (1996).

Kelly, Michael. *Ghosts of Halabja. Saddam Hussein and the Kurdish Genocide.* Praeger Security International. (2006).

Keyes, Geoffrey, ed. *The Apologie and Treatise of Ambroise Paré*. Chicago: University of Chicago Press. (1952).

Keynes, Milo. 'The Medical Health of Napoleon Bonaparte'. *Journal of Medical Biography* 4/2 (1996).

Kinzer. Steven. *Poisoner-in-Chief. Sidney Gottleib and the CIA Search for Mind Control* Henry Holt and Co. (2019).

Kizer, K.W. and Yorker, B.C. Health Care Serial Murder: A Patient Safety Orphan. *Journal of Quality of Patient Safety.* 36(4). (2010).

Klippel, Heike ; Wahrig, Bettina ; Zechner. Anke. *Poison and Poisoning in Science, Fiction and Cinema: Precarious Identities*. Springer. (2017).

Knight, Amy. Orders to Kill. The Putin Regime and Political Murder. Thomas Dunne Books. (2017).

Knight, Bernard. 'Ricin—a Potent Homicidal Poison'. *British Medical Journal* (1979).

Knoefel, Peter. 'Felice Fontana on Poison'. *Clio Medica* 15/ ½ (1980).

Kogon, Eugen. *Nazi Mass Murder: A Documentary History of the Use of Poison Gas.* Yale University Press (1994).

Kogon, Eugen; Langbein, Hermann and Ruckerl, Adalbert Eds. *The Theory and Practice and Hell: the German Concentration Camps and the System Behind Them.* Farrar, Strauss and Giroux. (2006).

Kolecki, Paul. 'The Poison Theory and Mozart's Death'. *Mithridata* XI/1 (2001).

Konkola, Kari. 'More Than a Coincidence? The Arrival of Arsenic and the

Disappearance of Plaque in Early Modern Europe'. *Journal of the History of Medicine and Allied Sciences* 47/1 (1992).

Kramer, Heinrich. Sprenger, James. *The Malleus Maleficarum*. New York: Dover Publications. (1971).

Krzyszczuk, Łukasz and Morta, Krzysztof Basilisk – The History of the Legend. *Alea.* Vol. 25/1 (2023), Kudlein, Rudolf. 'Medical Ethics and Popular Ethics in Greece and Rome'. *Clio Medica* 5 (1970).

Kuhn, Jens and Leitenberg, Milton. 'The Soviet Biological Warfare Program'. In *Biological Threats in the 21st Century*. Edited by Filippa Lentzos. London: Imperial College Press. (2016).

Kunreuther, Howard. 'Risk Analysis and Risk Management in an Uncertain World'. The Wharton Financial Institutions Center Working Paper. December (2001).

Kutulas, Judy. *After Aquarius Dawned. How the Revolutions of the Sixties Became the Popular Culture of the Seventies.* University of North Carolina Press. (2017).

Lane, Roger. *Murder in America: A History*. Columbus, OH: Ohio State University Press. (1997).

Lanfranchi, Pierluigi. *Foetor judaicus*: archéologie d'un préjugé. *Pallas.* Vol 104 No. 2 (2017).

Larkin, Thomas. *A Life of Suspicion and Distrust. Sino-American Relations, Racialized Anxieties, and Anglo Panic in Nineteenth-Century Hong Kong.* University of California Press. (2023).

Lascaratos, John. 'The Poisoning of the Byzantine Emperor Romanos III Argyros'. *Mithridata* V/2 (1995).

Lemay, Helen Rodnite, trans. *Women's Secrets: A Translation of Pseudo-Albertus Magnus' 'De Secretis Mulierum' with Commentaries*. Albany, NY: State University of New York Press. (1992).

Lentzos, Filippa, ed. *Biological Threats in the 21st Century*. London: Imperial College Press. (2016).

Lentzos, Filippa. 'Dual-use Biology: Building Trust and Managing Perceptions of Intent'. *The Nonproliferation Review* 27/4-6 (2020).

Lentzos, Filippa. 'How to Protect the World from Ultra-targeted biological Weapons'. *Bulletin of the Atomic Scientists* 76/6 (2020).

Lentzos, Filippa. Rose, Nikolas. 'Governing Insecurity: Contingency Planning, Protection, Resilience'. *Economy and Society* 38/2 (2009).

Lerner, Robert. *The Heresy of the Free Spirit in the Later Middle Ages*. Berkeley, CA: University of California Press. (1972).

Levey, Martin. Medieval Arabic Toxicology; the Book on Poisons of Ibn Wahshiya and its relation to early Indian and Greek Texts. *Transactions of the American Philosophical Society* v. 56, pt. 7. (1966).

Levey, Martin. Ibn al-Waḥshīya's "Book of Poisons," "Kitāb al-Sumūm": Studies in the History of Arabic Pharmacology II. *Journal of the History of Medicine and Allied Sciences,* Vol.18 (4) (1963).

Levey, Martin. Chemistry in the "Kitab Al-Sumum" ("Book of Poisons") by "Ibn Al-Wahshiya" *Chymia*, Vol.9. (1964).

Lewy, Guenter. *The Nazi Persecution of the Gypsies*. Oxford University Press (2000).

Lifton, Robert Jay. *Destroying the World to Save It: Aum Shinrikyo, Apocalyptic Violence, and the New Global Terrorism*. Metropolitan Books. (1999).

Lifton, Robert Jay. Aum Shinrikyo: The Threshold Crossed. *Journal of Aggression, Maltreatment & Trauma*, Vol.9 (1-2) (2004).

Lowe, Kate and McLaughlin, Eugene. 'Caution! The Bread is Poisoned': The Hong Kong Mass Poisoning of January 1857, *The Journal of Imperial and Commonwealth History*, 43:2, (2015).

Liu, Yan. 'Poisonous Medicine in Ancient China'. In *History of Toxicology and Environmental Health*, edited by Philip Wexler. Amsterdam: Elsevier. (2015).

Lucanie, Ralph. 'Papal Poisoning: Preferred Pastime of the Pontiffs'. *Mithridata* 8/1 (1998).

Maatouk, Ismael ; Moutran, Roy. History of Syphilis: Between Poetry and Medicine. *Journal of Sexual Medicine*. Vol.11 (1), (2010).

Machiavelli, Nicolo. *The Prince*. Independently Published. (2020).

Makra, Lazlo. 'Anthropogenic Air Pollution in Ancient Times'. In *History of Toxicology and Environmental Health*, edited by Philip Wexler. Amsterdam: Elsevier. (2015).

Manguel, Alberto. *Maimonides: Faith in Reason.* Yale University Press (2023)

Markush, R. E. 'Mental Epidemics: A Review of the Old to Prepare for the New'. *Public Health Reviews* 2/ 4 (1973).

Mari, Francesco. Polettini, Aldo. Lippi, Donatella. Bertol, Elisabetta. *Un giallo di quattro secoli fa. La morte di Francesco I de' Medici e della sua sposa Bianca Cappello.* Le Lettere. (2007)

Martin, Morag Sarah. 'Consuming Beauty: The Commerce of Cosmetics in France 1750–1800'. Unpublished PhD dissertation. University of California, Irvine. (1999).

Matyszak Philip. *Ancient Magic: A Practitioner's Guide to the Supernatural in Greece and Rome.* Thames and Hudson (2019).

Mayor, Adrienne. 'The Nessus Shirt in the New World'. *Journal of American Folklore* 108 (1995).

Mayor, Adrienne. *Greek Fire, Poison Arrows, and Scorpion Bombs*. New York: Abrams Press. (2003).

McKerrow, R. B., ed. 'Pierce Penilesse'. In *The Works of Thomas Nashe*. (1958).

Meharg, Andrew. *Venomous Earth: How Arsenic Caused the World's Worst Mass Poisoning*. London: Macmillan. (2005).

Meier, Mischa. 'The "Justinianic Plague": the Economic Consequences of the Pandemic in the Eastern Roman Empire and its Cultural and Religious Effects'. *Early Medieval Europe* 24/3 (2016).

Metraux, Daniel. 'Religious Terrorism in Japan: The Fatal Appeal of Aum Shinrikyo'. *Asian Survey* 35/12 (1995).

Middleton, Thomas. Webster, John. *Five Jacobean Tragedies*. Hertfordshire, England: Wordsworth Editions. (2001).

Mirzayanov, Vil. *State Secrets. An Insider's Chronicle of the Russian Chemical Weapon's Program*. Outskirts Press. (2008).

Mishan, Ligaya, What if the Powerful (and Paranoid) Starting Using Food Tasters Again. *New York Times*. 31 October (2018).

Moon, John Ellis van Cortland. 'Chemical Weapons and Deterrence: The World War II Experience'. *International Security* 8/4 (1984).

Moon, John Ellis van Courtland. 'Controlling Chemical and Biological Weapons Through World War II'. In *Encyclopedia of Arms Control and Disarmament*. New York: Scribner. (1993).

Moore, Rebecca, *A Sympathetic History of Jonestown: The Moore Family Involvement in the Peoples Temple*. The Edwin Mellen Press (1985).

Moore, Rebecca, *The Jonestown Letters: Correspondence of the Moore Family* 1970–1985, The Edwin Mellen Press. (1986).

Moore, Rebecca. *In Defense of People's Temple... And Other Essays*. Edward Mellen Press. (1988).

Moore, Rebecca. The Stigmatized Deaths at Jonestown. *Death Studies*, 35: 42–58, (2011).

Moore, Rebecca. Jonestown Forty Years On. *Nova Religio: The Journal of Alternative and Emergent Religions,* Volume 22, Issue 2. (2018).

Moore, Robert. *The Formation of a Persecuting Society*. Hoboken, NJ: Wiley-Blackwell. 2007.

Mueller, John. Mueller, Karl. 'Sanctions of Mass Destruction'. *Foreign Affairs.* (1999).

Muir, Edward. *Sex and Gender in Historical Perspective*. Baltimore, MD: Johns Hopkins University Press. (1990).

Muntner, Suessman. 'Maimonides' Book for Al-Fadil'. *Isis* 35/1 (1944).

Murakami, Haruki. *Underground: The Tokyo Gas Attack and the Japanese Psyche.* Vintage International. (2010).

Myers, Steven Lee. *The New Tsar: The Rise and Reign of Vladimir Putin.* Alfred A. Knopf. (2015)

Nakagawa, Tomomasa and Tu, Anthony. Murders with VX: Aum Shinrikyo in Japan and the assassination of Kim Jong-Nam in Malaysia. *Forensic Toxicology,* Vol.36 (2) (2018).

Nathwani, Amit. Polonium-210: A Firsthand Account. *The Lancet* Vol 388. Issue 10049. (2016)

Needleman, Lionel and Diane. 'Lead Poisoning and the Decline of the Roman Aristocracy'. *Classical Views* 4/1 (1985).

Noga, Maciej ; Jurowski, Kamil What do we currently know about Novichoks? The State of the Art. *Archives of Toxicology*, Vol.97 (3) (2023).

Nuland, Sherwin. *Maimonides*. New York; Schocken Books. (2005).

Olson, Kyle. 'Aum Shinrikyo: Once and Future Threat?' *Emerging Infectious Diseases* 5/4 (1999).

Olson, Paul J. "The Public Perception of 'cults' and 'New Religious Movements'." *Journal for the Scientific Study of Religion.* 45.1 (2006).

Ondaatje, Michael. *Running in the Family.* Toronto: McClelland and Stewart. (1982).

Organisation for the Prohibition of Chemical Weapons. "Incident in Salisbury." https://www.opcw.org/special-sections/salisbury-incident/ (2018)

Orfila, Mathieu. A General System of Toxicology. Legare Street Press. (2023).

Orwell, George. *Decline of the English Murder and Other Essays.* London: Penguin Books. (1978).

Oster, Emily. 'Witchcraft, Weather and Economic Growth in Renaissance Europe'. *Journal of Economic Perspectives* 18/1 (2004).

Owen, Robert. The Litvinenko Inquiry. The National Archives. January Available at https://webarchive.nationalarchives.gov.uk/ukgwa/ 20160613090324 /https://www.litvinenkoinquiry.org/report (2016)

Padover, Saul. Patterns of Assassination in Occupied Territory. *Public Opinion Quarterly* 7/4 (Winter 1943).

Palkki, David and Rubin, Lawrence. Saddam Hussein's role in the gassing of Halabja, *The Nonproliferation Review*, 28:1-3, (2021).

Palmer, Richard. 'Pharmacy in the Republic of Venice in the Sixteenth Century'. In *The Medical Renaissance of the Sixteenth Century*, edited by Andrew Wear, Roger French and Iain Lonie. Cambridge University Press. (1985).

Parachini, John, et al. *Diversion of Nuclear, Biological and Chemical Weapons Expertise from the Former Soviet Union. Understanding an Evolving Problem.* RAND. Santa Monica. (2005).

Parascandola, John. *The King of Poisons: A History of Arsenic.* Potomac Books. (2012).

Parks, George. 'The First Italianate Englishmen'. *Studies in the Renaissance* 8 (1961).

Pate, Jason. Cameron, Gavin. 'Covert Biological Weapons Attacks against Agricultural Targets: Assessing the Impact against US Agriculture'. BCSIA Discussion Paper 2001–9. John F. Kennedy School of Government. Harvard University. (2001).

Pennington. Hugh. Lethal Specks. *London Review of Books.* Vol 28 No. 14. (2006).

Pharr, Clyde. 'The Interdiction of Magic in Roman Law'. *Transactions and Proceedings of the American Philological Association* 63 (1932).

Phillips, Mark. *Francesco Guicciardini: The Historian's Craft.* Toronto: University of Toronto Press. (1977).

Plutarch. *The Fall of the Roman Republic.* London: Penguin. (1958).

Pollak, Otto. *The Criminality of Women.* New York: A. S. Barnes. (1961).

Pollard, Tanya. *Drugs and Theater in Early Modern England.* Oxford: Oxford

University Press, (2005).

Porter, Roy. 'Medicine and the Decline of Magic'. *Journal of the History of the Behavioral and Social Sciences* 41 (1988).

Price, Hereward. 'The Function of Imagery in Webster'. *Publications of the Modern Language Association* 70/4 (1955).

Price, Richard. 'A Genealogy of the Chemical Weapons Taboo'. International Organization 49/1 (1995).

Pushkin, Alexander. 'Mozart and Salieri'. Translated by Alan Shaw. *Russian Language Journal* 39/131 (1984).

Rach, JuliAnne. *Practicing Magic: An Evaluation of Magic, Gender and Power in Tacitus*. Dissertation at University of Arizona. (2022).

Ratzen, Richard. Ferngren, Gary. 'A Greek Progymnasma on the Physician-Poisoner'. *Journal of the History of Medicine and Allied Sciences* 48/2 (1993).

Reader, Ian. *A Poisonous Cocktail? Aum Shinrikyo's Path to Violence* (1996).

Reeds, Karen Meier. 'Renaissance Humanism and Botany'. *Annals of Science* 33 (1976).

Rees, M. 'Denial of Catastrophic Risks'. *Science* (2013).

Renehan, Robert. 'An Eighteenth Century Carved Piece with Medical Associations'. *Journal of the Warburg and Courtauld Institutes* 26/ 3/4 (1963).

'Report of the Commission Concerning the Events in Jordan September 1997'. Jerusalem Government Press Office. 17 February (1998).

Retief, F. P. Cilliers, Louise. 'Causes of Death among the Caesars (27 BC–AD 476)'. *Acta Theologica* 26/2. Supplementum 7. (2006)

Retief, F. P. Cilliers, Louise. 'Lead Poisoning in Ancient Rome'. *Acta Theologica* 26/2 Supplementum 7. (2010).

Ridley D. The Inquest touching upon the death of Dawn Kelly Sturgess. In: Wiltshire Council. https://www.wiltshire.gov.uk/media/3904/Scope-ruling-Sturgess-20-December-2019/pdf/Scope-ruling-sturgess-20-december-20191.pdf. (2019)

Rimmer, Abi. Salisbury novichok attack: five minutes with . . . Christine Blanshard. *BMJ (Online)*, Vol.366. (2019).

Roach, Mary. *Stiff: The Curious Life of Human Cadavers*. Norton. (2004).

Robb, George. 'Circe in Crinoline: Domestic Poisoning in Victorian England'. *Journal of Family History* 22/2 (1997).

Roberts, A. A. *Poison in Warfare*. William Heinemen. (1915).

Roberts, Adam. Guelff, Richard. *Documents on the Laws of War*. Clarendon Press. (1982).

Rolle, Renate. *The World of the Scythians*. London: B. T. Batsford. (1989).

Roper, Lyndal. *Witch Craze*. New Haven, CT: Yale University Press. (2004).

Rosenau, William. 'Aum Shinrikyo's Biological Weapons Program: Why Did It Fail?' *Studies in Conflict and Terrorism* 24/4 (2001).

Roth A and McCarthy T. 'It's got me': the lonely death of the Soviet scientist

poisoned by novichok. In: The Guardian. https://www. theguardian.com/
world/2018/mar/22/andrei-zheleznyakov-soviet-scientist-poisoned-
novichok. (2018).

Ruiz, Benjamin Salisbury, Novichok and the OPCW Loyola, LOJ
Pharmacology and Clinical Research, , Vol.1 (4) (2019).

Salazar, Christine. 'The Treatment of War Wounds in Graeco-Roman
Antiquity'. *Studies in Ancient Medicine* 21 (2000).

Sayer, Geoffrey Robley. *Hong Kong, 1841-1862.* Hong Kong: Hong Kong
University Press. (1980).

Scarborough, J. 'The Myth of Lead Poisoning among the Romans: An Essay
Review'. *Journal of the History of Medicine* 39 (1984).

Scarborough, John. "Early Byzantine Pharmacology." *Dumbarton Oaks
Papers* 38 (1984).

Scheidel, Walter. 'Emperors, Aristocrats and the Grim Reaper: Towards
a Demographic Profile of the Roman Elite'. *Classical Quarterly* 49/1
(1999).

Schmaltz, Florian. *One Hundred Years of Chemical Warfare: Research,
Deployment, Consequences.* Springer. (2017).

Schmidt. Ulf. *Secret Science: A Century of Poison Warfare and Human
Experiments.* Oxford University Press. (2015).

Shane, Scott. 'Poison's Use as Political Tool: Ukraine Is No Exception'. *New
York Times.* 15 December (2004).

Shapiro, James. *Shakespeare and the Jews.* New York: Columbia University
Press. (2016).

Silver, Eric. 300 West Bank Girls 'Poisoned.' *The Guardian.* 28 March (1983).

Sivak, Michale. Flannagan, Michael. 'Flying and Driving after the September
11 Attacks'. *American Scientist* 91/1 (2003).

Smith, George. The Recipe for Ricin. Part II. *GlobalSecurity.org.* Available at:
https://www.globalsecurity.org/org/nsn/nsn-040304.htm (2004)

Soleymani, Samaneh   and Zargaran, Arman. From Food to Drug:
Avicenna's Perspective, a Brief Review. *Research Journal of
Pharmacognosy,* Vol.5 (2), (2018).

Somerset, Anne. *Unnatural Murder: Poison at the Court of James I.* London:
Wiedenfeld and Nicholson. (1997).

Southon, Emma. *A Fatal Thing Happened on the Way to the Forum: Murder
in Ancient Rome.* Abrams Press. (2021)

Southon, Emma. *Agrippina: The Most Extraordinary Woman of the Roman
World.* Pegasus Books. (2019).

Spiers, Filippa. *Agents of War: A History of Chemical and Biological Weapons.*
Reaktion Books. (2021).

Stanislaw Andreski. *Syphilis, Puritanism and Witch-hunts.* Macmillan. (1989).

Stanley A. Moscow Journal; To the Business Risks in Russia, Add
Poisoning. *New York Times.* https://www.nytimes.com/1995/08/09/
world/moscow-journal-to-the-business-risks-in-russia-add-poisoning.

html?searchResultPosition=1 (1995).

Steindl, David ; Boehmerle, Wolfgang et al. Novichok Nerve Agent Poisoning. The Lancet (British edition), Vol.397 (10270) (2021)

Stern, Jessica. 'Dreaded Risks and the Control of Biological Weapons'. *International Security* 27/3 (2002/2003).

Stone R. How German military scientists likely identified the nerve agent used to attack Alexei Navalny. *Sci AAAS.* (2021).

Strickland, Debra Higgs. *Saracens, Demons, and Jews: Making Monsters in Medieval Art*. Princeton, NJ: Princeton University Press. (2003).

Sussman, George. 'Carriers of Cholera and Poison Rumours in France in 1832'. *Societas* 3/3 (1973).

Sweeney, John. *Killer in the Kremlin.* Transworld Digital. (2022).

Takahashi, Hiroshi et al. 'Bacillus anthracis Bioterrorism Incident, Kameido, Tokyo, 1993'. *Emerging Infectious Diseases* 10/1 (2004).

Tanimoto, Tetsuya ; Ozaki, Akihiko ; Harada, Kayo ; Tani, Yuta ; Kami, Masahiro Involvement of doctors in Aum Shinrikyo *The Lancet*, Vol.392 (10153), (2018)

Templer, Robert. Jonestown. *The Richmond Review*. Available at https://www.richmondreview.co.uk/temple02/ (1998).

Tenney, Frank. 'The Bacchanalian Cult of 186 BC'. *Classical Quarterly.* 21/3, 4 (June/Oct 1927).

Tezcür, Güneş Murat ; Horschig, Doreen. A conditional norm: chemical warfare from colonialism to contemporary civil wars. *Third World Quarterly.* Vol 42 (2) (2021).

The Invisible Enemy. The Economist. 16 November (2000).

The Sixth Voyage of Es-Sindibad of the Sea: Stories from 1001 Nights. New York: Harvard Classics. (1909)

The Middle East. When Poison Is Not Poison. *Journal of Palestine Studies* 13/1 (1983).

Thomas, D. M. *Alexander Solzhenitsyn: A Century in His Life*. New York: St Martin's Press. (1998).

Thomas, Keith. *Religion and the Decline of Magic Studies in Popular Beliefs in 16th and 17th Century England*. London: Penguin. (2003).

Thorndike, Lynn. 'Medieval Magic and Science in the Seventeenth Century'. *Speculum* 28/4 (1953).

Touwaide, Alain. Pietro d'Abano, De venenis: Reintroducing Greek Toxicology into Late Medieval Medicine in Toxicology in the Middle Ages and Renaissance, (2017).

Tribby, Jay. 'Cooking (with) Clio and Cleo: Eloquence and Experiment in Seventeenth-Century Florence'. *Journal of the History of Ideas* 52/3 (1991).

Tucker, Jonathan. *War of Nerves: Chemical Warfare from World War I to Al-Qaeda*. New York: Pantheon Books. (2005).

Urban, Mark. *The Skripal Files.* Macmillan. (2018).

Urdang, George. 'Pharmacopeias as Witnesses of World History'. *Journal of the History of Medicine and Allied Sciences* 1 (1946).

Ushiyama, Rin. *Aum Shinrikyo and religious terrorism in Japanese collective memory* (British Academy Monographs) (2023)

Vale, J. Allister ; Marrs, Timothy C. ; Maynard, Robert L. Novichok: a murderous nerve agent attack in the UK. Clinical toxicology (Philadelphia, Pa.), Vol.56 (11) (2018).

van Bergen, Leo. The poison gas debate in the inter-war years, *Medicine, Conflict and Survival*, 24:3. (2008).

Voigtländer, Nico and Voth, Hans-Joachim. Persecution Perpetuated: The Medieval Origins of Anti-Semitic Violence in Nazi Germany. *Quarterly Journal of Economics* 127. (2012).

Voiţă-Mekereş, Florica ; Delcea, Cristian ; Buhaş, Camelia Liana ; Ciocan, Veronica. Novichok Toxicology: A Review Study. Archives of Pharmacy Practice, Vol.14 (3) (2023)

Volkogonov, Dmitrii. Izvestiia. 'Stalin's Plan to Assassinate Tito'. History and Public Policy Program. Wilson Center (1993).

Volodarsky, Boris. *The KGB's Poison Factory From Lenin to Litvinenko.* Frontline Books. (2013).

Volodarsky, Boris. *Assassins. The KGB's Poison Factory 10 Years On.* Frontline Books. (2020)

von Hippel, F. Russian whistleblower faces jail. *Bulletin of Atomic Scientists.* (1993)

Walton, Calder. *Spies: The Epic Intelligence War Between East and West.* Simon and Schuster. (2023).

Warrick, J., and Troianovski, A. "Agents of Doubt. How a Powerful Russian Propaganda Machine Chips Away at Western Notions of Truth." *Washington Post*, December 10. https://www.washington post.com/graphics/2018/world/national-security/russian-propaganda-skripal-salisbury/?utm_term=.f14e4fb5de40 (2018).

Waters, Tim. 'The Fine Art of Making Poison'. *Discover* 13/8 (August (1992).

Watson, Emily. Homer. *The Odyssey.* W.W. Norton and Company. (2017).

Watson, Emily: Homer. *The Iliad.* W.W. Norton and Company. (2023)

Watson, Katherine. *Poisoned Lives: English Poisoners and Their Victims.* London: Hambledon and London. (2004).

Weber, Max. 'Politics as a Vocation'. In *Weber: Selections in Translation.* Edited by W. G. Runciman, translated by Eric Matthews. Cambridge: Cambridge University Press. (1978).

Weekly, Terry. 'Proliferation of Chemical Warfare: Challenge to Traditional Restraints'. *Parameters* 19/1 (1989).

Wein, Lawrence. Liu, Yifan. 'Analyzing a Bioterror Attack on the Food Supply: The Case of Botulinum Toxin in Milk'. *Proceedings of the National Academy of Sciences* 102/28 (2005).

Weitbrecht, Julia. "Thou hast heard me from the horns of the unicorns:" The

Biblical Unicorn in Late Medieval Religious Interpretation. *Interfaces* (Milano), 12 (5) (2018)

West, Clarence. 'The History of Poison Gases'. *Science* 49/1270 (1919).

Wexler, Philip. *History of Toxicology and Environmental Health: Toxicology in Antiquity II.* Academic Press (2015).

Whiteside, Thomas. 'Annals of the Cold War'. *The New Yorker*. 11 February (1991).

Whorton, James. *The Arsenic Century: How Victorian Britain Was Poisoned at Home, Work and Play*. Oxford University Press. (2010)

Wickens, G. M. *Avicenna, Scientist and Philosopher*. London: Luzac. (1952).

William of Malmesbury. *Gesta Regnum Anglorum or The History of the Norman Kings*. Translated by Joseph Stevenson. Dyfed, Wales: Llanerch Enterprises. (1991).

Williams, Neville. *Powder and Paint: A History of the Englishwoman's Toilet, Elizabeth I–Elizabeth II*. London: Longmans, Green and Co. (1957).

Williams, Steven. *The Secret of Secrets: The Scholarly Career of a Pseudo-Aristotelian Text in the Latin Middle Ages*. Ann Arbor, MI: University of Michigan Press. (2003).

Winslow, C. and Duran-Reynals, M. Jacme d'Agramont and the First of the Plague Tractates. Bulletin of the History of Medicine. Vol 22 No. 6. (1948).

Wolf, René. Judgement in the Grey Zone: The Third Auschwitz (Kapo) Trial in Frankfurt 1968, *Journal of Genocide Research*, 9:4. (2007).

Wolfgang, Marvin. *Patterns in Criminal Homicide*. Philadelphia: Pennsylvania University Press. (1958).

Wright, R. Putin, a Little Man Still Trying to Prove His Bigness. *The New Yorker*, March 20. https://www.newyorker.com/news/news-desk/putin-a-little-man-still-trying-to-prove-his-bigness (2018).

Xenophon, *Cyropedia.* George Bell. (1898).

Yardley E, and Wilson D. In Search of the 'Angels of Death': Conceptualising the Contemporary Nurse Healthcare Serial Killer. *Journal of Investigative Psychological and Offender Profiling*. 13(1). (2016).

Yorker, B.C., Kizer, K.W.et al. Serial Murder by Healthcare Professionals. *Journal of Forensic Science* 51(6) (2006).

Zariati, Parisa, et al. 'A Comprehensive Review: Toxicity of Nanotechnology in the Food Industry'. *Journal of Medical Discovery* 3/2 (2018).